Management of Safety, Health and Environment in South Africa

Management of Safety, Health and Environment in South Africa:

A Handbook

By

Theo C. Haupt

Cambridge Scholars Publishing

Management of Safety, Health and Environment in South Africa: A Handbook

By Theo C. Haupt

This book first published 2021

Cambridge Scholars Publishing

Lady Stephenson Library, Newcastle upon Tyne, NE6 2PA, UK

British Library Cataloguing in Publication Data
A catalogue record for this book is available from the British Library

ISBN (10): 1-5275-7370-2
ISBN (13): 978-1-5275-7370-3

TABLE OF CONTENTS

Foreword

'Prof Theo clearly did not sleep while putting this handbook together. A health and safety handbook of this magnitude will change the way that South African, and African health and safety practitioners see the subject. A very necessary addition to all practitioners' reference material when navigating the health and safety space. While it is set up for the South African health and safety industry, it can be used internationally. There are really well thought out supporting documents that can be adapted for all workplaces, ready to use. A mix of theory and a very practical approach that will assist in meeting good standards in the workplace for those just entering the field, or the seasoned practitioner who needs to brush up on their practices. A broad range of subject areas relating to health and safety are needed by a health and safety practitioner, and the handbook really delivers with the range needed. Areas generally not well managed, specifically in my field of occupational health will be extremely important in managing employee well-being, primary health care and limiting occupational diseases. Prof Theo has included his knowledge, and decades of experience across many countries into the textbook and enriched the South African health and safety industry in so doing.'

Dr Claire Deacon
Johannesburg
June 2020

PREFACE

The global performance track record of the management of safety, health and the environment (SHE) by industries has been persistently poor. I was inspired to conduct extensive research over many years by my interest in trying to understand the reasons for this poor performance and through that effort contribute to its improvement. My research interest led me to study at the University of Florida in the USA where I obtained a PhD. My doctoral dissertation examined a performance approach to construction safety and health given the dismal performance track record of this particular sector which regularly kills and maims a large cohort of people each and every year. My research activities in SHE over time have gained me both a national and international reputation of expertise in the SHE space. For example, about SHE I have addressed a wide range of audiences as an invited speaker, presented a large number of conference papers, published several journal papers, wrote a number of book chapters, served on the Occupational Health and Safety Advisory Committee advising the Minister of Labour and presented many SHE-related workshops across South Africa.

However, for many years I have had the desire to compile a comprehensive text or handbook on the management of safety, health and the environment (SHE) in South Africa considering that there very few if any such texts available. I wanted to use my expert knowledge and experience to fill this perceived vacuum. However, work on the actual text only began after presenting several iterations of a series of nine-day long SHE training workshops in KwaZulu-Natal in South Africa. The national lockdown due to COVID-19 in 2020 provided the space to accelerate progress on the final text. I used the workshop material and the various visual presentations to inform the content.

The handbook is divided into four parts, namely introduction to SHE management, management of safety, management of health, and management of the environment. It consists of 26 chapters that cover a wide range of relevant and current aspects of SHE management within the context of the South African legislative and regulatory framework and better practice. However, the principles of management are the same since the focus is not on mere compliance but on better practice which should be the goal of all SHE practitioners and all businesses irrespective of the sectors that they operate in. For example, the book covers the occupational health and safety legislative and regulatory framework, the need for SHE management, use of current technology, risk management, hazard identification and risk analysis, cannabis in the workplace, management of asbestos, COVID-19, disaster management and SHE management systems just to mention a few topics.

I believe the handbook is an important one that will help not only SHE practitioners, but anyone interested in wanting to know how better to manage SHE in the workplace. It is not intended to be exhaustive but rather to provide guidance on key aspects and principles that can be applied in any workplace in any industry. Writing the book was challenging given the dynamic nature of SHE management especially in the South African context to ensure that the content was current and user friendly. To remain abreast of the latest developments I consulted many online resources for detailed information, some of which are listed at the end of each chapter as references for further reading. In some cases, the references are extensive while in others they are not. Each chapter has a series of simple review questions which are useful for training and evaluation purposes. The book can be read cover to cover through each consecutive chapter from beginning to end or by referring to relevant sections in any order of preference.

I am thankful to Patric Pantelis of FTS Safety who insisted that I develop and present SHE training in the KwaZulu-Natal Province of South Africa and initially provided me with a collection of anonymously authored notes some of which I reworked, adapted and used. Where I used online sources, I attempted to contact the sources for permission to include them in this text. In every case I cited the source both intext and in the reference list to ensure that the contribution was duly acknowledged and fully recognized providing for readers to visit the sites themselves and familiarise themselves with the author/s firsthand. I acknowledge the University Extended Learning (UEL) unit of the University of KwaZulu-Natal in Durban, South Africa for allowing me to conduct several series of SHE management training courses. I am particularly thankful to my postgraduate students, Mariam Akinlolu and Mohlomi Terah Raliile who assisted me in critiquing and reviewing the text as I compiled it. I acknowledge that this handbook would not have been possible without part funding from the National Research Foundation (NRF) and that opinions expressed and conclusions arrived at are mine and are not necessarily to be attributed to UEL or the NRF.

Prof. Theo C Haupt
Durban, South Africa
June 2020

PART 1

INTRODUCTION TO SAFETY, HEALTH AND ENVIRONMENT MANAGEMENT

This section comprises 9 chapters that introduce key aspects of safety, health, and environment management in general industry in South Africa. Chapter 1 highlights the importance of effectively managing occupational safety, occupational health and the environment during general industrial activities. Chapter 2 outlines the general principles governing occupational health and safety. Chapter 3 discusses a systematic approach to managing safety, health, and the environment. Chapter 4 covers the importance of policy and the effective distribution of responsibilities for managing safety, health, and the environment at various levels. The critical role of communication and communication strategies as part of managing safety, health and the environment are discussed in Chapter 5. Chapter 6 deals with the importance of risk management and introduces various risk management concepts and approaches. Chapters 7 and 8 deal with broad legal concepts and principles that govern the South African legislative and regulatory framework and specifically addresses the various statutory requirements in the occupational safety, health, and the environment spaces. This section concludes in Chapter 9 with the operational management aspects of keeping and preserving records, reporting and control of important documents.

CHAPTER 1

THE NEED FOR SAFETY, HEALTH AND ENVIRONMENTAL MANAGEMENT

The effective management of safety, health and environment in all workplaces has many benefits. These benefits include, for example, the improved overall wellbeing of the workforce, increased productivity, reduced work-related accidents, injuries and fatalities, and a positive public impression of the organization and its management.

IN THE BEGINNING

Since the beginning of time man needed shelter and to survive and provide for his family – a notion captured by Maslow in his seminal Hierarchy of Needs. To be able to satisfy this very basic human need he had to work. However, he had to look for raw materials in his daily natural environment to use as 'tools' to perform tasks that he could not do entirely by himself. As a result, he was able not only to shorten the time it took to do some tasks but also do so much more in a much shorter time than he could without these 'tools. However, many of these rudimentary tools and implements could potentially hurt and even kill him because of their crude and rough shapes, heavy and awkward handling, and the type of material that they were made of. As man developed over time, he adapted his tools so that he could use them under safer and healthier conditions. Consequently, he became less ill from the activities he was engaged in and avoided unnecessary injuries. Some have suggested that these adaptations driven by his immediate needs for shelter and survival were the earliest beginnings of basic occupational health, safety and environment management.

Several key inventions over time influenced the development of man as a productive and industrious person at work. One such invention was the wheel which had a profound effect on the human economy and society. Since the introduction of wheels, it has become easier to travel and move about. A world with no wheels around is hard to imagine.

The invention of the wheel around 3,000 years ago, resulted in a revolt in the manufacturing industry. New tools and operating procedures were developed to cope with the demand created by new inventions and their use. As a means of efficiently carrying goods for long distances, vehicles, and modes of transport with wheels allowed for the broadening of trade networks. With access to a wider market, craftspeople could more easily specialize in specific crafts or trades and communities could expand if there was no longer any need to live close to areas where food was processed.

Man, slowly continued to develop additional new tools and machines, enabling him to achieve even more than what was previously possible and more efficiently. In this process he used existing tools to make and develop new ones. Certain specific trades were established and standards were set to promote uniformity. Safe practices became standard.

Further, the introduction of innovations generally taken for granted such as making running water available on tap and hygienic ablution facilities contributed substantially to the health and prosperity of man and society in general. Because of activities such as farming and mining, man was also starting to have a serious impact both directly and indirectly on the environment.

INDUSTRIAL REVOLUTIONS AND THEIR IMPACT

Before the Industrial Revolution began in 1760, it was normal to make a living through agriculture or the making and selling of products from the home. The pressure of the resultant new developments in machinery and manufacturing processes resulted in mass production and the establishment of the factory system. Urbanization increased as workers came to the cities seeking employment within these new mills and factories. This increase in the size of the available workforce and demand for work led to poor pay, hazardous factory working conditions and an increase in child labour. Although child labour offered a cost-effective solution because of the low wages that were paid, the lack of industrial experience resulted in serious accidents when working with foreign or unfamiliar machines or chemical substances. Working hours were invariably long and conditions dangerous for the workers, with many of them losing their lives at work. Further, a general lack of health and safety at work led to many of the child workers developing occupational diseases such as terminal lung cancer and early deaths.

Because of growing concerns relating to the effects of worker health and illness on productivity the Factory Acts in the UK was formed, for example. As the number of other work-related accidents increased, including mining fatalities, and the unknown effects of working with chemicals, the Government came under pressure to act against organizations who did not protect their employees effectively. Sir Robert Peel, who was a factory owner himself in Britain introduced the Health and Morals of Apprentices Act 1802, commonly known as the Factory Act which is generally regarded as the beginning of health and safety regulation and which required, for example:

- Sufficient windows and openings for ventilation
- Cleaning at least twice yearly with quicklime and water
- Limited working hours to no more than 12 hours a day (excluding time taken for breaks)
- Elimination of night-time work from 21h00 to 06h00
- Provision of suitable clothing and sleeping accommodation to workers

Consequently, various laws were introduced that included the significant appointment of factory inspectors initially to prevent injury and child labour. For the first time ever, an external set of rules was now governing the health and safety of workers in the workplace and being monitored. The main duty of factory inspectors was to prevent injury and overworking of child workers but also to formulate new regulations and laws to ensure the Factories Act could be suitably enforced.

The introduction of 'duty of care' provisions that were later introduced established that an employer was not responsible to ensure higher safety standards for an employee than he ensured for himself. A kind of 'treat others the way you would treat yourself' approach – the so-called Golden Rule.

As time passed mechanisms were introduced for workers to claim compensation for injury or death caused by a defect in equipment or machinery or negligence of a person given authority over the worker by the employer. Workers were eventually only required to prove that the injury occurred on the job, namely that it was work-related.

Progressively additional health and safety regulations were developed, introduced, implemented and enforced which included

- Requirements to provide safeguarding for machinery
- Raising the legal working age
- Increasing numbers of inspectors appointed across industries

Generally mass production, the use of machinery and the availability of cheap labour resulted in health and safety at work to unfortunately not be regarded as important. Sadly, it was commonly believed that it was cheaper to replace a dead or injured worker than to implement health and safety measures. After World War II, legislation was introduced to ensure employers provided their workers with a work environment free from hazards including exposure to toxic chemicals, excessive noise levels, mechanical dangers, heat or cold stress and unsanitary conditions.

A good and informative read is 'Two steps forward, one step back' A brief history of the origins, development and implementation of health and safety law in the United Kingdom, 1802–2014 by David Eves

It was important for employers to motivate their workforces to achieve higher levels of effectiveness, while gaining their trust and cooperation in the process. One way to achieve this level of performance on a sustainable basis was to improve the working and living conditions of workers to mutually negotiated and acceptable standards. This process included the improvement of health and safety standards of work activities themselves and the workplace and environment. Through ongoing research, health and safety standards have subsequently been implemented and adapted to meet industry demands.

Vast amounts of effort and money have been invested into the study of the effects and impacts of technological advancement on the environment. Research has shown that industrial activity has had devastating effects on the environment. Measures to protect the environment against the negative effects of human exploitation needed to be devised and adapted according to the ever-changing need to do so.

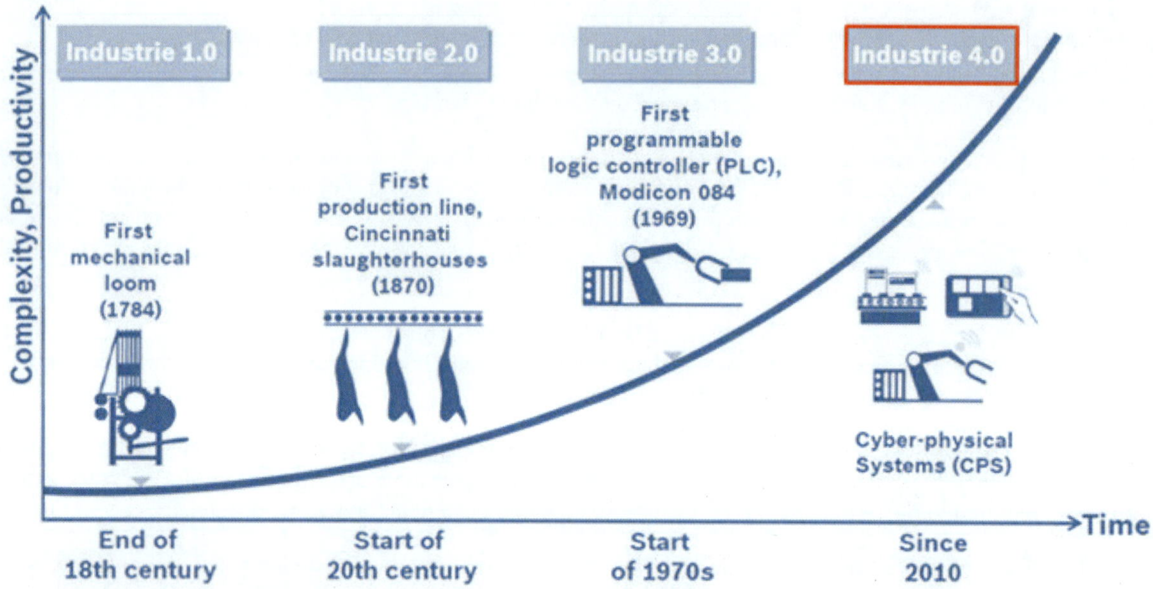

The Fourth Industrial Revolution also referred to as 4IR presents new challenges in the workplace. This new digital industrial technology makes it possible to gather and analyse data across machines, enabling faster, more flexible, and more efficient processes to produce higher-quality goods at reduced costs. This manufacturing revolution will increase productivity, shift economics, foster industrial growth, and modify the overall profile of the—ultimately changing the competitiveness of companies and regions. As a result, new threats to safety, health and the environment will emerge – some of the work activities which will present these threats are not even identifiable yet. This industrial revolution introduces challenging and interesting times ahead for entrepreneurs, employers, and workers as they adapt, survive, or die!

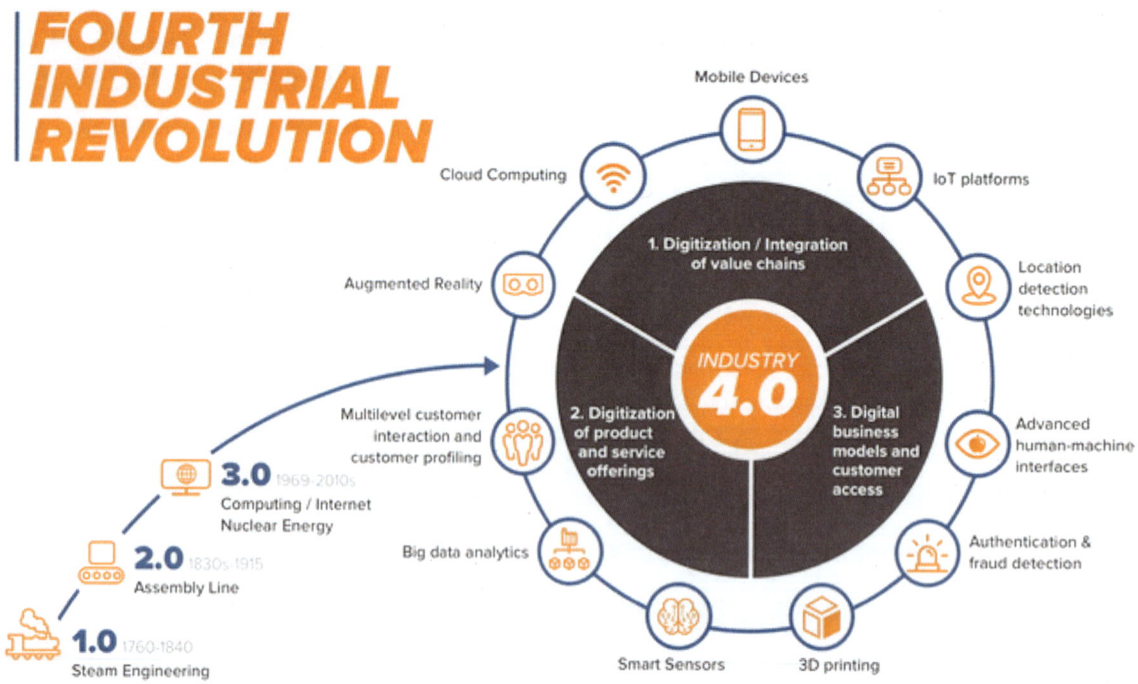

Source: https://cdn.shockoe.com

Advanced digital technology is already being used in manufacturing, but with 4IR, it will significantly transform product production and service delivery. It will lead to greater efficiencies and change traditional production relationships among all stakeholders in the supply chain that includes suppliers, producers, and customers—as well as between humans themselves and machines. It will challenge ideas of what being human actually means. Nine technology trends form the building blocks of Industry 4.0 (4IR) and are transforming industrial production. These are:

- BIG DATA AND ANALYTICS

In an Industry 4.0 context, the gathering and evaluation of data comprehensively from a large number of different sources such as production systems and equipment as well as the management systems of enterprises and customers will become universally standard to support real-time decision making.

- AUTONOMOUS ROBOTS

It is already possible for robots to interact with other robots and work safely alongside humans while learning from them. The benefits of this development are that these robots will be less expensive and have a greater range of capabilities than those already currently used in present manufacturing.

- SIMULATION

Simulations will be used more extensively than presently in plant operations to leverage real-time data. These simulations will effectively mirror the physical world in a virtual model. This model could include machines, products, and humans. This increased usage of simulations will allow operators to test and optimize the machine settings for the next product in line in the virtual world before the physical changeover, thereby driving down machine setup times and increasing quality, health, and safety. Simulations therefore provide an effective and efficient form of 'what if' scenarios.

- HORIZONTAL AND VERTICAL SYSTEM INTEGRATION

Within the context of 4IR, it is possible for organisations, departments, functions, and capabilities to become increasingly cohesive, due to the evolution of cross-company and universal data-integration networks and enable truly automated value chains.

- THE INTERNET OF THINGS (IOT)

The 4IR context means that more devices which may sometimes including unfinished products, will be enriched with embedded computing Consequently, this embedded computing will allow field devices to communicate and interact both with one another and with more centralized controllers. IoT will also decentralize analytics and decision making and enable responses in real-time.

- CYBERSECURITY

Given the increased connectivity and use of standard communications protocols that is facilitated within the context of 4IR, there will be an increased need to protect critical industrial systems and manufacturing lines from cybersecurity threats. Therefore, it is essential to have secure, reliable communications and sophisticated identity and access management of machines and users.

- THE CLOUD

4IR will result in production-related undertakings require high levels of data sharing across organizational sites and boundaries. Simultaneously, cloud technology performance will improve, achieving reaction times of mere milliseconds. Consequently, machine data and functionality will be deployed to the cloud increasingly, enabling more data-driven services for production systems.

- ADDITIVE MANUFACTURING

Additive manufacturing such as 3-D printing, is being increasingly used in organizations to prototype and produce individual components. With 4IR, these additive-manufacturing methods will be used more often to produce small batches of customized products to offer, for example, construction advantages, such as complex, lightweight designs.

- AUGMENTED REALITY

Augmented-reality-based systems support a variety of services. It is likely that organizations will use of augmented reality on a much broader basis in the context of 4IR to provide their workers with real-time information to improve their on-the-job decision making and work procedures.

Given the recent global challenges presented by the COVID-19 pandemic the opportunity is presented to expand research into technologies which will prevent or at least limit the exposure of all industry stakeholders to potential infection of not only the current COVID-19 but also increase preparedness for any future pandemic or global disaster. The efficacy of social distancing between workers in the workplace of around 2m places challenges in the way work activities are traditionally executed. Core work activities will be increasingly examined in the future and technologies will be developed that can execute them either without the physical involvement of workers themselves or involvement from a distance from other workers. Further, pre-workplace entry screening and detection technologies will be developed as early warning mechanisms to prevent possible infection at the point of entry to work premises. Protocols to promote off site and working from remote locations while still monitoring work activities and production progress will receive attention. Issues of cybersecurity will be addressed when using virtual communication platforms for meetings. Fully automated operator-less plant and equipment will be developed. Innovative logistical management of the supply chain and material delivery will be investigated and improved under the conditions of the 'new norm.' The legislative and regulatory framework that govern industries will be adapted to suit new contractual, working and labour employment conditions that include provisions for the use of new technologies, training of workers to use them as well as revisiting definitions of *force majeure.* The opportunities are endless given the vast amount of effort to improve overall health and safety performance using innovative and advanced technologies at worker, project, sector, national and global levels.

As a result, organizations will face formidable challenges in the adoption of these new technologies. To build and sustain competitiveness in the race to full implementation, they will need to broaden and deepen their practical knowledge about innovative digital and advanced technologies and their related uses. At the same time, they will need to manage SHE in this new and ever-rapidly-changing environment. The goal and principles of managing SHE will remain constant and the SHE operating framework will have to be adaptable enough to accommodate occupations that have not even been invented yet.

• REASONS FOR SHE PROGRAMMES

Legal considerations

In South Africa several laws influence and affect the employer/worker relationship. The Occupational Health and Safety Act, No. 85 of 1993 (OHS Act), the Mine Health and Safety Act, No. 29 of 1996 and the Compensation for Occupational Injuries and Diseases Act, No. 130 of 1993 (COID Act) are the pieces of legislation involved. It is imperative that SHE practitioners must familiarise themselves fully with all sections and regulations applicable to health, safety and the environment. Failure to comply with Acts, which regulate health and safety in industry and in the mines, can result in unnecessary delays in production, legal liability claims and possible criminal prosecution.

The following are, inter alia, other Acts regulating safety, health and the environment in South Africa, namely

- ✓ The Constitution,
- ✓ Basic Conditions of Employment Act,
- ✓ Explosives Act,
- ✓ Labour Relations Act,
- ✓ National Building Regulations Act,
- ✓ Hazardous Substances Act,
- ✓ Nuclear Act,
- ✓ Road Ordinance Act,
- ✓ Fire Act,
- ✓ Pollution Act, and
- ✓ Certain SABS codes of practices.

Moral and social considerations

Management and workers have, apart from their legal responsibilities, a moral and a social obligation towards other workers and themselves to ensure effective health, safety and environmental management and protection. Management also has a moral and social responsibility towards safeguarding the public, contractors and visitors who may be affected by the activities of their operations and the use of their products. Organizations must act responsibly regarding the hazards of the articles or substances they produce, process, use, handle, store, or transport. Social and moral obligations interlink with legal and financial considerations.

Economic considerations

According to Peter Drucker, the well-known Austrian-born American management consultant, educator, and author, "The first duty of an organization is to survive, and the guiding principle of business economics is NOT the maximization of profit, it is the avoidance of loss." The direct costs of incidents are according to the latest report of the Compensation Commissioner more than R450, 000, 000.00. The indirect costs can be as high as 50: 1 depending on the type of business and other factors.

If an organization experiences an annual incident loss of R200 000.00 and they sell their products at a 5% margin, then 20 times the amount must be sold to make up for the loss incurred. Business enterprises or public institutions must be financially viable by making profits to survive. In the case of the public sector, or other service orientated institutions and organisations, the profit refers to providing "optimum service levels at least cost".

- HEALTH AND SAFETY AS A VALUE

What an organisation believes to be morally correct and justifiable when questioned on the health and safety of its workers is a reflection on how it values its people. Many organizations make health and safety their number 1 priority. Sadly, priorities change from time to time. Instead, if health and safety was an organizational core value it is not subject to change. That is what organizations should be striving for.

- HEALTH AND SAFETY AS A COST FACTOR

Serious work-related injuries and diseases have an enormous cost implication on the economy of South Africa. The following are therefore important, namely:

✓ The financial implications of injuries or health problems for the worker.
✓ The potential suffering for the injured or sick worker and his / her family.
✓ The potential effect on the career and future employment opportunities of the worker.
✓ Loss in productivity associated with workers who are absent from work due to occupational injuries or illness.
✓ The loss in needed skills if the worker is no longer able to perform the duties, he / she previously performed.
✓ The cost of re-training workers if they are to be accommodated in an alternative position within the organization.
✓ The cost of recruitment and employment of replacement staff.
✓ Disruption of the normal flow of work and delivery of services if key staff are lost or if operations have to stop due to workplace accidents.
✓ The effect on the general morale of other workers when workplace accidents occur.

- HEALTH AND SAFETY AND THE WORKPLACE

Risks are accepted as a part of daily life. Working cannot be stopped just because there are risks associated with the work, but the risks and exposure to them can be effectively managed. There is no perfect workplace without risks. However, health and safety in the workplace is both a state of mind as well as a way workers work and act towards each other.

✓ Joint responsibility and self-regulation are necessary. All the role players in an organisation such as the employer, workers and trade unions, should work together and accept joint responsibility to make the workplace a safer place.
✓ Effective communication channels are critical. Communication channels should be available and regularly revised to ensure that health and safety issues and risks are communicated effectively. Communication should not only be focused on enforcing policies and procedures but also on reporting potential problems. Therefore, all workers should know to whom they must report these problems or improvement suggestions as well as the rules for reporting injuries sustained on duty.

Occupational health and safety legislation specifies that a workplace must ensure that persons other that those employed in the workplace and who are directly and indirectly affected by the activities of the workplace may not be exposed to health and safety risks. The implication is that, for example:

✓ A workplace may not pollute the natural environment or discard any material that may cause a health risk to others such as industrial toxic effluence and waste into rivers and canals.
✓ Persons who use products manufactured by the workplace must be informed of any health or safety risk they could be exposed to by using those products.

- HEALTH AND SAFETY AS A ZERO ACCIDENT VISION

It has long been argued in some quarters that accidents happen by chance. Consequently, accidents therefore cannot be influenced because one cannot affect chance. Because one cannot change chance trying to do so is wasting energy, time, and resources. Accidents are then perceived as an inevitable and integral part of life and work and caused by 'bad luck'. This argument and attitude create the situation where nothing is done, or a minimum effort is made to prevent accidents. The fact is that accidents are not fate or chance but unfortunate events that can be prevented. Everyone can do something about SHE. Prevention is the responsibility of everyone. Many organizations around the world have shown that having a profitable business does not require accidents to occur. These organizations have adopted a high standard SHE culture and the zero accident vision by accepting that accidents are preventable.

• GOVERNMENT OBJECTIVES

Government, like any business, sets objectives to be effective. These objectives are formulated and published. Measures are then implemented to achieve these objectives and the private sector is motivated to assist government in its efforts.

In terms of health and safety government may aim to increase the awareness of health and safety by training more people to a certain level of competence. To motivate trade and industry to participate in the effort, government may offer various incentives, such as, for example, subsidizing part of the health and safety training fee, and other tax incentives for having a consistent annual positive performance record.

• POLICY OF THE ORGANIZATION

The policy of the organization will determine its specific objectives regarding safety, health and environmental protection procedures. In addition to these objectives the organisation may set its own standards and procedures within the law. Notably, the Act and associated regulations prescribe the minimum requirements necessary to conform to the law.

For example, the law prescribes that the minimum depth of the tread on the tyres of a vehicle must be 1mm. The organisation may decide that, because of the distances travelled and the general condition of roads used the tyres on its vehicles should be replaced when the tread depth is 2mm, both to promote safety and to avoid the cost of a breakdown due to tyre failure.

• CORPORATE GOVERNANCE

Corporate governance has been defined as 'an internal system that includes the policies, processes and people, which serves the needs of shareholders and other stakeholders, by directing and controlling management activities with good business sense, objectivity, accountability and integrity.'

Key elements of good corporate governance principles include:

- ✓ honesty,
- ✓ trust and integrity,
- ✓ openness,
- ✓ performance orientation,
- ✓ responsibility and accountability,
- ✓ mutual respect, and
- ✓ commitment to the organisation.

Commonly accepted principles of corporate governance concerning health and safety in the workplace include:

- ✓ Interests of stakeholders, including workers: Organisations should recognise that they have legal and other obligations to all legitimate stakeholders.
- ✓ Integrity and ethical behaviour: Ethical and responsible decision making is a necessary element in risk management and avoiding lawsuits. Organisations should develop a code of conduct for their directors and executives that promotes ethical and responsible decision making.
- ✓ Disclosure and transparency: Organisations should clarify and make publicly known the roles and responsibilities everyone employed by the organisation and establish a level of accountability.

Compliance with health and safety legislation and regulations should form part of good corporate governance and not be the sole objective of OHS interventions.

• REVIEW QUESTIONS

1. Why should companies comply with health and safety legislation and regulations?
2. What is your view when it is claimed that accidents happen by chance?
3. How does health and safety save money?
4. Why would you encourage your organisation to have a SHE program?
5. How would you use the latest technology to improve SHE performance at your organisation?

• REFERENCES

Badria, Adel., Boudreau-Trudel, Bryan., Souissid, Ahmed, Saadeddine, Souissi. 2018. Occupational health and safety in the industry 4.0 era: A cause for major concern? *Safety Science.* 109, 403–411

BCG. 2020. *Man, Machine and Industry 4.0.* https://www.bcg.com/en-za/industries/engineered-products-infrastructure/man-machine-industry-4.0. Accessed July 25, 2020

Department of Labour (1993). Occupational Health and Safety Act No: 85 of 1993.

Heightsandsafety.co.za. (n.d.). Health and Safety Course. [online] Available at: http://www.heightsandsafety.co.za/health-and-safety-course.asp. Accessed July 25, 2020

Health and Safety Authority. (n.d.). Safety Statement and Risk Assessment. [online] Available at: https://www.hsa.ie/eng/Topics/Managing_Health_and_Safety/Safety_Statement_and_Risk_Assessment/. Accessed July 25, 2020

Kedproductivity. 2020. *Guiding Principles of Business economics and the avoidance of loss.* https://www.kedproductivity.com/articles/guiding-principle-business-economics-avoidance-loss. Accessed July 25, 2020

Kupper, Danial, (n.d.). *Embracing Industry 4.0 and Rediscovering Growth.* [online] Bcg.com. Available at: https://www.bcg.com/en-mideast/capabilities/operations/embracing-industry-4.0-rediscovering-growth.aspx. Accessed July 25, 2020

Lee, John, Cameron, Ian, and Hassall, Maureen. 2019. Improving process safety: What roles for digitalization and Industry 4.0? *Process Safety and Environmental Protection,* 132, 325-339

Olivier, Belinda. (n.d.). Corporate Governance. [online] Financialmarketsjournal.co.za. Available at: http://financialmarketsjournal.co.za/oldsite/8thedition/printedarticles/corporategovernance.htm. Accessed July 25, 2020

PwC. 2015. Industry 4.0: Opportunities and challenges of the industrial internet, Strategy, www.strategyand.pwc.com/Google Scholar. Accessed July 25, 2020

Schwertner, Krasimira., Zlateva, Plamena. and Velev, Dimiter. (2018). Digital technologies of industry 4.0 in management of natural disasters. *Proceedings of the 2nd International Conference on E-commerce, E-Business and E-Government - ICEEG '18.*

Staysafe. 2020. *History of workplace health and safety.* https://staysafeapp.com/blog/2019/12/09/history-workplace-health-safety/.Accessed July 25, 2020

The BCF Group. 2020. *History of health and safety in the workplace.* https://www.thebcfgroup.co.uk/health-and-safety-pages/history-health-safety-workplace.php. Accessed July 25, 2020

ThoughtCo. 2020. *The Invention of the wheel and wheeled vehicles.* https://www.thoughtco.com/wheeled-vehicles-history-practical-human-use-171870. Accessed July 25, 2020

CHAPTER 2

GENERAL HEALTH AND SAFETY PRINCIPLES

General health and safety principles can only be applied effectively if the following are also in place, namely

1. **COMMITTED AND INVOLVED TOP OR SENIOR MANAGEMENT** evidenced by:

   - ✓ visible, active commitment and involvement from the top or senior management of the organization
   - ✓ establishing effective ‘downward’ communication systems and management structures
   - ✓ integration of good health and safety management with business decisions.

2. **WORKER ENGAGEMENT AND INVOLVEMENT** evidenced by:

   - ✓ engaging the entire workforce in the promotion and achievement of safe and healthy conditions
   - ✓ effective 'upward' communication
   - ✓ providing high quality relevant training

THEORIES

• INTRODUCTION

Over the years there has been a paradigm shift from haphazardly managing health and safety on a reactive post-incident approach to also include other proactive disciplines, such as Managing of Environment, Safety and Health (MESH), or the Safety, Health and Environmental (SHE) approach including the Construction Regulations.

A SHE programme adds value to any business and is merely an extension of the normal management function with greater emphasis on increasing profits and minimising losses. An effective, well-managed programme should contribute to, and is a tool for improved safety, health, quality, productivity and cost control.

This chapter introduces several theories, concepts, principles and terminology of health and safety to inform the reader about work-related health and safety concerns, how to identify potential problems such as hazards and to participate in the SHE management programme of the organization.

There are many theories relative to how accidents are caused. Despite these theories, accidents have continued unabated. These theories have generally focused on the worker being the primary cause of accidents – a basic tenet of the behavioural safety approach espoused by Krause and Hidley and others and reliant on the psychological theory of behaviourism. The goal of behaviour-based health and safety is to change the behaviour of workers from ‘at risk’ behaviours to ‘healthy and safe’ behaviours. Skinner in his seminal work argued that Antecedents serve as triggers to observable behaviour. Consequences either enforce or discourage repetition of behaviours. According to Skinner rewards or positive consequences are supposed to keep workers healthy and safe whereas punishment or censure or negative consequences are supposed to discourage unhealthy and unsafe behaviours. Arguably, emphasis on individual failures results in depending on short-term solutions rather than identifying and dealing with more fundamental management or organisational problems. The proposed remedy typically targets a specific event or operative or worker, resulting in no effort to uncover the underlying cause or causes of the accident. Further, any changes merely at the direct level are highly unlikely to deliver the degree of change that is necessary, nor would the resulting potential improvement be sustained.

Some of the better-known older theories include, for example, the following

- Accident Proneness Theory;
- Goals-Freedom-Alertness Theory;
- Adjustment Stress Theory;
- Chain of events (Domino theory); and
- Distractions Theory.

Examples of the more recent accident causation theories include the following such as, namely

- Reason’s Framework for Accident Causation; and
- Constraint-Response Theory.

Accidents are preventable and should be regarded as failures of management or the management system. Put another way, accidents are caused by how the work system is managed. None of the theories comprehensively address these issues. However, in line with modern accident theory and thinking, the aim of organisations should be to shift the emphasis from errors on the part of the worker to the management and organisational errors that cause poor health and safety performance at the operational level.

• KEY CONCEPTS/DEFINITIONS

Accident - is an event or sequence of events that results in unintended harm in the form of injury, disease, death or property damage and possibly environmental pollution or degradation

Hazard – an unsafe condition or activity that, if left uncontrolled, can eventually contribute to an accident

Lost time accident - A work injury, including any occupational disease, illness or fatality, which arises out of and in the course of employment and which renders the injured worker or contractor unable to perform his/her regular/normal work on one or more full days or shifts (other than the day or shift on which the injury occurred), including Saturdays, Sundays and public holidays. (Normal work refers to any work where a person can perform his/her normal duties without restriction).

Risk – individual or combined assessments of probability of loss and potential of loss

Safety – the control of recognized hazards to obtain an acceptable level of risk; freedom from unacceptable risk of physical injury or of damage to the health of people, either directly, or indirectly as a result of damage to property or to the environment

• ACCIDENT VERSUS INCIDENT

The poor health and safety performance of most industries and the general lack of focus on the hazardous nature of activities is widely recognized and acknowledged. In discussing prevention, it is important to clarify the terms "accident" and "incident." The term "accident" implies that the event was not expected, foreseen or intended sometimes resulting from negligence that results in injury, loss or damage). Similarly, it is an unpleasant event that happens unexpectedly without advance planning and apparent cause and results in injury or damage.

By referring to the term "accident", an impression is created that it just happened, and that it was unexpected and unpreventable. It supports a mindset that a particular outcome was unavoidable. As a result, personal responsibility, and accountability are seemingly removed and can potentially create a means to avoid rectifying the situation, dodge scrutinising current methods, and yield to pressure to change.

On the other hand, an "incident" is defined as a minor happening, an event or an occurrence, whether predictable or not, that takes place because of something else. Generally, incident prevention commences with the acceptance that all injuries are preventable if the trigger events that led to the injury are avoided. Therefore, it is preferable to reduce the focus on accidents in favour of increasing the focus on incidents to reinforce the principle and culture of preventability. Further, this focus on incidents can potentially promote a better overall health and safety culture.

• RELATIONSHIP BETWEEN ATTITUDE, BEHAVIOUR AND CONDITIONS (ABC)

Herbert Heinrich analysed 75,000 industrial incidents and concluded that 90% of these incidents were preventable, with 88% of them due to unsafe acts and 10% to dangerous physical or mechanical conditions. The cause of the remaining 2% was not determinable. Therefore, he held that to improve health and safety the focus should be on changing the behaviour of the worker.

SHE practitioners are still directing most of their efforts at only 10% of the health and safety problem, namely the unsafe physical or mechanical conditions. If the inspection reports of the health and safety representatives are reviewed, they will notice this unfortunate trend.

It is argued by in particular proponents of behaviour-based safety that the basic or root problem should be identified and addressed, and 80% time allocated to 88% percent of the problem which involves people, and 20% of the time to the 10% problem – the conditions and work environment created by management and workers themselves. This focus has to date not yielded the desired overall improvement of SHE performance as it tends to ignore upstream causes and responds instead to trigger events which are downstream occurrences.

Consider, for example, the A, B, C of health and safety:

“A” - attitudes

Attitudes play an important role in human behaviour and may adversely affect a worker who would normally act in a manner that is safe without any threat to health or the environment. Attitudes develop over time and are not easily changed. Industry has recognised the effect that attitudes can have on, for example, production, morale, turnover, absenteeism, and health and safety. The measurement, development and changing of attitudes constitute a major problem for human resources practitioners and occupational psychologists and is of utmost importance for health and safety practitioners.

"B" - behaviour

There is a strong relationship between the attitude and behaviour of a person. Given that people differ, there are individual differences, and within this framework there are certain common patterns shared by all. Human behaviour in the work environment is manifested in, for example, individual differences in motivation, emotions, attitudes, conflicting motives, and job satisfaction. The management styles of supervisors influence the behaviour of workers in the work environment. If meaningful health and safety training is provided to workers with responsibilities and the appropriate accountability, a positive change in behaviour is likely to become evident.

"C" - conditions

If SHE standards are included at the beginning of a construction or installation project, for example, hazardous physical and mechanical aspects will be addressed from the conceptual phase to the decommissioning phase of the project at the end of its lifespan.

During a project the following phases are generally followed, namely:

- ✓ Conceptual phase;
- ✓ Design phase;
- ✓ Construction phase;
- ✓ Commissioning phase;
- ✓ Modification phase; and
- ✓ Decommissioning phase.

If the project is well planned and all the stakeholders are involved, the risks should be minimised, and hazards identified. Typical hazards include, inter alia:

- ✓ Mechanical hazards;
- ✓ Electrical hazards;
- ✓ Structural hazards;
- ✓ Chemical hazards (stresses);
- ✓ Physical hazards (stresses);
- ✓ Ergonomical hazards (stresses); and
- ✓ Biological hazards (stresses).

Because certain processes are hazardous, the necessary engineering, administrative and personal health and safety equipment (PPE) controls should be implemented using the hierarchy of controls with PPE being a measure of last resort.

It has been held, in particular, by behaviour-based health and safety practitioners that the ABC concept, if well planned, implemented and controlled, could be a tool for improved health, safety, environment quality, productivity and cost control. However, it does not cover all aspects such as management commitment and involvement and the multi-causal approach to accident, injury and near-miss (incident) prevention. It does not shift focus to fixing the system instead of fixing the blame for accidents.

• ACCIDENT OR INCIDENT CAUSATION

There is no agreement on the causes of accidents and many theories and models have been developed over the years. A theory or model of accident causation is required to examine the direct causes and the circumstances preceding the accident. These theories or models have been classified in several ways. For example, accident models could be classified as

- Sequential or event-based which describe accidents as the result of sequences of discrete events that occur in a specific temporal order;
- Epidemiological which describe accidents in analogy with a disease, namely the outcome of a combination of manifest and latent factors that happen to exist together in space and time; or

- Systemic which consider the performance of the system as a whole and describe accidents as emergent phenomena, or complex and interconnected network of events that exist coincidentally in a specific time and space, which are normal or natural in the sense of being something that must be expected.

Accidents and incidents could be classified in terms of accident causation models such as domino theory and multiple causation and human error theories such as behaviour models, human factor models and Ferrel theory which is shown in the following figure.

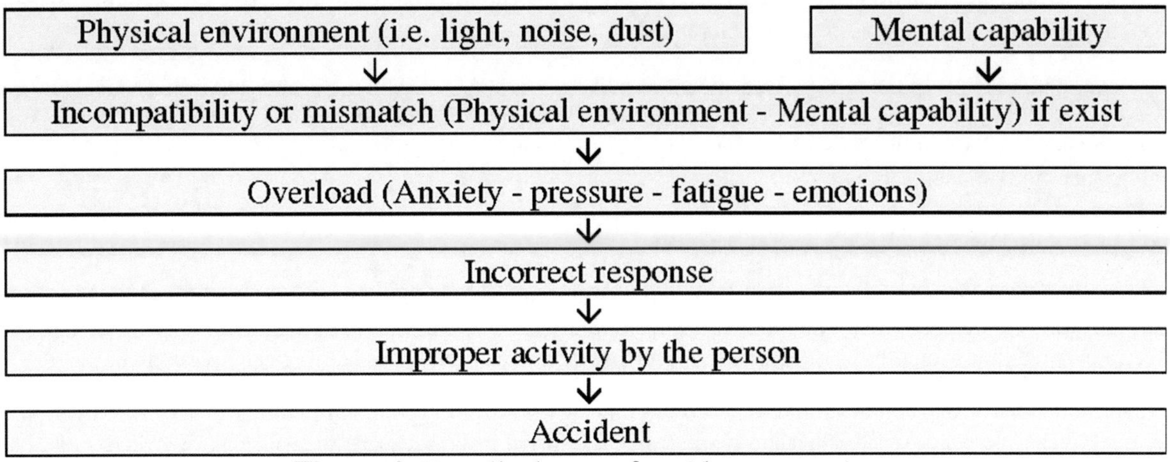

Figure 4: Ferrel's theory of accidents causation

No model currently provides an understanding of the causes of accidents sufficient to prevent the kinds of accidents that continue to occur and refer to

- behaviour models where the blame falls mainly on the human unsafe characteristic only; and
- human factors models where human error is the main cause of accidents but the design of the workplace and tasks that do not consider human limitations also contribute.

It is also argued that a systems view or model would focus on the activity level by looking at the production system comprised of interacting variables that generated risks and shape work behaviours. This type of approach suggests a reduction of task unpredictability and improvement of error management capabilities.

It is believed that the effective use of accident causation models or theories can identify the sources of accidents and ultimately reduce or eliminate accidents. Prevention efforts could then be directed at the root causes of accidents and not at the symptoms, resulting in improved accident prevention approaches. The theories provide theoretical frameworks to guide the retrospective analysis of accidents and the prospective identification of hazards. Typically, these theories have mostly focused on the worker being the primary cause of accidents – a basic tenet of the behavioural safety approach. The process of accident investigation typically involves the writing of a brief narrative description of what occurred with some information about the accident being recorded in a number of code categories - a somewhat backward-looking or retrospective approach designed to determine the facts surrounding accidents and identifying root causes and contributing factors possibly leading to them. The findings of these investigations inevitably tend to point the finger at failures or causes on the part of workers. The remedy targets a specific event or operative such that unfortunately no effort is made to uncover the underlying cause of the accident.

Accidents are preventable and should be regarded as failures of management or failure to effectively manage the work system. Few of the theories comprehensively address this view, largely because of the intrinsic litigation aspect of accident causation, which invited a retrospective approach to identify the guilty party or 'culprit' and distribute punitive measures. However, in line with the modern accident causation theories that examined both proximal accident and distal causal factors the aim of organisations should be to shift the emphasis from errors on the part of the individual to the management and organisational errors that caused poor health and safety performance in the first place. Modern theories represent a systemic approach to identify the underlying causes of accidents, taking into account decisions and actions upstream of the accident event. This forward-looking approach seeks to identify and eliminate causes and mechanisms of failure. Typically, accidents involved any combination of system design and technical flaws, operational or workforce failings, compromised organizational behaviours and management shortcomings, and possible deficient regulatory oversight.

There are many accident causation theories. It is beyond the scope of this text to discuss every theory and those that are actually discussed are not in detail either. It is important to note that as captured in the OSH Act input from both management and workers as partners is necessary if the overall SHE performance of the organization is to be improved.

• ACCIDENT PRONENESS THEORY

This common theory focuses on personal factors related to accident causation and is based on the assumption that when many persons are placed in similar working conditions some would be more likely than others to sustain an injury suggesting that accidents are not randomly distributed. There are two views, namely an older and newer view.

In terms of the older view, injuries happen to people who have a genetic predisposition to being injured. This view suggests that certain individuals have inherent characteristics that predispose them to a greater probability of being involved in accidents. This view is no longer favoured.

On the other hand, in terms of the new view, accident proneness is associated with the propensity of individuals by way of a personality trait to take risks or to take chances. Accident proneness might then change over time since risk taking is not a permanent or fixed trait. This view is more positive for health and safety, given that behaviour can be altered, through proper motivational techniques.

• GOAL-FREEDOM-ALERTNESS THEORY

This theory, which was first suggested in 1957 by Kerr, suggests that accidents are the result of unsafe behavior resulting from an unrewarding psychological climate that does not contribute to a high level of mental alertness. Accidents are therefore attributed to low-quality work behaviour occurring in an unrewarding psychological environment.

The theory holds that management should let a worker have a well-defined goal and give that worker the freedom to pursue that goal. The result will be that the worker focuses on the task that leads to that goal. Workers must have opportunities to participate in raising problems and solving them. Such participation would lead to alertness habits that would in turn promote high quality production, healthy and safe behaviours, and fewer accidents.

According to this theory, managers, and supervisors should be trained to make the work more rewarding for their workers. This intervention can be through managerial techniques, including participative management, clear work assignments, positive reinforcement, and goal setting.

• ADJUSTMENT-STRESS THEORY

A second theory, the adjustment-stress theory, was also developed by Kerr to explain accident causation. This theory states that health and safety performance is compromised by a climate or temporary conditions that impose stress which diverts the attention of workers. The adjustment-stress theory was developed to complement the goals-freedom-alertness theory, in that the goals-freedom-alertness theory states that the workers will be safe in a fulfilling or positive work environment, and the adjustment-stress theory spells out the conditions under which a worker will not be safe.

Any complications or negative stresses imposed on an individual either by the internal environment such as, for example, fatigue; lack of sleep; psychological stresses such as worry or personal problems; or by the external environment such as noise; temperature; or excessive physical strain will increase accident occurrence. If the worker cannot adjust to the stress – the attention of the worker is diverted during working hours - the chance of injury is escalated.

According to the theory, factors that divert attention and increase the probability of an accident may either be brought to the job or generated on the job.

Factors generated on the job

Practices and policies of managers are often the source of on-the-site job stress and are the easiest for management to minimise and control (Hinze, 2006). Such mental diversions and stress may arise from, *inter alia*:

- Unrealistic demands placed on workers which include pressure to keep costs below some level that may not be realistic;
- Pressure to meet an unrealistic deadline;
- Pressures related to threats to job security;
- Poor relationship with supervisor;
- Poor relationship with team members;
- Poor co-ordination;
- Excessive rework;
- Cost overruns;
- Long overtime commitments;
- Crowded work areas; and
- Workers being asked to work in an obviously unsafe environment (Hinze, 2006).

Factors brought to the job:

Stress on workers may be brought to the job and may affect the ability of the worker to remain healthy and safe in the workplace. These factors may be difficult for management to detect. Such examples include:

- Family: poor relationship with the spouse; death of a loved one; illness of family member; and financial problems; and
- Personal: substance abuse; poor health; fatigue and lack of sleep.

• DISTRACTIONS THEORY

The Distractions Theory of accident causation has a narrow site-based focus and correlates productivity with risk. It predicts that workers under stress working in an environment that is hazardous have a greater probability of being involved in accidents. The hazards distract workers while they are performing their work tasks and increase the probability of accidents and decrease their productivity. Health and safety and productivity are arguably compatible when hazards are removed resulting in a decrease in distraction. There are two types of distractions, namely

- Jobsite hazards
- Unsafe physical conditions

Productivity is compromised when the distraction due to hazards is high. Obviously, when the hazard level is high, it is expected that the worker will have a high level of hazard awareness. The attention paid to hazards and the avoidance of being injured become a distraction for the worker. Pressure to get the task done may cause the worker to be distracted and to ignore the hazard, resulting in an injury or accident.

- **Mental Distractions**

Workers will try to focus on the work to be done but may be distracted by worries caused by personal problems such as, for example, financial worries, divorce, and family disputes or job-related concerns such as, for example, tight deadlines and unduly close supervision. Even positive events can be contributing factors to mental distractions, such as parties, celebrations, paydays and going on holiday. Failure to be able to focus on the work increases the likelihood of being injured.

- **Mental Stresses**

Holmes and Rahe, two early psychological researchers examined the relationship between various life stresses and the onset of illness. Their work followed that of others who had provided evidence that stressful life events played a strong causative role in the onset of diseases.

Stress was evoked by both negative and positive events. Holmes and Rahe developed a scale of stressful events in which each type of event was given a value. The values of all the events that applied to a person were summed up. The higher the final score the higher the stressful condition and the higher the likelihood of contracting an illness. The scoring might be similarly applied to the distractions theory. The Holmes and Rahe Scale lists various potentially stressful events and the scores assigned to each.

Social Readjustment Rating Scale of Holmes and Rahe

Life Event	Value
Death of spouse	100
Divorce	73
Marital separation	65
Jail term	63
Death of close family member	63
Personal injury or illness	53
Marriage	50
Fired at work	47
Marital reconciliation	45
Retirement	45
Change in health of family member	44
Pregnancy	40
Sex difficulties	39
Gain of new family member	39
Business readjustment	39

Life Event	Value
Change in financial state	38
Death of close friend	37
Change to different line of work	36
Change in number of arguments with spouse	35
A large mortgage	31
Foreclosure of mortgage or loan	30
Change in responsibilities at work	29
Son or daughter leaving home	29
Trouble with in-laws	29
Outstanding personal achievement	28
Wife begins or stops work	26
Begin or end school	26
Change in living conditions	25
Revision of personal habits	24
Trouble with boss	23
Change in work hours or conditions	20
Change in residence	20
Change in schools	20
Change in recreation	19
Change in religious activities	19
Change in social activities	18
A Small Mortgage	17
Change in sleeping habits	16
Change in number of family get-togethers	15
Change in eating habits	15
Vacation	13
Christmas	12
Minor violations of the law	11

• H W HEINRICH'S CHAIN OF EVENTS (DOMINO THEORY)

H.W. Heinrich was an engineer working for a USA insurance company who conducted an analysis of 75,000 accident reports and attempted to develop a model that explains casual factors of an accidents that lead to injuries. He theorized that an accident is caused by one of five factors in a sequence that resulted in an accident, namely:

- *Ancestry and social environment.* This refers to the social background and influence of family that leads someone directly to…
- *Fault of person.* This refers to personal attributes such as greed, stupidity, and recklessness could be in the nature of a person or learned which leads directly to…
- *Unsafe act and/or mechanical or physical hazards or conditions.* This refers to the behaviour of workers that is unsafe and the unsafe conditions that may be because of the presence of mechanical or physical hazards which leads directly to…
- *Accidents.* These refers to events such as falls, collisions, etc., which leads directly to…
- *Injuries.* These could be sprains, lacerations, sprains, fractures and broken or dislocated bones etc.

Accidents are therefore characterized as occurrences that result from a sequence of events that occur in a specific order in that each event is consequently followed by another event. The theory is classified as a sequential accident or event-based model and was compared to dominos falling, that is if one event occurred, it would trigger the cause of the next event which will cause another event and so on. This theory portrays the concept of how accidents occur.

The last event is the accident itself and is invariably caused by a mechanical or physical hazard or the unsafe act of a person. The chain of events theory proposes that there are many causes to an accident rather than just a single cause. Heinrich's 'dominos' were as shown in the figure.

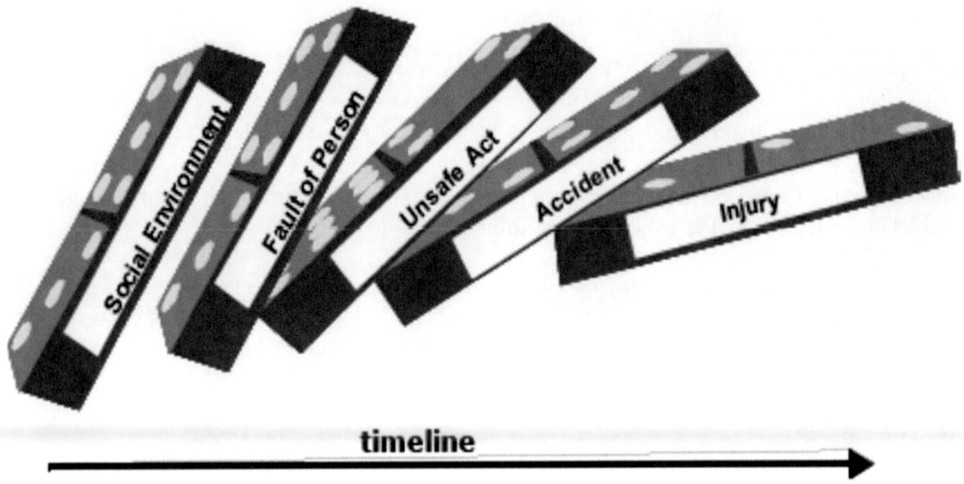

A more complete explanation of each of the factors is shown in the following table:

Incident Factors	Explanation of Factors
1. Ancestry and social environment	Recklessness, stubbornness, and other undesirable traits of character may be passed along through inheritance. Environment may develop undesirable traits of character or may interfere with education. Both inheritance and environment cause faults in person.
2. Fault in person	Inherent or acquired faults in person, such as recklessness, violent temper, nervousness, excitability, inconsiderateness, ignorance of safe practice, etc., constitute proximate reasons for committing unsafe acts or for the existence of mechanical or physical hazards.
3. Unsafe act and/or mechanical or physical hazard	Unsafe performance of persons, such as standing under suspended loads, starting machinery without warning, horseplay, and removal of save-guards, and mechanical or physical hazards, such as unguarded gears, unguarded point of operation, absence of rail guards, and insufficient light, result directly in incidents
4. Incidents	Events such as people falling, and striking of people by flying objects, are typical incidents that cause injury
5. Injury	Fractures, and lacerations, are injuries that result directly from incidents

In Heinrich's theory:

- ✓ A potential injury or the final domino occurs only because of an incident;
- ✓ An incident occurs only because of a personal or mechanical hazard;
- ✓ Personal or mechanical hazards exist only because of the faults of people; and
- ✓ Faults of people are inherited or acquired because of their social environment in which they were born, bred, and educated.

The domino theory asserts that, if there is an injury or damage to property, all five factors are involved. If the fifth domino falls, it is because the first domino fell, causing the others to fall in turn. If one of the factors in the sequence leading to an incident can be removed, the loss can be prevented. Elimination of an unsafe or unhealthy act or an unsafe or unhealthy condition makes the action of the preceding factors ineffective.

According to Heinrich 88% of accidents were caused by unsafe acts, and only 10% were caused by unsafe conditions. The domino theory held, therefore, that unsafe behaviour of people was the fundamental reason behind accidents. This theory is supported by many people who take the view that accidental injury is the consequence of the injured behaving in a way that disregards accepted behavioural norms. Many programs aimed at the prevention of occupational accidents still focus efforts to reduce or eliminate the immediate behavioural causes of accidents by attempting to change the behaviour of workers. Heinrich's Domino Theory has faced much criticism for focusing too extensively on the immediate causes or circumstances surrounding accidents and the role of human behaviour in accidents when it now widely recognised that unsafe acts and conditions have systemic and organisational causes. Accidents should not merely be attributed to either an unsafe condition or an unsafe act because most accidents are the result of a complex interaction of multiple contributing factors, causes and sub-causes. Therefore, the accident causation model developed by Heinrich as well as other behaviour models is over-simplified.

• AN UPDATED DOMINO THEORY

The Chain of Events (Domino) theory of Heinrich was later revised by Bird in 1974 by integrating the role of management in the accident causation process. Accident causation and the resultant losses such has injuries, fatalities and even the damage to property and wastage of the organisation's assets can be traced back to management negligence and lack of control. The revised sequence sets out the dominoes as follows:

- *Lack of control by management of root causes*: This leads to…
- *Basic contributing causes of accidents*: This refers to personal factors, such as lack of motivation, fatigue or insufficient health and safety knowledge. It could also be job factors like unrealistic work schedules and inadequate or insufficient resources, for example which lead to…
- *Immediate or direct causes of accidents*: This refers to sub-standard practices, conditions, hazards or errors, which lead to…
- *The accident*: This refers to falls, collisions, tripping, etc. which results in…
- *The loss or injury*: This can be minor, serious, or catastrophic and could lead to fatalities, loss of limb, damage to property, delays in project, increased costs etc.

In general, there are preceding series of events to every accident. If any one event or activity had been altered or done differently in the series of events, then the accident would not have possibly not occurred. Put another way, any one event in a chain that did not occur would break the chain or sequence and prevent the accidents. Each link in the chain may involve different people and be associated with different activities and only one link in the chain need to break to prevent the accident from occurring.

However, accidents are too often blamed on the event that occurs most proximal in time or place to the accident. This last event is typically some action performed by the worker. It is, therefore, common for many injuries to be blamed on worker behaviour. It is important to consider the other events in the chain, not just the final action of the worker who becomes injured. To simply blame the injured workers for many of their injuries and only identify corrective focused on eliminating the last event in the chain, is basically ignoring the roles that other parties play in influencing worker behaviour.

A study was conducted at the Naval Surface Weapons Centre in Silver Spring, Maryland in the U.S.A. where the concept of the chain of events theory was applied. The study focused on the role of management in accidents and was based on the premise that all accidents and hazards are indicators of management failures. The hypothesis for the study was "In the investigation of any accident, there can always be found some degree of management involvement or activity that might in some way have prevented the accident". Therefore, it was arbitrarily assumed that responsibility of causes of every accident will be for managements account, as well as for the existence of every hazard." Accidents were investigated to seek out the means that might have been available to management to change the course of events to avert the accident.

This Navy study included accident investigations with clear cut causes. Managerial opportunities to either minimise or avoid the hazards began to emerge. When each accident or hazard had been carefully evaluated, nearly every investigation revealed a failure of management. Once management recognised its failures, it was able to correct and prevent them. The effect of the corrective measures proved to be positive from an economic point of view, health improved at the facility, and there was higher productivity. In the long term, this acceptance of responsibility by management resulted in overall reductions in the costs of the Navy operations.

The traditional focus on direct or immediate causes of accidents in accident prevention involves treating the symptoms of the problem rather than their underlying root causes. Accidents will continue to occur if long term solutions do not focus on management control and errors of management which is the first domino in the sequence of events.

• DR WILLIAM HADDON'S TEN STRATEGIES

Dr William Haddon as part of his efforts to understand accident causation suggested the following **TEN** strategies designed either to suppress conditions that produce incidents or to enhance conditions that retard incidents from occurring, namely:

1. Prevent the creation of energy/hazards
   Examples: prevent the generation of thermal, kinetic, or electrical energy, or ionising radiation. Prevent the production of plutonium, thalidomide, or LSD.

2. Reduce the amount of energy/hazards
   Examples: reduce speed of vehicles, lead content of paint, mining of asbestos, make less beverage alcohol — a hazard and its consequences, such as drunken driving.
3. Prevent the release of energy/hazards that do exist
   Examples: bolting or timbering mine roofs, impounding nuclear wastes, the undermining of cliffs and the encapsulation of Polychlorinated Biphenyl (PCBs).
4. Modify the rate or spatial distribution of release of the energy/hazard from its source
   Examples: brakes, shutoff valves, reactor control rods. The third strategy is the limiting case of such release reduction but is identified separately because in the real world it commonly involves substantially different circumstances and tactics.
5. Separate energy/hazards in time or space from that which is to be protected
   Examples: isolation of persons with communicable diseases, walkways over and around hazard, evacuation, the phasing of pedestrian and vehicular traffic, whether in a work area or on a city street, the banning of vehicles carrying explosives from areas where they and their cargoes are not needed. This strategy, in a sense is also concerned with rate-of-release modification, has as its hallmark the elimination of intersections of energy and susceptible structure — a common and important approach.
6. Separate energy/hazards and that which is to be protected by interposition of a material barrier
   Examples: the use of electrical and thermal insulation, shoes, safety glasses, barrier cream, safety helmets, shield, lead aprons, childproof poison container closures, vehicle air bags.
7. Modify relevant basic qualities of energy/hazard
   Examples: to modify appropriately the contact surface, sub-surface or basic structure, as in eliminating, rounding, and softening corners, edges and points with which people can, and therefore sooner or later do, come in contact. This strategy is widely overlooked in architecture, resulting in many minor and serious injuries. Altering a pharmacological agent to reduce side-effect. Using a smaller grid to manufacture machine guarding, thus preventing fingers from reaching dangerous nip points.
8. Strengthen the structure that might be damaged by the energy/hazard
   Examples: immunization/vaccines such as those for polio, yellow fever and smallpox, are analogous strategies in the close parallel set to reduce losses form infectious diseases. Making structures more fire resistant, making motor vehicles more crash resistant.
9. To begin to counter the damage caused by environmental hazard/energy
   This strategy in loss reduction applies to the damage not prevented by measures under the eight preceding measures – to move rapidly in the detection and evaluation of damage that has occurred or is occurring, and to counter its continuation and extension – this is where the emergency preparedness organisation should be mobilised. Examples: extricating trapped miners, sprinkler and other suppresser responses, fire doors, fire alarms, emergency medical care, emergency transport.
10. Stabilise, repair, and rehabilitate
    It encompasses all the measures between the emergency period following the damaging energy exchange and the final stabilisation of the process after appropriate intermediate and long-term reparative and rehabilitative measures. Examples: post-traumatic cosmetic surgery, physical rehabilitation of injured after serious injuries, rebuilding after fire losses or bomb blasts.

• MULTIPLE CAUSATION MODEL

The theory of multi causality states that there are more than one event or even multiple events that precede every accident. Contributing causes also play a role in accident causation as they combine randomly resulting in an accident. Using as an example of a common accident scenario of a man falling off a defective stepladder; present accident investigation approaches will most likely identify it only one act (climbing a defective ladder) and/or one condition (a defective ladder). Corrective action for this accident would be to discard the defective ladder to prevent it from being used again. If the Domino Theory was used, this would be the typical response to the investigation but by using Multiple Causation Theory, the surrounding factors such as lack of supervision would have been evident.

Accident investigation processes or methods that attempt to uncover only the unsafe act or the condition is only successful with mitigating the symptoms of the problem, as they may be the "immediate/proximate causes" and not "root causes". To have permanent improvement and to prevent the same accidents from reoccurring the root causes must be identified. Root causes often relate to the first event in the chain of events which relate to errors of management and may be due to, for example, management policies, procedures, supervision, effectiveness, and training.

• REASON'S EPIDEMIOLOGICAL MODEL

This theory of accident causation spans the entire accident sequence from organisational to individual levels. It follows the more current approach by seeking causal factors that were removed in both time and space from the onset of the incident. Previously, accident investigations tended to highlight the role of the frontline operator as the most obvious and immediate instigator of the accident. Accidents were particularly prone to such interpretations – incidents regularly occurred to individuals acting alone. Reason's theory incorporated an organisational level analysis that considered the input of management and decision-makers – Swiss cheese model of accident causation. Most health and safety systems have several layers or plates. Due to human error these plates had holes which allowed for the potential incident to pass through the plate or layer of the health and safety system. Often, the next plate or layer intercepted the potential incident and prevented its occurrence. However, when holes lined up the incident was more likely to occur. The model divided failures into two categories, namely active and latent failures as shown in the following figure.

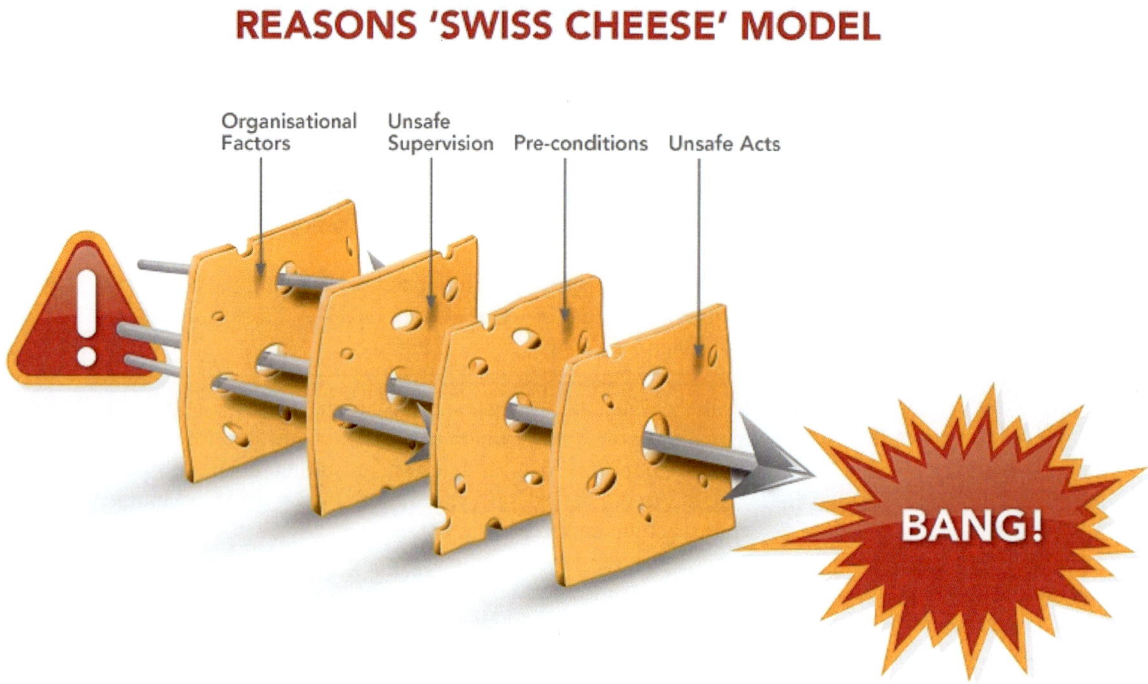

Reason's Swiss cheese model (Source:http://www.ohsa.com.au/wp-content/uploads/2017/08/OHSA-ICAM-Lead-Incident-Investigation-Course-Reasons-Swiss-Cheese-Model.jpg)

Active failures were those errors and violations that had an immediate adverse effect. These were generally associated with the activities of 'front-line' operators" which corresponded to, for example, the activities such as contacting overhead power lines or failure to wear Personal Protective Equipment (PPE).

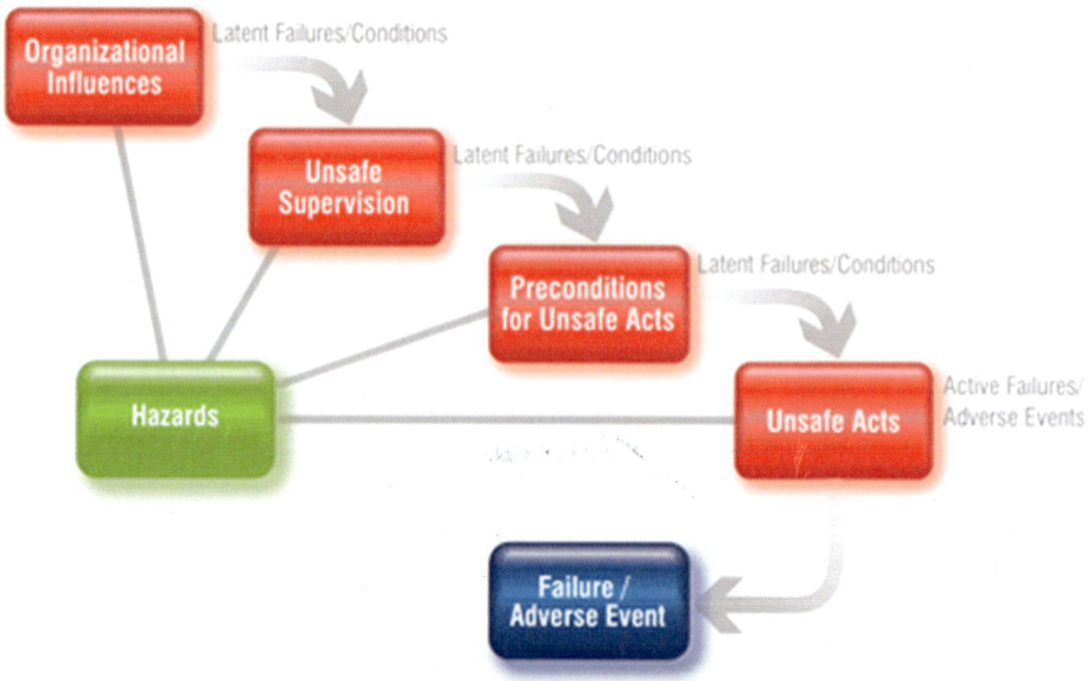

Latent and active failures
(Source: http://media.noria.com/sites/magazine_images/201003/Exponent_Figure2.gif)

Many health and safety interventions targeted the general operative such as, for example, programmes to encourage the wearing of hard hats or instituting health check campaigns. However, there are an infinite number of unsafe acts that could precipitate accidents in the workplace. Most of these are unforeseeable and occasionally quite bizarre. Therefore, attempts to reduce the number of unsafe acts would only have limited value. It would be more beneficial to aim at the level of latent failures.

Latent failures corresponded to errors at the head office and site management levels. These are decisions or actions, the damaging consequences of which may lie dormant for a long time, only becoming evident when they combine with local triggering factors. Their defining feature is that they are present in the system well before the onset of a recognisable accident sequence. Research by the Health and Safety Executive (HSE) in the UK found that many of the preconditions of unsafe behaviour originated in poor management decisions or an organisational culture in which health and safety goals might have been considered subordinate to production goals. This research also noted that violations were known to occur more frequently in situations where responsibilities were ambiguous or ill-defined, training poor and time pressures high.

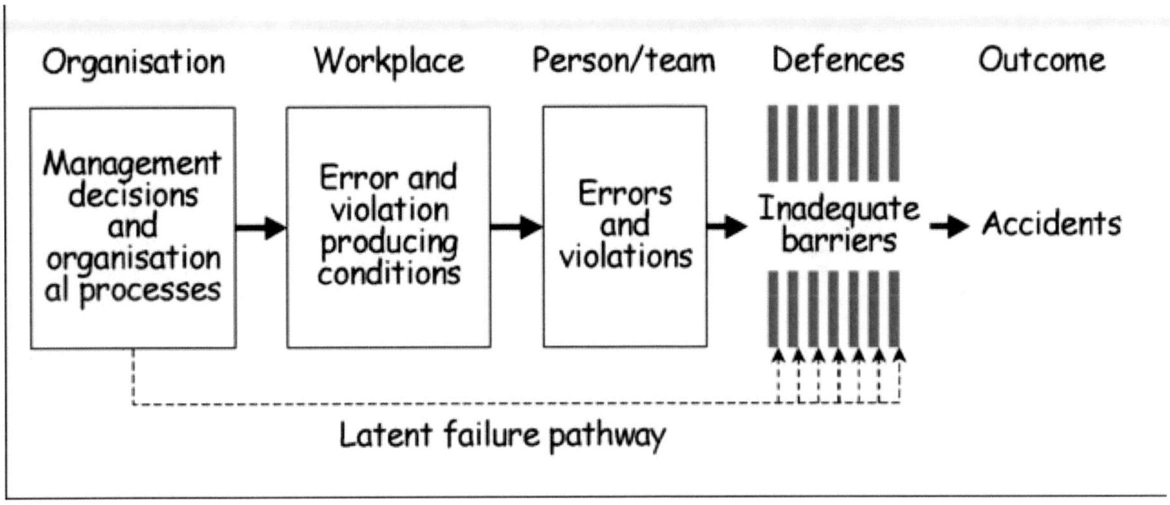

A model of organisational accident causation showing impact of latent failures
(Source: Hollnagel, 1999)

• CONSTRAINT-RESPONSE THEORY FOR CONSTRUCTION ACCIDENTS

Suraji, Duff and Peckitt of University of Manchester Institute of Science and Technology (UMIST) and the Health and Safety Executive (HSE) in the United Kingdom, developed a causal model specific to construction accidents. They cited Reason's model as a theoretical description but noted the lack of specific detail necessary to guide practical investigation and intervention. The effective mitigation of causal factors required better knowledge of which factors were most influential, who may reasonably be expected to control those factors and how such control may most effectively be achieved.

Like Reason's model, the Constraint-Response Model extended the scope of the accident causation process to include management and organisational aspects. It mapped the potential contributions of all participants within the project organisation to the accident causation process. The model classified two types of factors, namely distal and proximal, equivalent to latent and active failures in Reason's configuration as shown in the following diagram.

Distal factors were at management level and included

- Project conception constraints;
- Project design constraints;
- Project management constraints;
- Construction management constraints;
- Sub-contractor constraints; and
- Operative constraint.

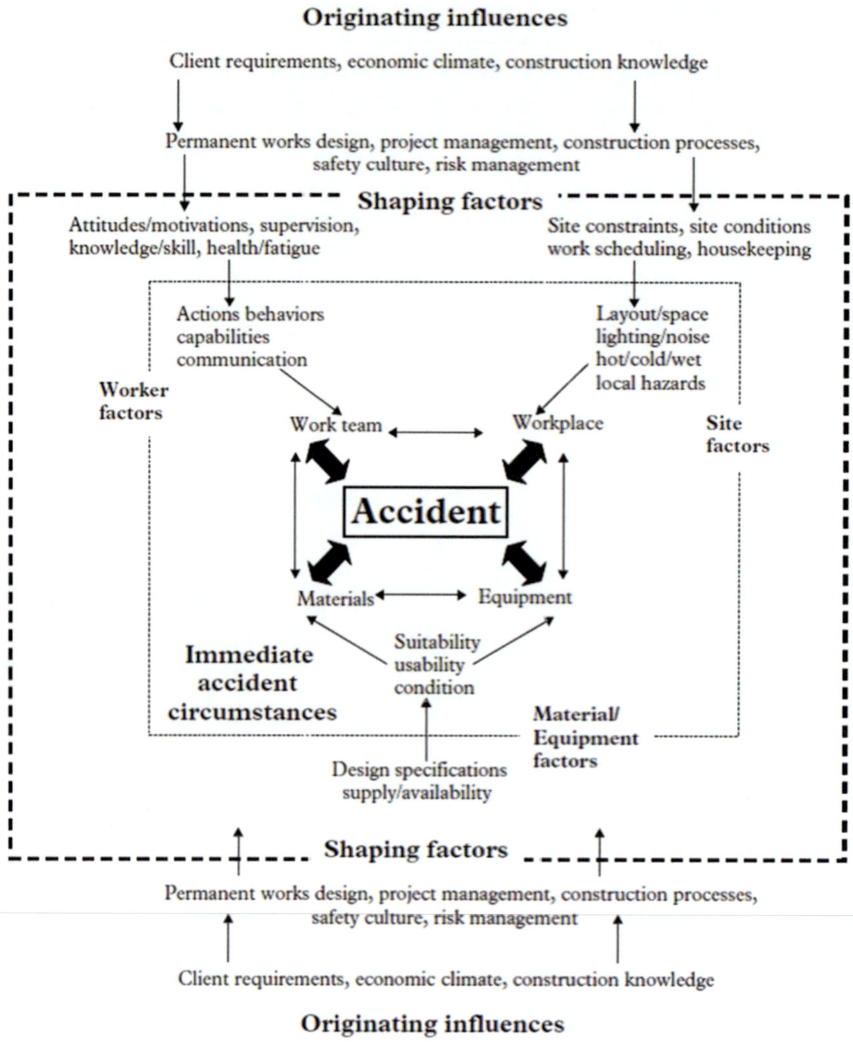

Loughborough Construction Accident Causation model (Haslam *et al.*, 2005)

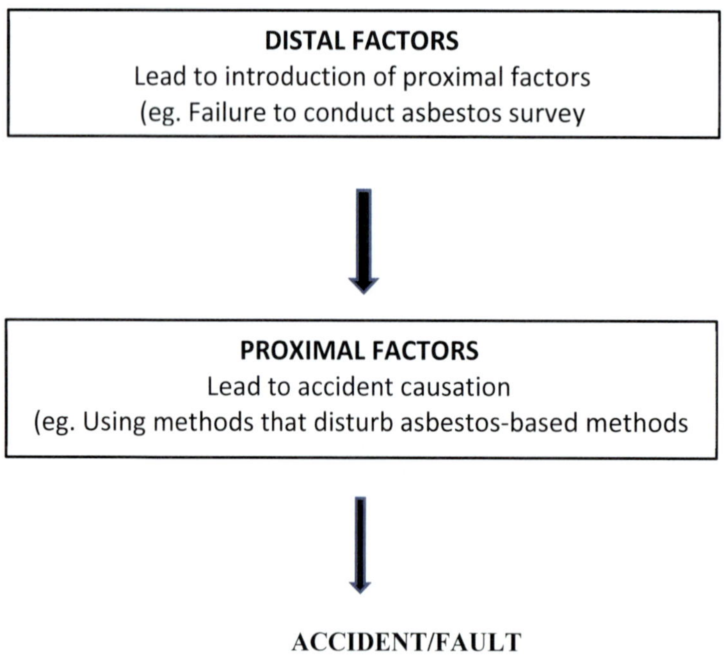

These precipitated potentially unsafe responses by clients, designers, the project team of clients, principal contractors, sub-contractors, and operatives. These constraints and responses included the influence of management and organisational factors; environmental factors such as economic, legislative, political, and social; and individual participant factors.

Proximal factors, on the other hand, operated at site management and injured person levels and included

- Inappropriate construction planning;
- Inappropriate construction control;
- Inappropriate site condition;
- Inappropriate construction operation; and
- Inappropriate operative action.

These could be identified as immediate causes of construction accidents. The theory was premised on each participant in the construction process experiencing constraints on their particular activity which could be, for example, a client typically facing difficulties in obtaining funding. The responses of the client would in turn create a set of constraints for other participants in the process. A client may have decided to reduce the project budget resulting in the project design team being constrained by an inadequate design budget. The design team may have reduced design resources for the contract. Additionally, the project management team may, in turn, then be constrained and affected by the late delivery of the design details, and so on throughout the project chain. This unfortunate sequence of constraints and responses would ultimately create situations where the proximal factors were manifest such as 'unsuitable existing topography' (inappropriate site conditions) or inadequate supervision of operative work (inappropriate construction control).

- HUMAN ERROR THEORIES

Human error theories were best captured in behaviour models and human factor models. Behaviour models identified the workers as being the main causes of accidents.

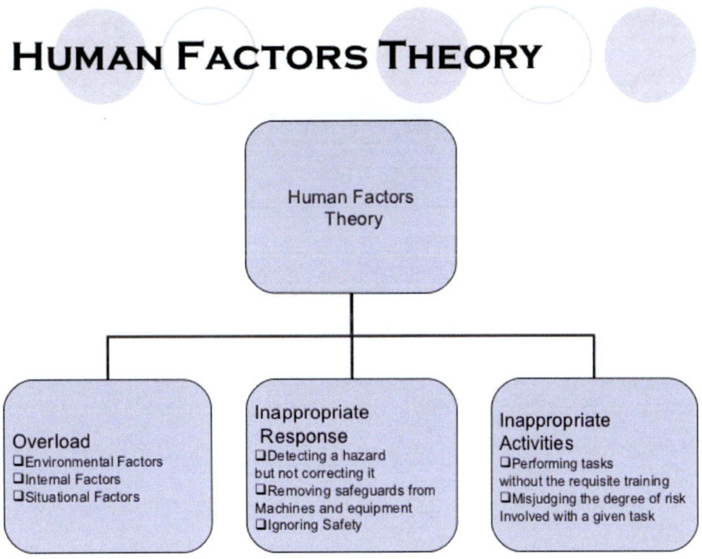

(Source: https://image.slidesharecdn.com/theoriesofaccidentcausation-100315052440-phpapp01/95/theories-of-accident-causation-11-728.jpg?cb=1268630759)

This approach studied the tendency of humans to make errors under various situations and environmental conditions, with the blame mostly falling only on the human or unsafe characteristics. Human error was defined as any one set of human actions that exceeded some limit of acceptability. Many researchers have devoted much time and effort to defining and categorising human error.

Like the behavioural models, the human factors approach held that human error was the main cause of accidents. However, the blame not only fell on the human unsafe characteristics, but also on the design of the workplace and tasks that failed to consider human limitations that could have harmful effects. In other words, the overall objective of the human factors approach was to arrive at better designed tasks, tools, and workplaces, while acknowledging the limitations of human physical and psychological capabilities.

• SYSTEMIC ACCIDENT MODELS

A system in the context of construction is not static but dynamic and continually changing to achieve scheduled milestones while reacting to a changing environment. Systemic accident models describe an accident as a complex and interconnected network of events and consider the performance of the entire system. They view accidents as emergent phenomena that arise from interactions among system components such as people, social and organizational structures, engineering activities, and physical system component). They do not happen in isolation but are part of a wider system of causal factors or spheres of influence. It is this foundation that forms the basis of the systems approach to understanding and managing risk, in which all events or activities are considered within the context of a wider system of influence, such that no one event can be viewed in isolation from its surrounding context. These causal factors may be human, technical, and environmental.

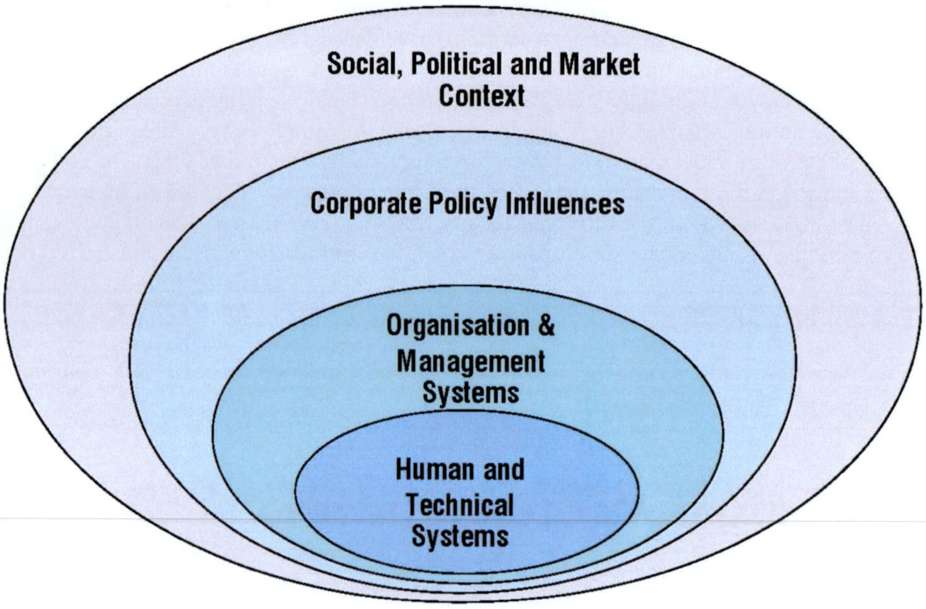

A graphic representation of examples of spheres of influence

Examples of these spheres of influence that influence the performance of people and hardware in a hazardous situation are shown in the diagram. These influences can either be internal to the organization or external to the organization. For example,

- Direct level influences would include, inter alia, the following, namely
  - Level of competence;
  - Motivation or morale;
  - Fatigue or level of alertness;
  - Equipment operability;
  - Situational awareness or risk perception;
  - Inspection and maintenance;
  - Level of information or advice available; and
  - Workplace environment.
- Organizational level influences would include, inter alia, the following, namely
  - Training;
  - Planning;
  - Management and supervision;
  - Health and safety culture; and
  - Inspection and maintenance.
- Policy level influences would include, inter alia, the following, namely
  - Contracting policy;
  - Organisation's culture;
  - Organizational structure;
  - Health and safety management;
  - Labour relations; and
  - Profitability.

- Environmental level influences would include, inter alia, the following, namely
  - Political;
  - Regulatory framework;
  - Market conditions; and
  - Societal.

The latent conditions that lead to accidents can arise from poor organisational practices, regulatory policies and societal influences that have played a significant part in creating the corporate culture that manifests within organisations. From this context, unacceptable health and safety acts within the workplace are seen as the consequences of organisational failures rather than the causes of accidents. While the immediate causes are important in understanding the circumstances surrounding an accident, it is the underlying aspects that make the most significant contribution to the causes of accidents. Due to the persistence and lack of overt effect of the more remote factors, it is these aspects that need to be targeted if health and safety is to be assured over time.

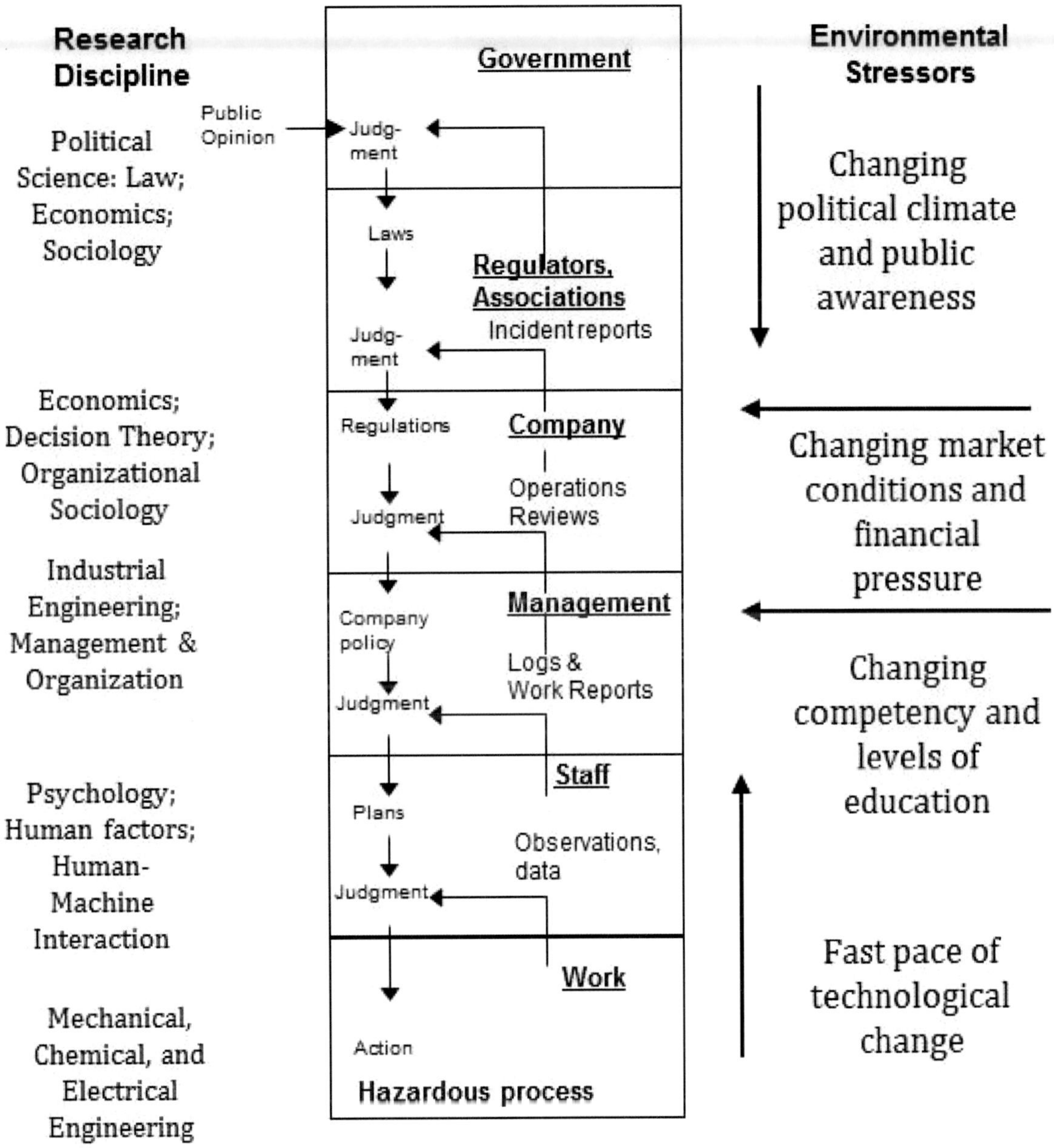

Rasmussen's (1997) sociotechnical model
(Source: https://www.researchgate.net/publication/315771917_Sociotechnical_systems_theory_and_the_regulation_of_safety_in_high-risk_industries_-_White_paper)

Rasmussen's hierarchical socio-technical framework endeavours to model the dynamics of complex sociotechnical systems. Health and safety depend on the control of work processes in the context of pressures and constraints in the operational environment. Several hierarchical levels are involved.

The organizational and management decisions made at higher levels should transmit down the hierarchy whereas processes at lower levels should propagate up the hierarchy to form a closed loop feedback system. This model suggests that accidents are caused by decisions and actions by decision makers at all levels and not just by workers at the process control level. The various levels are also impacted by external disruptive forces which are unpredictable, rapidly changing and influence the entire system.

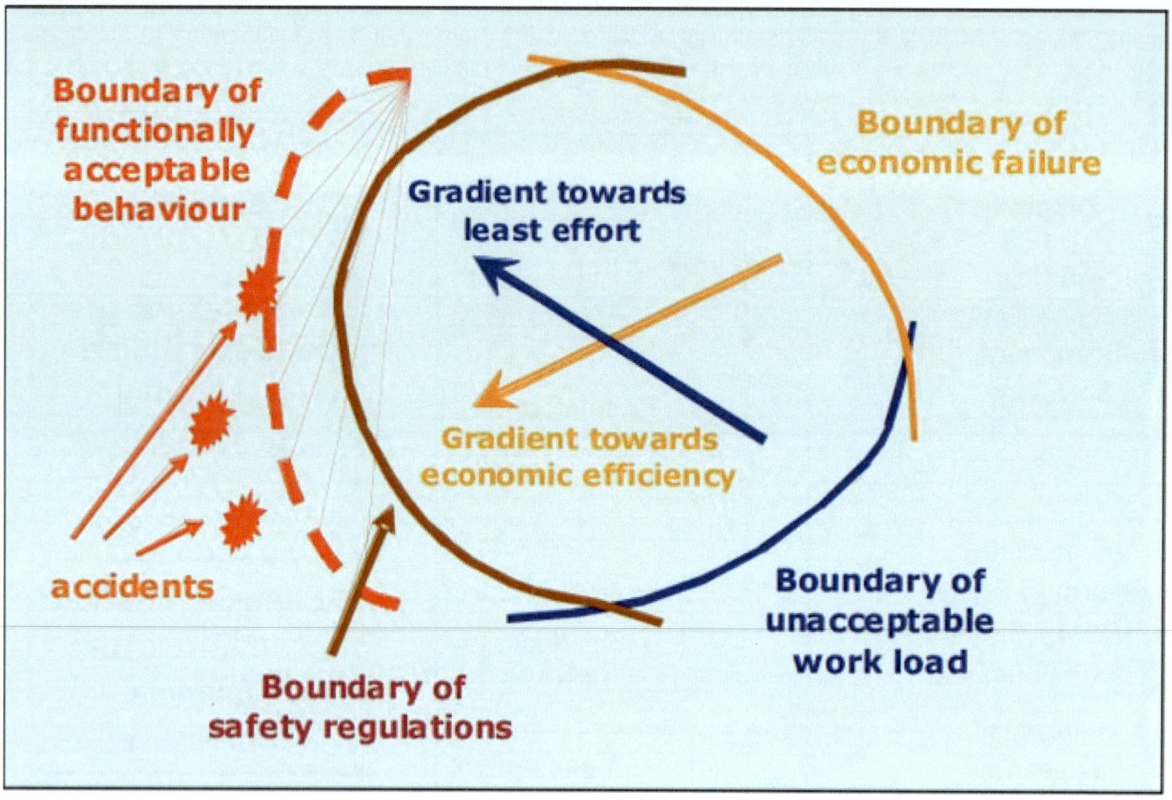

Boundaries of Safe Operation (Rasmussen 1997)
(Source: https://www.researchgate.net/publication/228683461_A_review_of_accident_modelling_approaches_for_complex_socio-technical_systems/)

The behaviour of workers is context dependent and is shaped by the dynamic conditions in the work environment. To be able to analyse the health and safety of a work domain or environment it is necessary to identify the boundaries of health and safety operations and the dynamic forces that may cause the system to migrate towards or cross these boundaries. It is evident as shown in the figure that dynamic forces can cause a system to modify its structure and behaviour over time. The safe space of performance within which workers and management can navigate freely is contained within three boundaries, namely individual unacceptable workload; financial and economic constraints; and health and safety regulations and procedures. As people are subjected to dynamic forces, they adapt their behaviour and cross the boundaries that might lead to an accident if control is lost at the relevant boundary. Therefore, there is a need to identify the boundaries of health and safety operations, make these boundaries visible to everyone and create opportunities to control behaviour at the boundaries.

The technical, human, and organizational factors in complex socio-technical systems are considered in the Systems-Theoretic Accident Model and Processes (STAMP). Accidents are caused by inappropriate or inadequate control or enforcement of health and safety-related constraints rather than by discrete events and are therefore viewed as a control problem. Health and safety are therefore managed by a control structure embedded in an adaptive socio-technical system.

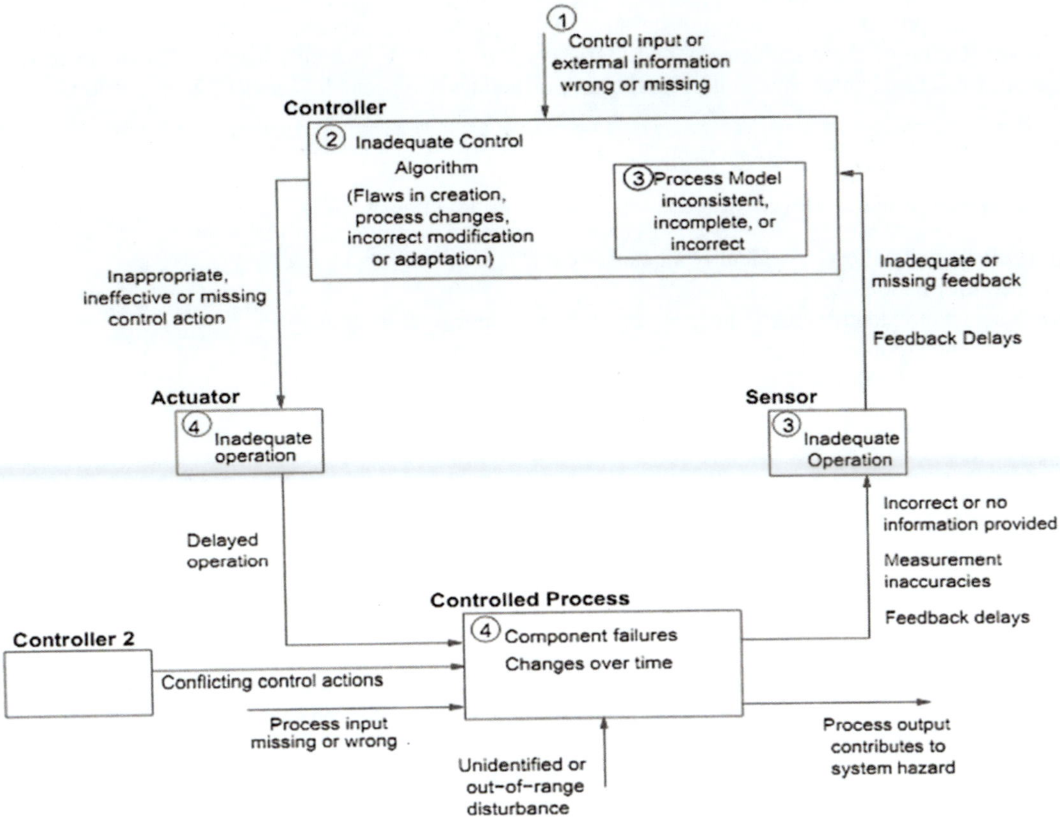

Flaw classification

The causes of accidents can be understood in terms of flaws in control loops between components during design, development, and implementation operations. These flaws have been classified as shown in the following list.

Flaw 1: Inadequate Enforcement of Constraints (Control Actions)

- Unidentified hazards;
- Inappropriate, ineffective, or missing control actions for identified hazards
  - Design of control algorithm (process) does not enforce constraints
  - Process models inconsistent, incomplete, or incorrect (lack of linkup)
  - Inadequate co-ordination among contractors and decision makers (boundary and overlap areas)

Flaw 2: Inadequate execution of control action

- Communication flaw
- Inadequate actuator operation
- Time lag

Flaw 3: Inadequate or missing feedback

- Not provided in system design
- Time lag
- Inadequate sensor operation (incorrect or no information provided)

Each control loop is investigated at every level of the socio-technical control structure and its contribution to inappropriate health and safety behaviour evaluated. As a result of this approach attention is not only on the proximate events of accidents but also on the influence of other organizational factors and role players.

• INCIDENT INVESTIGATION

Every year companies all over the world suffer huge losses because of both minor and major incidents which occur within their organizations. These incidents cause occupational injuries, illness, environmental damage or damage to property, materials and equipment. To survive and expand in these competitive world markets, losses need to be prevented as far as possible or be reduced to a minimum. Each worker has a responsibility to reduce any risk of exposure to health and safety hazards within their organization.

Minor daily incidents provide early warning of possible major incidents, which could have serious consequences if not addressed. Investigation of these incidents is necessary to prevent future incidents, namely "Learn from the past to improve the future". This chapter only addresses the basic principles and needs for incident investigation.

• AIMS OF INVESTIGATIONS

The aims of an effective incident investigation and prevention programme are to:

1. Objectively and collectively identify all causes contributing to all incidents.
2. Collect information.
3. Analyse data to identify contributory factors and their inter-relationships.
4. Determine the root or real causes of events contributing to the incident.
5. Formulate an appropriate action plan and prevent further losses.
6. Create ownership among workers by making them part of the investigation process – worker engagement.
7. Demonstrate that the organisation is concerned about health and safety – involvement of top management.

Together these steps combine to prevent the re-occurrence of incidents from the identified root problems or root causes. For any programme to be successful, all steps must be carried out efficiently. Incident investigation must focus on fact finding rather than fault finding. The objective is to put preventative measures and controls in place – the intention is not to blame or 'find the culprit'. If ignored, today's "small incident" may become tomorrow's fatality.

• THE INVESTIGATION PROCESS

Incident investigation should be a systematic and pre-planned exercise. To ensure this, a systematic process must be followed.

Phase 1 — situation analysis

The investigator(s) should arrive at the scene as soon as possible after the incident has taken place and establish what should have happened and what happened.

Phase 2 — collect information

Ask questions; observe the general scene and activities and view documentation.

Phase 3 — analyse data

Root causes must be established from the data analysis.

Phase 4 — remedial action plans

Final conclusions and recommendations may be drafted and submitted to the appropriate persons. When evaluated and agreed upon, action plans may be implemented.

Phase 5 — audit

The audit procedures may be assessed and reviewed to ensure that they are appropriate and effectively address the risk factors agreed on as acceptable.

This process should not be a linear process — namely, as a once-off step-by-step process. The phases in the process must be interactive and iterative.

NOTE: All phases must be incorporated during every investigation. The depth and scope of this coverage will depend on the nature of factors identified during active investigation.

No incident investigation can be considered complete or effective if the following actions are not physically concluded. Assumptions are neither scientific nor acceptable – aspects of each of the following are necessary for the formulation of recommendations:

- ✓ The incident investigators visited the scene of the incident.
- ✓ Direct and immediate causes have been identified.
- ✓ The root causes were identified.
- ✓ Positive and practical recommendations have been made to prevent or minimize the recurrence of similar incidents.
- ✓ Specific persons have been made responsible for the implementation of remedial action according to target dates.

- ✓ Follow-up actions have been carried out to audit the remedial action after implementation in order to determine if such actions are effective and practical.

Root causes

Fundamentally, the root causes of incidents are the absence or neglect of or deficiencies in management systems. These normally include, but are not necessarily limited to:

- ✓ A weakness in the integrated implementation or administration of the SHE management system.
- ✓ Lack of total participation or commitment to the maintenance of the programme.
- ✓ Highlight areas which require more management control.
- ✓ Highlight areas where competency-based training, monitoring, and mentoring are required.
- ✓ The inter-linking of several factors namely "a small stone left untouched will not result in a landslide".

Remedial action plans

Recommendations should always address improvements to the system which are aimed at addressing the incident's root causes. It can therefore be said that recommendations address failures in the overall SHE management system. These recommendations are the outcome of an investigation. Recommendations are therefore the tool used to accomplish the main objectives and rationale of incident investigation, namely, to prevent the recurrence of the same or similar incidents and minimise the losses incurred.

Criteria for successful recommendations

Do the recommendations:

- ✓ Take into consideration the options for reducing the frequency, minimizing personnel exposure and reducing the consequences of one or more root causes?
- ✓ State the intended purpose of all recommendations clearly and precisely?
- ✓ Require practical, feasible and achievable interventions?
- ✓ Improve inherent health and safety by reducing the intensity and severity of incidents, providing for change or substitution?
- ✓ Reduce inventories if and where appropriate?
- ✓ Increase the number of events necessary to generate an incident?
- ✓ Correlate with findings derived from the data analysis and facts?
- ✓ Illustrate compatibility and cohesion with the SHE system?
- ✓ Reduce the possibility of actual implementation giving rise to other new risks?

Incident costing

The outcome of an incident can often be determined in costs, whether loss in lives or financial loss. Costs are usually classified as direct (insured) and indirect (uninsured) costs. Some examples of each include:

Direct/Insured costs

- o Medical expenses

The cost of medical treatment for work-related injuries/illnesses is covered in terms of the requirements laid down by state compensation laws. Medical expenses that qualify to be reported to the Compensation Commissioner or insurer include any emergency room or hospital expenses, medical or other treatment costs, and rehabilitation expenses.

- o Worker's Compensation, Disability Payments and Death Benefits

Compensation paid to disabled worker or next of kin from compensation insurance fund provider and required by law.

Indirect/Uninsured costs

- o Cost of training new worker(s)

If a replacement worker is employed, the cost of training that worker including materials and supervisory time needs to be considered as a cost of the incident. For example, if due to a lack of experience or knowledge a replacement worker is only half as productive as the injured worker, half of the salary of the replacement worker should be apportioned to incident costs.

- Cost of investigation time

Time spent by management and other workers on investigating an incident, processing documentation, forms, claims, and attending hearings, for example, should be included in the direct costs of the incident.

- Uninsured medical costs

These costs are usually the medical service, or first aid provided by the plant or organisation's nurse or doctor. This service should be considered as a reduction in the direct medical care costs of incidents, which will result in the overall lowering of expenses for this operation.

- Legal expenses

Any cost to the organisation or the individual because of legal expenses including witnesses' traveling expenses, attorney, and court costs as well as fines imposed.

- Image of the organization

Incidents may be covered by the news media or may be discussed by workers with the public or trade union organization. This negative publicity can affect customers' perception of the quality of products and may influence the public and shareholders' perception of the integrity of the organisation. This may result in poor image of the organisation, raise questions on ethical codes of practice and lead to a loss in sales, customer confidence and product loyalty.

- Inspections

A government-initiated inspection or prohibition order may affect production levels as well as result in fines or penalties against the organisation.

- Time lost from work by injured

This includes any wages paid to the injured worker that are not covered by Compensation Commissioner.

- Time lost from work by uninjured persons

Time lost when workers stop work to assist, view or discuss the incident. This would also include lost production time used to repair equipment, clean up the incident site, as well as any additional time that future tasks may take without the input of the injured worker.

- Damage to tools, equipment, products, material, buildings, and production

Incidents may result in damage to equipment, property, or buildings. These also include any costs to move or reorganize material or equipment, special repairs to equipment or replacement parts, costs of lost or damaged parts as well as costs of delayed or stopped production. For example, an incident could damage a pallet of finished goods, which may have to be scrapped or reworked. A stop in production could impact on the completion of a product run, possibly resulting in missing a shipping date, incurring a penalty, or even losing a specific order or an entire customer account.

- Loss of sales

Damaged product needs to be re-worked to fill orders. Client may place order elsewhere.

- Loss of client base

Clients may remain loyal to new supplier who was selected as a result of inability to deliver on time.

- Loss of income/profit

Incidents may result in a short- or long-term reduction in efficiency. For example, a process may take longer if workers are afraid to use the equipment, or if they must take more time to check equipment set-up before running the line.

✓ Extra cost of overtime

The difference between normal wages and overtime wages needed to counteract lost production time, as well as any other services necessary to make up for the interruption of production.

✓ Lost time of supervisors

Time supervisors and management spend away from normal responsibilities as a result of an incident.

✓ Decreased output

Incidents result in decreased production if a worker cannot complete the scheduled work quota if a replacement worker is necessary, or if there is any unscheduled downtime.

• REVIEW QUESTIONS

1. How effective are worker-centred accident causation theories or models in improving health and safety performance?
2. Why are indirect/uninsured costs of accidents more difficult to quantify?
3. What should be the aim and objectives of an effective incident investigation process?
4. Why should an incident investigation process be systematic and comprehensive?
5. What is the difference between an accident and an incident?

• REFERENCES

Abdelhamid, Tariq. S. and Everett, John G. 2000. Identifying root causes of construction accidents, *Journal of Construction Engineering Management,* 126 (1), 52-60

Anonymous. Nd. Loose notes on SHE management

Arboleda, Carlos. A. and Abraham, Dulcy. M. 2004. Fatalities in Trenching Operations – Analysis using Models of Accident Causation, *Journal of Construction Engineering and Management,* March/April, 273-280

Asanka, Waga, Arachchige and Ranasinghe, Malik, (2015). Study on the impact of accidents on construction Projects, 6th International Conference on Structural Engineering and Construction Management, Kandy, Sri Lanka, 11Th - 13th December 2015

Austin, John. 2006. An Introduction to Behavior-Based Safety, *Stone, Sand & Gravel Review,* January/February, 39

Bird, Frank, E. 1974. *Management guide to loss control,* Institute Press, Atlanta

Dahlbäck, Olof. 1991. Accident-proneness and risk-taking, *Personality and Individual Differences,* 12 (1), 79-85

Durnwald, D.C. (2012). Incident Investigating and Reporting Procedure. BP Oil Refinery, Major Incident Announcement, 3

Farmer, E. and Chambers, E. G. 1929. A Study of Personal Qualities in Accident Proneness and Proficiency. *I. H. R. B. Report No. 55*

Farmer, E. and Chambers, E. G. 1939. A Study of Accident Proneness Among Motor Drivers. *I. H. R. B. Report No. 84*

Farmer, E. and Chambers, E. G. 1933. Tests for Accident Proneness. *I. H. R. B. Report No. 68*

Geller, E. Scott. 2001a. *Working safe: How to help people actively care for health and safety (2nd ed.).* Lewis Publishers, Inc.

Geller, E. Scott. 2001b. *Keys to Behavior-based Safety.* Government Institutes, Rockville

Gibb, Alistair, Lingard, Helen, Behm, Michael and Cooke, Tracy, (2014), 'Construction accident causality: learning from different countries and differing consequences', Construction Management and Economics, vol. 32, no. 5, pp. 446-459.

Goh, Yang. Miang., Brown, Helen. and Spickett, Jeff. (2010), Applying systems thinking concepts in the analysis of major incidents and safety culture, Safety Science, 48, 302-309

Hämäläinen, Paivi., Takala, Jukka., and Saarela, Kaija, Leena, (2006), Global estimates of occupational accidents, Saf. Sci., vol. 44, pp. 137–156

HaSPA (Health and Safety Professionals Alliance). (2012). The Core Body of Knowledge for Generalist OHS Professionals. Tullamarine, VIC. Safety Institute of Australia.

Harvey, Michael. (1984). *"Theories of accident causation".* [online] Archive.org. Available at: https://archive.org/stream/theoriesofaccide00harv/theoriesofaccide00harv_djvu.txt.

Haupt, Theo. and Harinarain, Nishani. A multi-causal approach to construction accident investigation. In Haupt, T.C. and Fester, F.C. (Eds.) The Built Environment 7, Proceedings of the 7th Built Environment Conference, Cape Town, July 28-30, pp. 272-286, ISBN 978-0-620-55984-3

Health and Safety Authority. 2002. *Fatal Accidents in the Irish Construction Industry 1991-2001: A Survey of Contributory Factors.* Marie Dalton, HSA, internal document

Health and Safety Executive. 2001. *Improving Health and Safety in Construction. Phase 1: Data Collection, Review and Structuring.* Contract Research Report No. 387/2001

Heinrich, Herbert, William. 1931. Industrial accident prevention: A scientific approach. New York. McGraw-Hill.

Heinrich, Herbert, William. 1959. *Industrial Accident Prevention,* McGraw-Hill, New York

Heinrich, Herbert, William, Petersen, Dan. and Roos, Nestor. 1980. *Industrial Accident Prevention,* McGraw-Hill, New York

Hinze, Jimmie. 1996. The distraction theory of accident causation, *Proceedings of the 1st International Conference on Implementation of Safety and Health on Construction Sites, CIB Working Commission W99: Safety and Health on Construction Sites,* Alves Diaz, L.M. and Coble, R.J. eds., Balkema, Rotterdam, 357-384

Hinze, Jimmie. W. 2006. *Construction Safety.* Alta Systems, Inc., Gainesville, Florida

Hollnagel, Erik. 2002. Understanding Accidents – From Root Causes to performance Variability, *IEEE 7th Human Factors Meeting, Scottsdale, Arizona*

Holmes, Thomas. and Rahe, Richard. H. 1967. The Social Readjustment Rating Scale, *Journal of Psychosomatic Research,* 11, 213

Hopkins, Andrew. (1995), Making safety work: Getting management commitment to occupational health and safety, Allen & Unwin, St Leonards

Hosseinian, Seyyed Shahab and Torghabeh, Zahra, Jabbarani, (2012), "Major theories of construction accident causation models: A literature review, International Journal of Advances in Engineering & Technology, Vol. 4, Issue 2, pp. 53-66, ISSN: 2231-1963

Howell, Gregory. A., Ballard, Glenn., Abdelhamid, Tariq. Sami. and Mitropoulos, Panagiotis. 2002. *Working near the edge: A new approach to construction safety*, In Formoso, CT., Ballard, G. (Eds) Proceedings of 10th annual conference of the International Group for Lean Construction, Gramado, Brazil, 1-15

Hudson, Patrick., 2014. *Accident Causation Models, Management and the Law,* Journal of Risk Research, Vol 7 Issue 6, page 749-764

Katsakiori, Panagiota., Sakellaropoulos, George. and Manatakis, Emmanuel., (2009), Towards an evaluation of accident investigation methods in terms of their alignment with accident causation models, Safety Science, Vol 47, p1007-1015

Kerr, Willlard. 1957. *Complementary Theories of Safety Psychology*, The Journal of Social Psychology, 45 (1), 3-9

Klumb, Petra. L. 1995. Cognitive failures and performance differences: Validation studies of a German version of the cognitive failures questionnaire, *Ergonomics,* 38 (7), 1456-1467

Krause, Thomas. R. 2000. Moving to the Second Generation in Behavior-Based Safety, *ASSE Professional Development Conference and Exposition, June 25 - 28, 2000, Orlando, Florida*, American Society of Safety Engineers

Krause, Thomas. and Hidley, J.H. 1990. *The Behavior-Based Safety Process*, Global Professional Publications, Irvine

Krause, Thomas. and Hidley, John. (2009). *Taking the lead in patient safety*. Hoboken, N.J.: John Wiley & Sons.

Lutchmiah, Trinisha. 2018. An analysis of the causes of construction accidents in South Africa: a Case Study Approach. An unpublished M.Sc. dissertation. University of Kwa-Zulu Natal

Manu, Patrick., Ankrah, Nii., Proverbs, David. and Suresh, Subashini. 2010. An approach for determining the extent of contribution of construction project features to accident causation, *Safety Science,* 48, 687-692

Marks, Eric., Teizer, Jochen. and Hinze, Jimmie. (2014). Near Miss Reporting Program to Enhance Construction Worker Safety Performance, Construction Research Congress

Marsden, Eric. (2017). *The Heinrich/Bird safety pyramid: Pioneering research has become a safety myth.* [online] Risk Engineering. Available at: https://risk-engineering.org/concept/Heinrich-Bird-accident-pyramid.

Mitropoulos, Panagiotis. Abdelhamid, Tariq. S. and Howell, George.A., 2005. Systems Model of Construction Accident Causation, *Journal of Construction Engineering and Management,* July, 816-825

OSHA (2015). *Incident Investigations: A Guide for Employers*. [online] Osha.gov. Available at: https://www.osha.gov/dte/IncInvGuide4Empl_Dec2015.pdf.

Peckitt, S.J., Glendon, A.I. and Booth, R.T. 2004. Societal Influences on Safety Culture in the Construction Industry, *Construction Safety Management Systems,* Rowlinson, S. ed., Spon Press, London, 17-54

Petersen, Dan. 1971. *Techniques of safety management,* McGraw-Hill, New York

Qureshi, Zahid. H. 2008. *A review of Accident Modelling Approaches for Complex Socio-Technical Systems*, DSTO Defence Science and Technology Organization, Edinburgh, South Australia

Rahe, Richard. H. 1968. Life-change Measurement as a Predictor of Illness, Proceedings of the Royal Society of Medicine, 61(11 Pt 1), 1124–1126

Rasmussen, Jens. 1997. Risk Management in a Dynamic Society: A modelling problem, *Safety Science,* 27 (2), 183-213

Reason, J. 1990. *The Contribution Of Latent Human Failures To The Breakdown Of Complex Systems.* Phil. Trans. R. Soc. Lond. B. 327, 475-484

Researchomatic. Nd. *James Reason's Swiss Cheese Theory*, https://www.researchomatic.com/James-Reasons-Swiss-Cheese-Theory-129350.html (Accessed July 26, 2020).

Saleh, Joseph. H. and Pendley, Cynthia. C. 2012. From Learning from accidents to teaching about accident causation and prevention: Multidisciplinary education and safety literacy for all engineering students, *Reliability Engineering and System Safety,* 99, 105-113

SAMTRAC, (2017), HSE Terminology: A Glossary. [ONLINE] Available at:http://blog.samtrac.com/hse-terminology. [Accessed 22 January 2018].

Shavell, Steven. 1979. Risk sharing and incentives in the principal and agent relationship. *The Bell Journal of Economics,* 10 (1), 55-73

Shaw, L. and Sichel, H.S. 1971. *Accident proneness: Research in the occurrence, causation, and prevention of road accidents.* Pergamon, Oxford

Smallwood, John. and Haupt, Theodore. C. 2001. "Safety and Health Team Building," In Coble, Hinze and Haupt (eds), *Construction Safety and Health Management,* Prentice-Hall, New Jersey, 115-144

Suraji, Akhmad., Duff, A. Roy. and Peckitt, Stephen. J. 2001. Development of Causal Model of Construction Accident Causation. *Journal of Construction Engineering and Management*, 127(4), 337-344

Swuste, Paul. (2008), You will only see it, if you understand it" or occupational risk prevention from a management perspective, Human Factors and Ergonomics in Manufacturing, 18, 438-453.

Taylor, Geoff., Easter, Kellie. and Hegney, Roy., (2004), Enhancing Occupational Safety and Health, Elsevier Butterworth-Heinemann.

Tjiger's Health and Safety. 2020. Health and Safety Theories. https://tijgersi.co.za/abc-theory/ Accessed July 26, 2020

Toole, T.Michael. 2002. Construction Site Safety Roles, *Journal of Construction Engineering and Management*, 128(3), 203-210

Van den Broek, Karla. and Op De Beeck, Rik. (2016). *Accidents and incidents: OSHwiki*. [online] Oshwiki.eu. Available at: https://oshwiki.eu/wiki/Accidents_and_incidents.

Vicente, Kim. J. 1999. *Cognitive work analysis: Towards safe, productive and healthy computer-based work,* Lawrence Erlbaum, Mahwah, New Jersey

Whittington, C., Livingston, A. and Lucas, D.A. 1992. *Research into Management, Organisational and Human Factors in the Construction Industry.* Health and Safety Executive Contract Research Report No. 45/1992

Wijne, Morne. (2018). *Health and Safety Theories - Tijgers i.* [online] Tijgers i. Available at: https://tijgersi.co.za/abc-theory/.

Zeller, D. 1986. Heinrich revisited, *Professional Safety,* 31 (10), 40-4

CHAPTER 3

SYSTEMATIC MANAGEMENT OF SAFETY, HEALTH AND THE ENVIRONMENT

WHY SHE MANAGEMENT IS NECESSARY

According to recent International Labour Organization (ILO) statistics on occupational safety and health, every 15 seconds, a worker dies from a work-related accident or disease. Every 15 seconds, 153 workers have a work-related accident. Every day, 6,300 people die because of occupational accidents or work-related diseases – more than 2.3 million deaths per year. About 317 million accidents occur on the job annually; many of these resulting in extended absences from work. The human cost of this daily adversity is vast, and the economic burden of poor occupational safety and health practices is estimated at 4% of global Gross Domestic Product (GDP) each year.

These are shocking statistics and pose a heavy burden for society. Implementing a strong occupational SHE management system will help organizations reduce accidents and ill health, avoid costly prosecutions, reduce insurance costs as well as creating a culture of positivity in the organization.

SHE MANAGEMENT SYSTEM ELEMENTS

An effective SHE system or programme generally includes the following elements, namely

• SHE POLICY STATEMENT

The SHE policy statement includes the following:

A written policy document that has been signed by the CEO or most senior manager in the organization. It is important that this policy document is widely displayed and contains a review date.

• KEY PERSON OR STAKEHOLDER RESPONSIBILITIES

It is also important to document the responsibilities of each key person or stakeholder in the organization. In particular, the following could be included where applicable in the organization, namely:

- ✓ Responsibilities of workers
- ✓ Responsibilities of supervisors and first-line supervisors (OHSAct, Section 8(2)(i) appointees)
- ✓ Responsibilities of senior management (OHSAct, Section 16(1) and 16(2) appointees)
- ✓ Supervision of machinery (General Machinery Regulations (GMR) 2(1))
- ✓ Shifts man appointment (GMR 4(3))
- ✓ Responsibilities of health and safety coordinators/representatives/officers
- ✓ Emergency Controller
- ✓ Fire Fighting Co-ordinator
- ✓ Fire Fighter
- ✓ Fire Team Member
- ✓ Fire Fighting Equipment Inspector
- ✓ First Aid Coordinator
- ✓ First Aider
- ✓ Incident Investigator
- ✓ Stacking Supervisor
- ✓ Ladder Inspector
- ✓ Hazardous Chemical Substance Coordinator
- ✓ Ergonomics Survey Officer
- ✓ Health and Hygiene Coordinator
- ✓ Pollution Surveys
- ✓ Work Permit Survey Officer
- ✓ Air Power Tools Inspector
- ✓ Explosive Power Tool Controller

- ✓ Explosive Power Tool Operator

• HEALTH AND SAFETY REPRESENTATIVES AND COMMITTEES

The policy should include the following, namely:

- ✓ Appointment of representatives per the GAR 6 & Section 17 & 18 of the Act
- ✓ Establishment of health and safety committees per Section 19 & 20 of the Act
- ✓ Appointment of Health & Safety Committee Members
- ✓ Appointment of Chairperson of Health & Safety Committee
- ✓ Monthly meetings

• RISK ASSESSMENT

The following risk assessments should be done as required, namely

- ✓ Baseline risk assessments
- ✓ Issue-based risk assessments
- ✓ Continuous risk assessments

• SAFE WORK OR OPERATION PROCEDURES

Safe work or operation procedures that are known by the relevant or affected workers and physically available at each workstation are critical elements of any effective SHE management policy and plan. The following are essential, namely:

- ✓ After the critical area activities have been identified, the standards for the activities must be determined to establish safe and healthy work procedures
- ✓ Written standard work or operating procedures (SWPs or SOPs) must be compiled for each critical activity
- ✓ Workers need to be properly trained to execute the work activities in accordance with the SWPs or SOPs

• WORKER ORIENTATION

Worker orientation or induction will include training about SHE awareness on topics such as, for example:

- ✓ Emergency procedures
- ✓ Nomination and introduction of first aider/s
- ✓ Ensure workers know the location of first aid stations
- ✓ SHE responsibilities of the worker
- ✓ Details about responsibilities as determined by legislation
- ✓ Information about how injuries should be reported
- ✓ Explanations of unsafe conditions and acts
- ✓ Use of personal protective equipment
- ✓ Right to refuse hazardous work
- ✓ Identifying hazards, including those outside their own work area
- ✓ Explanation of the reasons for each health and safety rule
- ✓ Specific roles and responsibilities of each worker
- ✓ Scope of the authority of workers as determined by job descriptions that includes when they have the right to stop work activities

• TRAINING

It is essential to develop training programmes regarding the following aspects:

- ✓ Knowledge and understanding of the SHE programmes, rules and procedures of the organization as well as the specific roles and responsibilities of each worker
- ✓ Systematic programme of induction or orientation and ongoing training for workers and those who may be transferred between divisions, jobs or tasks
- ✓ Training in the handling of risks, hazards and dangers, precautions to be taken and procedures to be followed especially in emergencies
- ✓ Training in hazard identification, risk assessment, mitigation and control
- ✓ Training for all persons who may manage workers, contractors, suppliers, and manufacturers
- ✓ Training of top or senior management in their important roles and responsibilities
- ✓ Training and awareness programmes for contractors, temporary/casual workers and visitors according to the level of risk that they could be exposed to

- ✓ Training in the correct inspection and reporting procedures
- ✓ Training in proper incident/accident investigation procedures
- ✓ Training in the effective monitoring of the quality and effectiveness of the SHE programme

• WORKPLACE INSPECTIONS

An effective SHE policy and management plan will include the following inspections:

- ✓ Those done by Health and Safety Representatives
- ✓ Those done by line managers and supervisors
- ✓ Where prescribed by regulations and codes of practice
- ✓ Regular, planned workplace inspections
- ✓ Frequent and routine equipment inspections
- ✓ Special inspections as required

Note that standard inspection lists could be useful tools. However, care should be taken to avoid checklist tick fever

• CERTIFICATIONS

Where applicable the following certifications may be necessary by an Approved Inspection Authority (AIA) for:

- ✓ Ventilation surveys
- ✓ Light surveys
- ✓ Dust surveys
- ✓ Noise surveys
- ✓ Lifting equipment
- ✓ Pressure vessels

• MEDICALS

Procedures are necessary for various medicals various such as, for example:

- ✓ Procedures for pre-medicals, baseline, periodic and exit medicals where prescribed by a regulation or code of practice
- ✓ Baseline and regular periodic medicals where the workplace has exposures to noise, heat, dust, chemical exposure, asbestos, lead, cannabis, and coronaviruses

• REPORTING, RECORDING, AND INVESTIGATING ACCIDENTS

Reporting procedures are necessary according to legislation such as, for example:

- ✓ Section 24 - Reportable incidents (WC.I2)
- ✓ Section 25 – Occupational deceases (WC.l1)

• EMERGENCY PROCEDURES

Emergency procedures in a SHE policy and management plan are to include, inter alia:

- ✓ Building construction details
- ✓ Access to and egress from all areas of the facility or plant
- ✓ Emergency exits that are clearly sign posted
- ✓ Emergency lighting
- ✓ Fire precautions and procedures
- ✓ Firefighting appliances that are sufficient, appropriate, and properly maintained
- ✓ Fire drills and alarm checks to ensure that the procedures are familiar to all workers in the plant
- ✓ Proper storage of flammable/combustible materials such as gas cylinders, fuel oils, wood, and paper waste
- ✓ Identification of fire / explosion risk areas and instructions for isolating power, fuel, and gas should this become necessary
- ✓ Evacuation procedures and responsibility for roll calls
- ✓ Worker training in procedures and general fire safety practices
- ✓ Other emergencies and similar procedures to those for accidents and fire are required for emergencies to cover events such as, for example:
  - o Gas leaks
  - o Explosions

- Pressure vessel rupture
- Building collapse
- Chemical leaks and spillage
- Bomb threats

• FIRST AID

The SHE policy and management plan should also include:

- ✓ Training of first aiders
- ✓ Provision of sufficient stocked first aid boxes
- ✓ Inspection and control of first aid boxes to ensure levels and adequacy of contents

• SHE CULTURE

To be effective the SHE policy must be aimed at developing a positive and good SHE culture in the organisation through the management systems, policies, working procedures, and communication flow of the organization and their levels of effectiveness should be measured by appropriate audits of the SHE management systems. It is possible to address behavioural aspects in terms of how workers act in the workplace and can be measured through self-reported measures, outcome measures and peer observations. However, the focus in any SHE program should not only be on the downstream behaviours of workers but include upstream aspects of management and the management system. The psychological component must also be addressed that relates to worker norms, values, attitudes, and perceptions of SHE in the workplace which are commonly measured in the form of a SHE climate survey.

It is important to distinguish between a SHE culture and a SHE climate. A SHE culture is the way in which SHE is managed in the workplace. It is the combination of beliefs, perceptions, and attitudes that workers have toward the health and safety of workers and the overall health and safety of the work environment. It is therefore concerned about shared attitudes, values, goals, and practices. On the other hand, SHE climate is the perceived value placed on SHE in an organization at a particular point in time. These perceptions and beliefs can be influenced by the attitudes, values, opinions, and actions of other workers in an organization, and they can change over time and across different circumstances. Therefore, taking a measure of the SHE climate in the organization is indicative of the state of the SHE culture in it. If the climate is positive, then the SHE culture is probably strong. If it is negative, it is necessary to intervene to foster a stronger culture of safety among the workforce and management.

• CONTRACTOR MANAGEMENT

The way contractors that service the organization will be managed must also be included in the SHE policy and should cover the following:

- ✓ Contractor agreements
- ✓ Validity of COID registration in the form of valid letters of good standing (LOGS)
- ✓ Medical certificates of workers
- ✓ Current worker certifications and qualifications to demonstrate their competence
- ✓ Machinery and equipment checks and certifications
- ✓ Risk assessments for contractor activities

• MATERIAL AND MACHINERY HANDLING

The following aspects of material and machine handling are important for effective SHE management, namely:

- ✓ Handling of Hazardous Chemical Substances and the provisions of Material Safety Data Sheets (MSDS's)
- ✓ Handling of Hazardous Biological Agents
- ✓ Material handling rules
- ✓ Lock out and tag out procedures
- ✓ Hot work permits
- ✓ Maintenance programs
- ✓ Vehicle safety rules
- ✓ Personal protective equipment requirements
- ✓ Engineering standards and procedures
- ✓ Purchasing standards and procedures
- ✓ Preventive maintenance

- HEALTH AND SAFETY AUDIT

The following audits are necessary for effective SHE management, namely:

✓ Internal audits
✓ External audits

OHSAS 18001:2007	ISO 14001:2015	BS 8800
1,1 OHEALTH AND SAFETYMS requirements	2,1, EMS requirements	3,1 Initial status reviews
1,1,1 General requirements	2,1, 1General requirements	3,1,1 General requirements
1,1,2 OHEALTH AND SAFETY policy	2,1,2 Environmental policy	3,1,2 OHEALTH AND SAFETY policy
1,1,3 Planning	2,1,3 Planning	3,1,3 Organizing
1,1,4 Hazard identification, risk assessment and determining controls	2,1,4 Hazard identification, risk assessment and determining controls	3,1,4 Hazard identification, risk assessment and determining controls
1,1,5 Objectives and programme(s)	2,1,5 Objectives, targets and programme(s)	3,1,5 OHEALTH AND SAFETY objectives
1,1,6 Legal and other requirements	2,1,6 Legal and other requirements	3,1,6 Legal and other requirements
1,2, Implementation and operation	2,2 Implementation and operation	3,2 Implementation and operation
1,2,1 Resources, roles, responsibility, accountability and authority	2,2,1 Resources, roles, responsibility and authority	3,2,1 Resources, roles, responsibility and authority
1,2,2 Competence, training and awareness	**2,2,2 Competence, training and awareness**	**3,2,2 Competence, training and awareness**
1,2,3Communication, participation and consultation	2,2,3 Communication	3,2,3 Communication of relevant information
1,2,4 Documentation	2,2,4 Documentation	3,2,4 Documentation
1,3,1 Control of documents	2,3,1 Control of documents	3,3,1 Control of documents
1,3,2 Control of records	2,3,2 Control of records	3,3,2 Control of records
1,2,2 Competence, training and awareness	**2,2,2 Competence, training and awareness**	**3,2,2 Competence, training and awareness**
1,3,3 Emergency preparedness and response	2,3,3 Emergency preparedness and response	
1,3,4 Operational control	2,3,4 Operational control	3,3,4 Operational control
1,3,5 Checking	2,3,5 Checking	3,3,5 Checking
1,3,6Internal audit	2,3,6 Internal audit	3,3,5 Internal audit
1,3,7 Evaluation of compliance	2,3,7 Evaluation of compliance	3,3,6 Evaluation of compliance
1,3,8 Performance measurement and monitoring	2,3,8 Monitoring and measurement	3,3,7 Periodic status review
1,3,9 Nonconformity, corrective and preventive action	2,3,9 Nonconformity, corrective and preventive action	
1,3,10 Management review	2,3,10 Management review	3,3,10 Management review

IMPLEMENTING SHE MANAGEMENT

- INTRODUCTION

The success of the SHE system is dependent on how it is introduced into and operationalized in the organisation. The development of a standard for the organisation for each of the elements is an ongoing process. The standard will need to be upgraded or improved and the new standard will be implemented. There are many variations to the Plan, Do, Check and Act continuous improvement loop also known as PDCA. Some have four steps while others have 6 steps. However, they all aim to achieve the same objective overall continuous improvement.

- THE IMPROVEMENT LOOP

Initially the improvement process commences by entering the loop at the planning and development stage usually after conducting a needs assessment and analysis. Once the SHE programme has been implemented the loop can be entered at any stage. The improvement loop is the key to the search for continuous improvement in SHE performance.

- PLAN AND DEVELOP

This stage refers to the process that will lead to the development of new goals and targets based on the data gathered from the needs assessment and the means to realize these goals and targets. New ideas and suggestions are implemented because of this thought process.

- IMPLEMENT

During this stage the standards to improve the SHE performance in the organization are implemented. The standards become the new and improved standards or the way of doing.

- MAINTAIN

This phase is sometimes referred to as the monitoring or checking phase. The standards that are in place should be improved on an ongoing basis. The target dates or periodic testing frequency could, for example, be increased. Zero defects should be the outcome of ongoing and continuous improvement.

- EVALUATE

The effectiveness of the implementation and maintenance phases are measured during the evaluation phase. Evaluation should be done on an ongoing basis. As soon as a deviation from the standard is detected corrective action should be taken. Avoid evaluating at fixed periodic intervals as this would lead to adjusted action before the evaluation. The evaluation will therefore not give a true reflection of the situation which will most likely result in inappropriate adjustments.

- DEVELOPING STANDARD DOCUMENTATION

- WRITING A STANDARD DOCUMENT

Standards in accordance with the risks in the organisation must be developed for implementation in the organisation. These should be captured in a written form.

- IMPORTANCE OF DOCUMENTATION

Documentation is essential for the implementation of the SHE management system. The objective of the documentation is:

✓ To standardise the requirements of the system.
✓ To integrate health and safety into the management of the system.
✓ To provide standard records that will control operating procedures throughout the organisation.

Various documents need to be developed that will be used to implement, for example, the element of ventilation and could include:

✓ Ventilation study
✓ Proof of action taken with study
✓ Velocity measurements/tests
✓ Copy of maintenance / service records
✓ Reports of monitoring points

Some of these documents will be combined to avoid repetition.

Element ventilation: natural and artificial ventilation study (internal)

ORGANISATION/DEPT: ____________________ DATE: ___________				
NAME: ____________________________ SIGNATURE: _____________				
Ventilation should be adequate in all areas so that the air breathed by workers does not endanger their health. Where a health hazard exists the employer shall provide adequate extraction systems to ensure the safety of workers and if this is not possible, he shall provide respiratory equipment of a type that reduces the exposure of the worker to a safe level.				
ITEM OBSERVED	**OBSERVATION/ COMMENT**	**RECOMMENDATION**	**ACTION BY**	**DATE**
IRRITATING FUMES				
1. Bad odours?				
2. Process odours?				
3. Welding fumes?				
4. Chemicals fumes				
• DUSTS				
1. Hazardous dust?				
2. Red eyes etc.?				
3. Dust explosion?				
• MISTS OR VAPOURS				
1. Solvent vapour?				
2. Acid ?				
3. Alkaline?				
1. Ammonia				
2. LPG				
3.Flammable/Explosive				
Ventilator openings				
Air conditioners				
Air filters and ducts				
Regular inspections				
Worker complaints				
Other observations				

- IMPLEMENTATION PHASE

- OBTAINING COMMITMENT

Any SHE system can only be implemented with the commitment and involvement of the top or senior management team of the organization. This commitment may be obtained by increasing the awareness of SHE within the organization. The following information is most likely to get the commitment and involvement of senior management, namely:

- ✓ Statistics and details of Injuries/incidents/deaths
- ✓ Inquiries from trade unions
- ✓ Pressure from customers
- ✓ Financial losses
- ✓ Moral obligations towards workers

- MAKING A PROPOSAL TO MANAGEMENT

Top management would almost certainly require a proposal which should include the following information:

Current Status

- ✓ Number of workers injured in the past year.
- ✓ Types of injuries sustained.
- ✓ Number of incidents such as, for example, trips, slips and falls.
- ✓ Financial loss incurred in the past year.
- ✓ Fines/litigation - current, pending and potential.

Proposed Status

- ✓ Health and safety policy.
- ✓ Number of injuries objective.
- ✓ Time schedule for these.
- ✓ Decrease in injuries sustained.
- ✓ Curbing financial losses.

- SHE ORGANISATION

For the implementation of a SHE system allocation of responsibilities is necessary to workers in the organisation. These responsibilities will be delegated to every person in the organisation and not only to management. The Chief Executive Officer (CEO) will have responsibilities and so will the elected Health and Safety Representatives.

Some of these are legal requirements as evidenced from the element of ventilation.

SHE STANDARDS

ELEMENT: VENTILATION: NATURAL AND ARTIFICIAL AUDIT GUIDELINES			
ITEM UNDER REVIEW	**TOTAL**	**ITEM UNDER REVIEW**	
STUDY MADE TO DETERMINE IF ANY PROBLEM EXISTS		**ENVIRONMENTAL STRESSORS (Occupational Health Hazards)**	
* All areas identified, and study done.		* Workers educated on relevant stressors.	
* Results measured against standards and deviations actioned.		* Programme to provide for: Identification of chemical stressors:	
* Worker complaints investigated and actioned appropriately.		Toxicity of substances determined.	
Medical surveillance to include:		Toxicological effects per substance identified.	
* Health risk-assessment on vapours, mist, gases, fumes, solvents, dusts.		Absorption established.	
* Exposure limits verified annually.		* Identification of physical (mechanical stressors)	
* Programme incorporates biological monitoring.		* Hazards determined.	
* Analysis carried out by accredited laboratories.		* Effects identified.	

VENTILATION SYSTEMS INSTALLED AND MAINTAINED		* Established exposure.	
* All hazardous areas identified and provided with ventilation systems.		• Medical surveillance programme.	
* Ventilation systems adequate.		***Lighting and vision:***	
* Velocity measurements taken and recorded at pre-determined intervals.		* Is visual ability per job category identified prior to job placement.	
* Supplementary protection provided where needed.		* Lighting inadequacies pertaining to ageing addressed.	
* Ventilation systems included in planned maintenance.		* Occupational hazards involving eyesight (vision) identified.	
* Air conditioner humidifier systems tested for bacteria or other pathogenic organisms.		* Heat stress / Cold stress	
ENVIRONMENTAL MONITORING/SAMPLING (Micro)		* Extremes of temperature (heat/cold)	
* All monitoring points identified and recorded.		* Areas identified per job category above 30° wet/dry bulb reading	
* Monitoring conducted according to standards with appropriate and calibrated equipment.		* Occupational hazards pertaining to heat/cold stress identified.	
* Monitoring carried out by competent persons.		* Action plan for heat/cold exposure.	
* Radiation		**Psychological Stressors**	
* Ergonomics		* Psychological stresses i.e. high risk activities, stressed working relationships and job dissatisfaction to be addressed.	
Biological Stressors		* Workers to be made aware of resources available regarding assistance / support / outside services.	
* Occupational hazards involving bacteria, viruses, fungi and parasites to be determined.			
* Action plan for deviations.			
		TOTAL	

Associated documents

1.	Ventilation survey
2.	Results of airflow (velocity) measurements.
3.	Inspection checklists for air conditioners.
4.	List of identified monitoring points — if applicable.
5.	Certificate of competency — person who carries out monitoring.
6.	Certificate of competency — approved inspection authority who carries out occupational hygiene survey.
7.	General Administrative Regulation — Annexure 1: Material Safety Data Sheet.
8.	Certificate of fitness — workers exposed to cold stress / heat stress.

ACTION/PROCEDURE STANDARDS ARE MET WHEN THE CRITERIA OR ACTIONS BELOW ARE IMPLEMENTED AND MAINTAINED	▪ *RESPONSIBILITY*
1. A ventilation survey is conducted every two years.	
2. Artificial ventilation is provided and maintained where it is required.	
3. Air conditioners and extraction fans are inspected on a regular basis (3 or 6 monthly)	
4. Welding areas have sufficient cross ventilation.	
5. Where welding is carried out in a confined space precautions are taken to ensure that welding fumes are extracted, and/or welders are protected against inhaling these fumes.	
6. Where air contaminants are produced during the process, these are monitored by a competent person at specific intervals, and no person is exposed beyond the TLV exposure limit.	
7. An internal ventilation — air contaminant survey is conducted annually.	
8. Material safety data sheets need to be completed for toxic substances.	
9. Physical stressors: Vibration studies on sources identified.	
10. Lighting and vision: Worker job specification provide for visual abilities per job category.	
11. Cold stress / heat stress workers identified and to undergo individual medical tests to be issued a Certificate of Fitness.	
12. Biological stressors: Areas where these can be found identified and action plan in place to prevent the spreading of contaminants.	
13. Psychological stressors: Environmental Awareness Programme (EAP) in place.	

COMMITMENT TO IMPLEMENT AND ENFORCE/UPHOLD ABOVE STANDARD			
NAME	**DESIGNATION**	**SIGNATURE**	**DATE**

- SUGGESTED LEGAL APPOINTMENTS

To have a complete SHE system the following list of suggested legal appointments might be considered and should be used as standard documents, namely:

APPOINTMENT	LEGAL REQUIREMENT	NAME OF APPOINTEE
Person responsible for building work	GSR 11	
Person responsible for pollution control	ISO 14001	
Person responsible for stacking operations	GSR 8	
Person responsible for ladder inspections	GLS	
Person responsible for scaffolding inspections	GSR 13	
Excavation Inspector	GSR 13	
Person responsible for lifting equipment inspections	DMR 18	
Goods Hoist Inspector	DMR 17	
Builder's Hoist Inspector	DMR 19	
Pressure Vessel Inspector	VUP	
Boiler Inspector	VUP	
Portable Gas Container Inspector	VUP	
Inspector of Refrigeration Installations	SABS 0174/DMR 15	
Hazardous substances co-ordinator	OHSAct Sec. 10	
Person responsible for management of radio-active sources	Act 53, 1992 (Amendment of Haz Sub Act) and Reg 247 of 26/2/1993)	
Radio-active Sources Coordinator	Act 53, 1992 (Amendment of Haz Sub Act) and Reg 247 of 26/2/1993)	
Motor Transport Officer	RTQS	
Earth Leakage Relays Inspections	GLS/EMR 9	
Accredited Person - Electrician	OHSAct	
Appointment: Supervisor of Machinery	GMR 2	
Person responsible for alarm system	GLS	
Fire teams	GLS	
First-Aid team	OHSAct	
Services team	OHSAct	
APPOINTMENT	**LEGAL REQUIREMENT**	**NAME OF APPOINTEE**
Emergency co-ordinator	GLS	
Fire Prevention and protection co-ordinator	GLS	
Incident investigator(s)	GAR 10	
Manager responsible for health and safety	OHSAct Sect. 16 (2)	
Person responsible for machinery	GMR 2	
Health and Safety Co-ordinator	GLS	
Health and Safety Representatives	OHSAct Sect. 17	
Health and Safety Committee Members	OHSAct Sect. 19	
First Aiders	GSR 3	
Management Self - Audit Team	GLS	
Person responsible for issuing Safe Work permits	GLS	
Environmental Programme Co-ordinator	GLS	

• PROGRAMME MAINTENANCE

Given that a set of standard documents has been implemented in the organization with implementation sheets that will ensure that the SHE of the organisation is carried out under controlled conditions, lack of constant control could result in failure to achieve the SHE objectives and targets. The implementation of the program needs to be monitored.

• ANNUAL PLANNER AND CONTROL

The SHE programme should be set out on an annual basis and all elements should be covered. The individual organisation would according to the risks/hazards develop their own operational control programme. Making use of an annual planner is a useful tool to ensure that no elements in the programme are missed or overlooked. The following is an example of an annual planner showing various elements that need attention and the suggested frequency with which these must be done. The list is not exhaustive and should be tailored to the specific organization and its activities and services.

EXAMPLE OF SHE ANNUAL PLANNER

(2=every two years; Q=quarterly; M=monthly; A=annually; D=daily; W=when necessary/as needed)

ORGANISATION: ______________________ YEAR: _______

ELEMENT	ACTION BY	FREQ.	JAN	FEB	MAR	ARR	MAY	JUN	JUL	AUG	SEP	OCT	NOV	DEC
Light Survey		2												
Ventilation Study		2												
Air Conditioner Inspection		Q												
Extraction Systems Inspection		M												
Hygiene Facility Inspection		M												
Pollution Study		A												
Stacking Inspection		Q												
ELEMENT	**ACTION BY**	**FREQ.**	**JAN**	**FEB**	**MAR**	**ARR**	**MAY**	**JUN**	**JUL**	**AUG**	**SEP**	**OCT**	**NOV**	**DEC**
Machine Guarding		A												
Ladders (Portable)		Q												
Scaffolding		W												
Lifting Gear		Q												
Oxy/Acetylene Equipment		Q												
Vessels under Pressure		2												
Refrigeration		Q												
Hazardous Substances		Q												
Vehicles/Forklifts		D												
Drivers' Licences		Q												

Portable Electrical Equipment		D												
Earth Leakage Relays		Q												
General Electrical		A												
Hand tools		D												
Ergonomics		A												
Respirators		M												
Noise Survey		2												
Hearing tests		A												
Fire Equipment Service		A												
Fire Equipment Inspection		M												
Fire Training		2												
Fire Alarm Test		1												
Evacuation Drill		M												
Health and Safety Talk		M												
Update Posters		M												
Health and safety Newsletter		Q												
Health and safety Noticeboard		M												
Health and safety Report		A												
Health and safety Training		A												
Medical Examinations		A												
Health and safety Inspections		M												

Plant and equipment Inspection		M												
Health and safety Self Audit		A												
Job health and safety Observations		A Yearly												

• EVALUATION OF THE PROGRAMME

The regular and systematic evaluation of the SHE programme is necessary, and it suggested that it be done according to the steps of the audit process such as, for example:

Step 1: Planning and Preparation
Step 2: Pre-Audit
Step 3: Incident Experience Review
Step 4: Risk Assessment
Step 5: Systems Audit
Step 6: Compliance Audit
Step 7: Evaluation of Findings
Step 8: Feedback
Step 9: Report
Step 10: Post Audit Activities

• STEP 1: PLANNING AND PREPARATION

This step is necessary to enable the auditor to use the time available effectively and productively at the organization being audited. The following will form part of this exercise, namely:

- ✓ Formal notification of the audit.
- ✓ Information about the organization being audited possibly in the form of a questionnaire.
- ✓ Audit scope.
- ✓ Audit team selection.
- ✓ Previous audit review.
- ✓ Incident statistics review.
- ✓ Legislation and standards review.
- ✓ Audit agenda.
- ✓ Confirmation by the organization being audited.

• STEP 2: PRE-AUDIT SCHEDULE

Items in the pre-audit schedule could include:

1. Arrival and Introduction.
2. Opening Meeting.
3. Organizational information.
   - 3.1. Organisational structure
     - 3.1.1. Organogram.
     - 3.1.2. Levels of education.
     - 3.1.3. Staff's complement.
     - 3.1.4. Staff turnover.
   - 3.2. Production process
     - 3.2.1. Plant layout.
     - 3.2.2. Process flow.
   - 3.3. Responsibility
     - 3.3.1. Contractors.
     - 3.3.2. Transportation.
     - 3.3.3. Off-site contracts.
     - 3.3.4. On-site externally owned.
     - 3.3.5. Outside activities.
   - 3.4. Energy sources
     - 3.4.1. Type.
     - 3.4.2. Consumption.
     - 3.4.3. Storage.
4. Participation

- ✓ Team
- ✓ Names
- ✓ Responsibilities

- STEP 3: INCIDENT EXPERIENCE REVIEW

The procedure to be followed should be:

- ✓ All Compensation Commissioner (CC) forms will be scrutinized such as first and final medical reports to ensure that the classification of injuries and diseases have been done correctly.
- ✓ All CC forms must be available for the last 12 months. The first and final medical reports of the medical practitioner are used for this purpose as these are the only official documents that will indicate lost time or days booked off.
- ✓ In certain instances, the number of compensation claims will be clarified with the Compensation Commissioner. For example, when no first medical report is available. The calculation of man-hours will also be verified based on the assumption that an average person works 167 man-hours per month or 2,000 man-hours per year.

Calculation of Disabling Incident Frequency Rate (DIFR):

The DIFR is calculated as follows:

total number of disabling incidents = injuries + illnesses + diseases

<u>Number of Disabling Incidents x 200 000 (factor)</u>

Total number of man-hours worked

Basis for using this factor:

- ✓ 100 workers work approximately 200,000 man-hours per year.
- ✓ DIFR is therefore an indication of the percentage of workers per year disabled by an incident.
- ✓ The DIFR calculation is applied internationally, and therefore the formula provides for international comparison.

A disabling incident is defined as any incident which arises out of and in the course of duty, resulting in any occupational illness, injury or disease; and giving rise to any related temporary or permanent disablement as determined by a medical practitioner.

- STEP 4: RISK ASSESSMENT

It is important to anticipate, identify and evaluate hazards and to give advice on the avoidance, elimination or control of those hazards, in order to reach a state where the residual risks are regarded as acceptable. It is also important to identify and give priority to those hazards which will result in major harm or damage and impact on the business should there be exposure to them.

Hazard

A hazard is defined as any substance or force with the potential danger to injure, negatively affect health or negatively impact the environment. The hazard (danger) potential is determined by the physical or chemical characteristics of such hazard, which pose a threat to the health, and safety of the worker, property or environment. Workers and the environment are exposed to hazards under various circumstances and for varying lengths of time. Some hazards have a greater potential to cause severe damage, while others have less.

Risk

Risk is the presence of uncertainty whether an event, incident, activity, condition or situation will happen, and is measured in terms of the variability in frequency and severity of the expected outcome/consequences. The number of workers and the length of time that they would be exposed to this situation are also measured.

Note: The biggest hazard is not necessarily the biggest risk, but there tends to be a correlation.

The following is an example of two cases to illustrate the difference in factors that will influence the way the exposure will be managed and mitigated:

CASE 1	CASE 2
✓ Highly concentrated acid. ✓ Small quantity (50 ml). ✓ Closed, durable container. ✓ Used in laboratory. ✓ Used by a formally qualified worker	✓ Medium concentrated acid. ✓ Large quantities (5,000 litres). ✓ Open glass containers. ✓ Carried up and down slippery steps. ✓ Handled by a worker with no qualifications

Formal Risk Surveys

The surveys should be available at the beginning of the audit. The auditor will spend quite some time on scrutinizing this list of surveys to assess, for example:

- ✓ Light intensity.
- ✓ Safety glass.
- ✓ Ventilation.
- ✓ Noise.
- ✓ High risk substances.
- ✓ Ergonomics
- ✓ Pollution.
- ✓ Fire.
- ✓ Emergency.
- ✓ PPE.
- ✓ Hazardous task identification.
- ✓ Work permits.

Assessment Criteria

For each of the surveys the following criteria will be used in the assessment process, namely:

- ✓ Report format.
- ✓ Competency level of assessor.
- ✓ Instrumentation specifications and calibration.
- ✓ Assessment procedures.
- ✓ Coverage of all risks
- ✓ Frequency of measurements.
- ✓ Identification of measurement points for future reference.
- ✓ Standards applied.
- ✓ Existing conditions reflected.
- ✓ Assessment reflects the actual severity of conditions.

Incident Records and Statistics

Incidents will be assessed in accordance with the incident records of the organization.

The auditor will take each of the risk surveys, read through it and will then check it according to the assessment criteria.

- STEP 5: SYSTEMS AUDIT

A SHE system provides a framework to organizations. Detailed management systems will depend on various factors such as, for example, organizational culture, technology level, regulations and standards applicable, inherent production process hazards, and educational level of workers. It is difficult to evaluate the SHE programme of an organization if its management systems are not understood and the soundness of those systems tested.

A thorough assessment of precisely how the management systems intend to meet legislative, regulatory and risk requirements, will also provide the basis for developing verification strategies for gathering the evidence needed to achieve the objective of the audit.

SHE management systems are the framework for guiding, measuring, and evaluating SHE performance. Generally, these management systems consist of:

Organization. The internal structure that establishes roles and responsibilities, accountabilities and reporting relationships

Guidance. Plans, policies, procedures, directives and standards that provide instructions as to how activities and functions are to be carried out.

Controls. Inspections, reviews, etc., built into facility operations to ensure that performance is consistent with objectives and requirements.
Communication. Mechanisms for collecting, handling and reporting information.

Management systems inform about "what, who, why, where, when, how it is done and monitored."

The various segments of the overall management system will need to be reviewed to evaluate compliance with specific and general legal requirements and inherent risks. These include, for example

- ✓ SHE policy of the organization.
- ✓ SHE organization.
- ✓ Training and communication practices.
- ✓ Critical equipment inspection and maintenance procedures.
- ✓ Inspection and audits records and procedures
- ✓ Incident investigation procedures.
- ✓ Permit to work procedures.
- ✓ Colour coding and demarcation standards.
- ✓ Lock-out and tag out, safe work procedures and emergency procedures.
- ✓ Medical examinations.
- ✓ Switches, isolators and valves identification standards.
- ✓ SHE specifications and associated engineering controls.
- ✓ Construction
- ✓ Quality

SHE Policy

The auditor will examine the SHE Policy and determine the level of management involvement and commitment to the policy. This could be done using a balanced scorecard approach.

SHE Organization

The auditor will examine the existing organizational structure from the CEO or most senior manager to SHE Representatives to determine its effectiveness.

Medical Examinations

The auditor will examine and test the procedures regarding medical examinations, particularly those that are mandatory, such as, for example,

- When they are done?
- Who should be examined?
- How often?
- Why?
- Who keeps records?

The auditor needs to know how the SHE system is managed. From this a testing plan will be developed and used to test compliance.

- STEP 6: COMPLIANCE AUDIT

The compliance audit is done to assess the degree of compliance with the internal management systems of the organization. This is the stage when specific documents or equipment will be inspected. It is important to inspect sufficient documents or equipment to provide a statistically defendable and reliable finding. For example, the following situation illustrates the point:

- ✓ At the organization there are 10 forklifts, the auditor will then inspect all 10. The minimum that he should inspect is 30% or 3.
- ✓ At the organization there are 100 fire extinguishers. The auditor would inspect 5% or 5. The minimum that he should inspect is 3% or 3.
- ✓ The auditor must now decide on which 3 fire extinguishers would be inspected?

- SHE TRAINING AND AWARENESS

It is well-established that training and awareness are important if improved performance is to be achieved. Worker involvement and engagement are necessary in those aspects of the work which directly affect them and provide and environment where they are encouraged to make suggestions on working methods, equipment, materials, safety, and health, for example, so that their ideas and opinions can be considered. Any decisions that incorporate the suggestions from workers are more likely to be accepted by the workforce.

A typical example could involve the design of a guard for a machine. If the ideas of the workers who work with the machine are incorporated into the design of the guard, it is more likely to be accepted and used. Extended worker involvement and engagement builds confidence and trust in the organization because it demonstrates that management regards workers and their views as important. Consequently, everyone in the organization senses that they are working together to achieve shared and understood organizational objectives.

When workers are considered, they generally can make substantive contributions in hazard identification, in proposing solutions and applying those solutions. SHE programmes are obviously more effective if workers have "bought into them". Their involvement and engagement must be more than just lip service. Rather, they must be meaningful and include decision-making capability.

- SHE PUBLICITY AND PROMOTION

SHE publicity forms an important part of the education and training process within the SHE space in an organization. It is not necessarily confined to campaigns as such and may include public relations and low-key educational programmes designed specifically to inform rather than to persuade or change behaviour. For example, workers returning to work after the national lockdown due to COVID-19 will need to be informed about the 'new norm' in their workplaces since changes will have had to be made that would affect them both directly and indirectly. SHE publicity can be an ideal way of promoting the SHE image of the organization while also being an effective tool to prevent accidents and incidents. Generally, SHE publicity can take one or both of two forms, namely:

✓ Promotions (what an organization does to promote SHE behavior and practices within its workplace).
✓ Campaigns (what it does to publicize SHE behavior and practices to works, families and areas serviced by the organization).

Well promoted programmes involve both management and workers in SHE activities.

- INDICATIONS AND NEED FOR A PROGRAMME

The following are indicative of the possible need for a programme:

✓ Consequences of a national disaster such as the COVID-19 pandemic.
✓ An increased rate of injuries and incidents may signal that a programme is needed.
✓ If housekeeping is deteriorating, protective equipment is not used and guards are not replaced, it is time to improve supervisory SHE awareness and to promote renewed and increased interest in SHE.
✓ Incomplete or missing accident reports also indicate decreasing supervisory interest and perhaps even failure of workers to report minor incidents and injuries.

- OBJECTIVES AND BENEFITS

A well-planned programme, although it cannot be expected to do everything, may help workers to become more committed to SHE and as such it may, for example,

1. Help to develop positive SHE work habits and attitudes especially when large-scale changes have had to be introduced into the workplace that will affect how they conduct themselves at work.
2. Focus attention on specific causes of accidents.
3. Supplement SHE training.
4. Give workers a chance to participate in accident prevention activities.
5. Provide a channel for communication and co-operation between workers and the management of the organization.
6. Improve worker, customer, and community relations because it is evidence of the commitment and involvement of the management of the organization.

The following need to be considered, namely:

- ✓ The policy and experience of the organization.
- ✓ Available budget or funding and facilities.
- ✓ Types of operations and services of the organization.
- ✓ Types and categories of workers who work in the organization.
- ✓ Basic levels of human interest.

The following forms of media and forums provide opportunities, namely:

- ✓ SHE committees and observers.
- ✓ Meetings.
- ✓ Contests and competitions.
- ✓ Posters and displays.

Other promotional methods that can be used to create and maintain interest in SHE could include:

- ✓ Special campaigns.
- ✓ Short talks, training courses and demonstrations.
- ✓ Publications and newsletter.
- ✓ Public address systems.
- ✓ Suggestion systems.

A SHE training programme is particularly needed under the following conditions:

- ✓ For all new or reassigned or transferred or redeployed workers.
- ✓ When new equipment or processes are introduced.
- ✓ Whenever procedures are revised or updated.
- ✓ When new information is available or required.
- ✓ When worker performance needs improvement or to provide worker growth and development.
- ✓ When worker interest in SHE, efficiency, and productivity needs a boost.

Certain symptoms may indicate that training is needed such as, for example:

- ✓ Proportionately higher accidents or injury rates than normal or other organizations doing the same work.
- ✓ Increasing accident rates and/or insurance rates.
- ✓ High worker turnover.
- ✓ Excessive waste or scrap material.
- ✓ Recent organization expansion or procedural changes.
- ✓ Increased or high level of absenteeism.
- ✓ Changing regulatory requirements.
- ✓ Need to upgrade or update knowledge or responsibilities of workers.
- ✓ Poor job satisfaction.

Several modes of training could be adopted such as:

- ✓ On the job training.
- ✓ Group techniques.
- ✓ Conference method.
- ✓ Individual techniques.
- ✓ Computer-assisted training.

• CONTRACTOR CONTROL

Any person who enters the premises or job site of an organization with the intention of performing any work must be considered as a contractor. The Act specifically provides that the employer should ensure the health and safety of all visitors on its premises or sites. Many incidents occur when contractors are not informed of the specific hazards on the premises both generally and specifically during their time on the premises or in the plant or do not abide by the SHE rules and regulations on the specific site on which they are working.

• CONTROL

Before contractors can enter any site or workspace, the following must be ascertained:

- ✓ Status of workman's compensation registration.
- ✓ Responsible person (person in charge).
- ✓ Extent to which workers have been trained in SHE
- ✓ Possession by workers of the necessary PPE

Areas where contractors are performing their tasks should be demarcated and placed under the responsibility of the responsible contractor. The primary contractor should ensure that work is performed according to legislative requirements, as well as the SHE rules of the organization.

A contract should be signed by the contractor prior to the commencement of work on site which clearly indicates all the SHE requirements as well as the rules and regulations applicable to the work which is to be performed. The contractors and their workers should also be inducted in the SHE rules of the organization as well as the PPE which is required to be used during work performed.

All special requirements and arrangements must be made quite clear to the contractor.

Control measures are necessary to ensure that contractors and the responsible person abide by agreements and regulations as stipulated in the applicable contract.

• BUDGETING FOR SHE

Many organizations have a specific budget for carrying out the SHE policy of the organization. A system of budgetary control establishes various budgets, which set out in financial terms the responsibilities of management in relation to the requirements of the overall policy of the organization. There should be a regular comparison of actual results with budget forecasts both to ensure that the objectives of the SHE policy are met and to form a basis for any revision of such policy.

The budget sets clear financial limits to component parts of the organization. Therefore, activities are planned in line with the policy of the organization and within the financial limits.

• PREPARATION OF BUDGETS

The following steps are typical of those taken in the preparation of individual budgets and the master budget for any organization. A SHE manager or similar person responsible for SHE at various locations, such as a group SHE manager, would be responsible for his own master budget. This would then form part of the overall master budget of the organization. The individual budgets are based on the overall SHE objectives of the organization. Forecasts are made regarding the prevailing general economic conditions and the conditions in the industrial sector. These individual budgets could include, inter alia:

- ✓ A SHE budget prepared based upon the forecasts that highlight key task areas for action.
- ✓ A production budget prepared in conjunction with these conditions and considering all the materials and other resources required to carry out those key SHE task areas.
- ✓ The administrative cost budget for each area of activity.
- ✓ A capital expenditure budget covering anticipated changes in legislation or purchase of specialized equipment or other modifications.

An example of audit guidelines for a hazardous chemical substances (HCS) standard in an organization:

ITEM UNDER REVIEW RECORD OF ALL HCS	TOTAL	ITEM UNDER REVIEW RADIO-ACTIVE SOURCES	TOTAL
* Risk analysis conducted on all HCS and record of all substances including products, by-products and waste from the manufacturing process available.		* Appointment of the responsible person and alternate up to date.	
* Inventory of raw materials and all intermediate mixtures available to OHS.		* Letter of the current legal authority in order.	
* Records kept up to date.		* Register of sources in order and up to date.	
* All products/substances containers clearly labelled.		* All sources secured/locked and barred against unauthorized access.	
PERSON DESIGNATED TO CO-ORDINATE HCS		* All personnel trained in the correct procedures for handling, maintaining and using of sources.	
* Competent person designated to co-ordinate all HCS.		* Worker health surveillance to be conducted prior to, during and on termination of service.	
MATERIAL SAFETY DATA SHEETS (MSDS) AVAILABLE AND USED IN DEPARTMENTS, BY BUYER, FIRST-AIDERS/OHS		* Workers made aware of health hazards related to radiation exposure measures.	
* All MSDS written in compliance with legislated/ international standards.		* Emergency plan includes radiation exposure measures.	
* MSDS available in all user departments, buying office, OHS or first-aid post and stores.		* OHS/First Alders trained.	
* All users/handlers of HCS trained and made aware of hazards, procedures and personal protective equipment.		**IONISING RADIATION**	
* System to cater for the training of clients on occupational health hazards.		* Study available indicating levels of radiation exposure monitoring (i.e. dosimeter, dose exposure, time exposure).	
* MSDS kept up to date with the latest information on the product/substance.		* Transfer of radiation exposure records during job changes maintained.	
* Authorised bulk storage volumes displayed.		* Qualifications of occupational health staff conducting monitoring.	
* MSDS available to transportation contractors where appropriate.		* Availability of decontamination process.	
* Transportation requirements/legislation complied with.		**NON-IONISING RADIATION**	
* MSDS includes emergency procedures including spill containment and clean-up procedures.		* Processes utilising non-ionising radiation identified.	
* Transportation staff/contractors trained in hazards, handling methods and these procedures (HAZCHEM signs).		* Health risks pertaining to non-ionising radiation identified.	
* Clients provided with MSDS as appropriate.		* Health and safety risk awareness pertaining to non-ionising radiation.	
		TOTAL	

ASSOCIATED DOCUMENTS

1.	Appointment: Hazardous Chemical Substances Controller or similar and proof of competency.
2:	Alphabetical list of all substances on site – HCS risk analysis.
3.	Any specific HCS training programme and attendance register.
4.	Material Safety Data Sheets (MSDS) for all HCS.
5.	Standards and SWPs or SOPs relating to HCS control.
6.	List of applicable legislation and regulations.
RADIO-ACTIVE MATERIAL	
1.	Appointment of responsible person and alternate.
2.	Letter of authority from responsible agency.
3.	Register of radioactive sources.
4.	Handling and firefighting procedures for radioactive material.
5.	Any specific training programme on radio-active material management and handling and attendance register.
IONISING/NON-IONISING RADIATION	
1.	Certificate of competency of person responsible for conducting monitoring.
2.	Emergency action plan and remedial action in the event of exposure.
3.	Any specific training programme on ionising and non-ionising radiation and attendance register.

IMPLEMENTATION

ACTION/PROCEDURE/STANDARDS ARE MET WHEN THE FOLLOWING CRITERIA OR ACTIONS ARE IMPLEMENTED AND MAINTAINED	RESPONSIBILITY
1. The appointed Hazardous Chemical Substances Controller is:	
2. An alphabetical list of all substances/chemicals is available and kept up-to-date.	
3. Material Safety Data Sheets in an easily readable form are available for all identified hazardous chemical substances	
4. A copy of the alphabetical list and material safety data sheets are available in all user departments, buying office, OHS or first-aid post and stores.	
5. All products/chemicals are clearly and correctly labelled as per the list of substances.	
6. Any specific HCS and MSDS training programme and attendance register	
7. Users of hazardous chemical substances are trained in the use of the chemicals.	
8. Bulk storage of hazardous chemical substances is indicated in an appropriate way.	
9. Where hazardous chemical substances are transported by contractors — they are supplied with the necessary MSDS's.	
10. All the applicable legal requirements are complied with at all times during the transportation of hazardous chemical substances.	
11. All transportation staff are trained on the hazards, methods, and spillage containment procedures of the substances that they transport (HAZCHEM signs posted on vehicles).	
RADIOACTIVE MATERIALS	
1. The person responsible for control of radioactive material is:	
2. The alternative person for control of radioactive material is:	
3. The co-ordinator for control of radio-active material is:	
4. A letter of the current legal authority is available and in order.	
5. A letter of the Department of Health is available and in order.	
6. A register of radioactive sources is available and up to date.	
7. All sources are secured/locked and barred from unauthorised access.	
8. A formal fire procedure in case of fire in the presence of radioactive material is available.	

9. All persons working with sources are trained in the fire procedure.	
10. All personnel working with sources have been trained in the correct procedures for handling, maintaining and using the sources.	
11. Qualifications of occupational health staff conducting monitoring available	
12. Radiation exposure badges worn by affected personnel	
13. Emergency procedures and action plans in place in the event of an exposure	
14. Procedure for the decontamination of substances / objects/ persons in the event of an exposure	
15. Any specific training programme on ionising and non-ionising radiation and attendance register.	

COMMITMENT TO IMPLEMENT AND ENFORCE / UPHOLD THE HCS STANDARD			
NAME	**DESIGNATION**	**SIGNATURE**	**DATE**

AN EXAMPLE OF AN APPOINTMENT OF A HAZARDOUS CHEMICAL SUBSTANCES CONTROLLER

Attention of:

I, ________________________, having been appointed in terms of Section 16(2) of the Occupational Health and Safety Act (85 of 1993), to ensure full compliance with the Act, hereby appoint you

____

as Hazardous Chemical Substances Controller for ______________________________________

____

Your duties as a Hazardous Chemical Substances Co-coordinator are as follows:

1. Keep an updated alphabetical list of all Hazardous Chemical Substances.
2. Review the alphabetical list on a quarterly basis by completing the quarterly review form.
3. Obtain the required hazard information (General Administrative Regulations, Annexure 1) for all identified Hazardous Chemical Substances from the supplier or manufacturer of the product.
4. Exercise control over the safe storing, labelling, issue and handling of all Hazardous Chemical Substances.
5. Ensure that all hazardous chemical substance containers are safely disposed of if they are not returned to the supplier.

Please confirm your acceptance of this appointment by signing and returning the duplicate copy of this letter to the undersigned.

____________________ __________

SIGNATURE DATE

MANAGER

I understand the implications of this appointment and confirm my acceptance thereof.

____________________ __________

SIGNATURE DATE

HAZARDOUS SUBSTANCES CO-ORDINATOR

AN EXAMPLE OF A HAZARDOUS CHEMICAL SUBSTANCES (HCS) CONTROL TRAINING PROGRAMME:

INSTRUCTOR: ______________________________ DATE: _____________

THE FOLLOWING ITEMS WERE REVIEWED FOR EACH OF THE FOLLOWING SUBSTANCES:

1. Material name and its ingredients/contents.
2. Hazard data referring to health hazard.
3. Basic physical data e.g. flash point, solubility in water etc.
4. Fire extinguishing media.
5. Reactivity data.
6. Spill, leak and disposal procedure.
7. Personal protection information.
8. Special precautions, i.e. storage, handling, transportation etc.

		HAZARD SHEET
1.		
2.		
3.		
4.		
5.		
6.		
7.		
8.		
9.		
10.		

IIAZARDOUS SUBSTANCES TRAINING—ATTENDANCE REGISTER

WORKER NAME	**DEPARTMENT**	**SIGNATURE**

HAZARDOUS CHEMICAL SUBSTANCES (HCS) CONTROL

SAMPLE LETTER: OBTAINING HAZARD INFORMATION FROM SUPPLIER

TO: ____________________ FROM: ________________________

________________________ ________________________

________________________ ________________________

DATE: ____________________

Dear: ____________________

As part of our SHE programme, we require information on the possible toxicity, nature of potential health and safety hazards, and safe handling and storage precautions, of the following products/chemicals which you supply us.

Would you kindly provide us with this information as soon as possible using the enclosed General Administrative Regulation, Annexure 1 form included?

If you do not manufacture the substances mentioned below yourself, will you kindly obtain and let us have the information from the manufacturer who supplies you.

1. ______________________ 2. ________________________

3. ______________________ 4. ________________________

5. ______________________ 6. ________________________

Regards,

NAME: ______________________

DESIGNATION: ________________

HAZARDOUS CHEMICAL SUBSTANCES (HCS) CONTROL

QUARTERLY REVIEW OF ALPHABETICAL LIST OF HAZARDOUS CHEMICAL SUBSTANCES

1. Ensure that the alphabetical list is complete.
2. Ensure that each substance that requires a Material Safety Data Sheet has one.
3. Indicate the action to be taken by making use of the following key.

A List complete, review again in three months.

B List complete, but MSDS must still be obtained for certain substances. Immediately initiate action to obtain the required MSDS.

C List incomplete, and MSDS incomplete. Immediately initiate action to update list and to obtain the hazard information

DATE REVIEWED	REVIEWED BY	ACTION REQUIRED	BY DATE	SIGNATURE

• REVIEW QUESTIONS

1. Why is an organizational SHE policy necessary?
2. What strategy is necessary to convince the management of the organization to introduce a SHE management program?
3. What core elements should be included in a SHE management system?
4. Why is it necessary to have a documented SHE organizational structure?
5. What should the goals of a SHE propaganda program be?

• REFERENCES

Anonymous. Nd. Loose notes on SHE management

Ccohs (n.d.). *Effective Workplace Inspections*. [online] Ccohs.ca. Available at: https://www.ccohs.ca/oshanswers/prevention/effectiv.html. Accessed July 31, 2020

Department of Labour (1993). Occupational Health and Safety Act No: 85 of 1993.

HSA (n.d.). *Safety and Health Management*. [online] Health and Safety Authority. Available at: https://www.hsa.ie/eng/Topics/Managing_Health_and_Safety/Safety_and_Health_Management_Systems/. Accessed July 31, 2020

HSA (n.d.). *Workplace Safety and Health Management* (online) https://www.hsa.ie/eng/Publications_and_Forms/Publications/Safety_and_Health_Management/Workplace_Safety_and_Health_Management.pdf. Accessed July 31, 2020

ILO (n.d.). *World Statistic*. [online] Ilo.org. Available at: https://www.ilo.org/moscow/areas-of-work/occupational-safety-and-health/WCMS_249278/lang--en/index.htm. Accessed July 31, 2020

ISO (2015). *ISO14001 ENVIRONMENTAL MANAGEMENT SYSTEMS.* [online] Iso.org. Available at: https://www.iso.org/files/live/sites/isoorg/files/archive/pdf/en/pub100329.pdf. Accessed July 31, 2020

OHSAS 18001:2007 (2007). *OHSAS 18001:2007 OCCUPATIONAL HEALTH AND SAFETY ASSESSMENT SERIES.* [online] Producao.ufrgs.br. Available at: http://www.producao.ufrgs.br/arquivos/disciplinas/103_ohsas_18001_2007_ing.pdf. Accessed July 31, 2020

OSHA (2002). *Materials Handling and Storage*. [online] Osha.gov. Available at: https://www.osha.gov/Publications/osha2236.pdf. Accessed July 31, 2020

OSHA (2015). *OSHA Safety and Health Program Management Guidelines*. [online] Osha.gov. Available at: https://www.osha.gov/shpmguidelines/SHPM_guidelines.pdf. Accessed July 31, 2020

Safeopedia. What is the difference between a safety culture and a safety climate and why does it matter. https://www.safeopedia.com/what-is-the-difference-between-a-safety-culture-and-a-safety-climate-and-why-does-it-matter/2/8059. Accessed July 31, 2020

Lin, Shu-Chiang., Mufidah, Llma. and Persada, Satria, Fadil. 2017. Safety culture exploration in Taiwan's metal industries: identifying the workers' background influence on safety climate. *Sustainability 9*(11), 1965, https://doi.org/10.3390/su9111965

CHAPTER 4

POLICY AND RESPONSIBILITIES

SHE POLICY

A policy is defined as "a set of ideas or a plan of what to do in particular situations that has been agreed to officially by a group of people or a business organization." This definition should be kept in mind when forming the SHE policy, as many organizations manage to go off on a tangent when constructing a policy, either omitting elements that should be included, or including elements that are not specifically relevant to the policy itself, or perhaps belong elsewhere.

A SHE policy should be based on the primary aim and vision of the organization which is to maximise the profit incentive of the business through the manufacture and sale of the products or services that it provides. The SHE policy should make it clear that the intention of the organization is to make a profit without putting workers, or others, at any risk of their safety, health, and physical wellbeing.

The SHE policy should be public, openly visible, and acceptable by all workers across the organisation. It can be brought to the attention of the workers through, for example:

- Displaying it on staff notice boards
- Handing out a copy of it to all workers
- Including it into the induction program of new workers
- Giving training sessions to existing workers when new policy is introduced or when existing policy changes
- Having a section in the corporate newsletter dedicated to the SHE policy

Available information, facilities and arrangements in the organisation should support the policy. With these structures in place, it ensures that the objectives of the policy are achievable. Successful implementation does depend to a large degree on the commitment, involvement and attitude of management towards SHE.

A successful SHE policy will result in high standards of compliance with physical and environmental requirements, carefully designed and observed safe and healthy systems of work, workers with high standards of cleanliness and housekeeping, the maintenance of safe means of access will be kept intact and toxic, corrosive and flammable substances will be controlled where it is relevant or necessary to do so. Furthermore, it will contribute to effective consultation for SHE and positive participation by workers. Training programmes will be well developed and be running on a continuous basis on all levels with high participation in them.

CONTENTS OF THE SHE POLICY

The three most common challenges with written SHE policies are that:

- The policies exist, but few people are aware of them;
- The amount of detail required may generate a mass of paperwork; and
- The policies are simply a restatement of the legal duties and obligations of the organization.

The SHE policy statement should show:

- The necessity for active leadership from top and senior management, direct participation and involvement of management and enthusiastic support of the entire organisation
- The intention to integrate loss control into all operations, including compliance with all applicable standards
- The visible and tangible commitment and involvement of the management of the organization in occupational health and safety and management of the environment
- The importance that the organization places on the health and safety of workers.
- The emphasis placed by the organisation on efficient operations, with a minimum of incidents, accidents and losses

The SHE policy statement should, inter alia, include the following elements:

- Statement of the basic objectives and detailed roles and procedures to cater for specified hazards and provide a framework for setting SHE goals and objectives
- The commitment and involvement of top management to a safe and healthy workplace, the alignment of SHE principles with the strategic direction of the organization, planning objectives, and measurement of the results
- Defined duties and extent of responsibility for SHE of specified line management levels while also clearly identifying who has overall responsibility for SHE in the organization
- Clear definition of the functions of various SHE personnel and their relationship to line and senior management
- A commitment to grow stakeholder and worker engagement and commitment through meaningful consultation on critical SHE aspects in the organization including the SHE policy itself
- A statement confirming the responsibility of all workers and stakeholders to uphold the SHE principles or the organization and the maintenance of a safe and healthy workplace
- The identified and prioritized hazards as well as the precautions to be taken to mitigate them including the roles of workers, visitors, and contractors
- Details of how the SHE policy will be reviewed and improved where necessary
- The system for supervising SHE performance and how information about such performance will be communicated including what information system will be used to monitor the implementation of the SHE system and identify needs that arise
- A description of how the SHE activities will be resourced and funded by the organization including budgeting process
- A statement of how compliance with legislation, regulations, codes of practice and standards and compliance will be achieved

Further, an effective SHE policy is generally made up of three major sections or elements, namely:

- **Statement of Intent including specifying objectives**
  - This part outlines the overall philosophy of the organization in relation to the management of SHE.
- **Organisational details** (details of the people and their allocated duties with respect to SHE involvement)
  - Clearly indicates who is responsible to whom for what in the form of a management structure diagram or organogram
  - Demonstrates how accountabilities are determined
  - Specifies how and when policy implementation is to be monitored, evaluated and reviewed
  - Indicates the functions and responsibilities of SHE committees and representatives
  - Shows how SHE and associated accountabilities are addressed in the job descriptions of workers
- **Organizational SHE systems and procedures**
  - This element gives practical details of how the overall SHE policy will be implemented in the organization
  - Includes details of, for example, procedures that cover COVID-19 protocols such as screening, testing, referrals, distancing, and working remotely, SHE training, audits, accident reporting and investigation, safe systems of work (SWPs or SOPs), permit to work systems, noise and environment control, utilization of personal protective equipment, fire health and safety, liaison with contractors, manufacturers and suppliers, machine guarding, emergency procedures and medical and worker welfare considerations.
  - This section is regarded as the primary part of the SHE policy as it spells out at operational level how SHE is to be maintained at all levels throughout the organization

All departments should prominently and visibly always display SHE procedures applicable to their department to their workers to ensure that the workers are familiar with these.

• ORGANIZATION OF THE SHE POLICY

The SHE policy details the structure and framework in which the organisation functions with respect to managing safety, health, and the environment. It is also true that having a policy guide that stipulates the how's, when's and why's is not enough to achieve excellent performance. Top or senior management must set the example by not only being committed but also be visibly involved and provide through their actions the motivation to comply and implement the policy and ensure its continued smooth execution. The role of management extends to the coordinating of various activities so that the organization works as a cohesive team and contributes together to the success of the organisation through the effective management of SHE.

All workers in the organization with SHE responsibilities and duties need to have clear and specific plans of action. A useful tool is the use of a management structure or organogram that allocate jobs and determines the roles, responsibilities, and functions of all workers in the organization. This approach shows clearly the functions of everyone in the workforce which can be expanded in the form of written explanations on job descriptions.

- ACCOUNTABILITY AND RESPONSIBILITY IN THE SHE POLICY

For the effective implementation of the SHE policy in an organisation, SHE management within the organisation should have the following characteristics, namely:

- A clear indication in both written and pictogram forms of the operation of uninterrupted and logical delegation of duties through line management where risks arise and before incidents happen.
- It should identify key personnel, by title that are accountable to top or senior management for ensuring that SHE systems and procedures for the working conditions of the organization are developed, utilised and maintained.
- It should define the roles of line and functional management through specific job descriptions
- It should provide visible and tangible support through functional management in the form of, for example, SHE advisers, medical advisers, designers, chemists, and engineers.
- Nomination of competent people to measure and monitor SHE performance.
- Stipulation of the censure for failing to meet the requirements of the SHE policy.
- Transparent information on the total resources available for SHE includes the budget development and approval processes.

The extent to which the SHE policy is exercised and integrated into the culture of the organisation will determine its effective management.

- HOW TO DRAFT A SHE POLICY

It is important to be aware that a SHE policy becomes a legally binding document for all parties in the organization from top or senior management to workers. Unfortunately, policy statements have historically been lengthy and longwinded. A word of caution is necessary. Since the organization will be held accountable for the commitments that are made in the policy statement, it should not make any commitments that it has no intention of adhering to or that it cannot live up to. Aim to have the SHE policy statement fit on a credit card sized business card which can be carried in the shirt pocket by a worker. It should be concise and short enough for the worker to be able to recite the policy statement of the organization. Evidence of a positive SHE culture is when every worker knows and can repeat the SHE policy of the organization when requested to do so.

Note that:

- The Act does not require employers to have an OHS or SHE Policy unless directed to by the inspectorate of the Department of Labour. This generally happens if the employer is working in a particularly high risk industry.
- The Mines Health and Safety Act does require the mine to have a OHS Policy.

The policy requirements of, for example, OHSAS 18001:2007 (being replaced with ISO 45001:2018) and ISO 14001:2017 provide excellent guidance on the formulation and content of SHE policies.

NOTES

- Top management for the defined organization should be involved in defining the policy statement.
- Ensure that the SHE Policy Statement is aligned with the SHE objectives and targets of the organization. In other words, the organization may not have a policy statement without corresponding objectives and targets, nor may objectives and targets be formulated that are not referenced in the policy statement.
- The intention of the standards is that the policy statement may only be generated once all the hazards that are typically encountered in the daily operations of the organization have been identified and assessed.

As a minimum the following need to be addressed in the policy:

- CLEARLY ARTICULATED MANAGEMENT COMMITMENT TO AND INVOLVEMENT IN THE FULL LIFE OF THE SHE MANAGEMENT SYSTEM
  - Develop, document, implement, maintain, and improve the SHE management system
  - Availability of resources
  - Allocation of roles, responsibilities, and authority of persons within the overall management structure
- MANDATORY COMMITMENT STATEMENTS
  - Commitment to reviewing and the continual improvement of SHE performance
  - Commitment to preventing environmental pollution, ill health, injury and fatalities
  - Commitment to complying with the prevailing legislative and regulatory framework and other requirements
  - Commitment to effective and transparent communication to people working on behalf of or for the organization

- o APPROPRIATENESS AND RELEVANCE OF THE POLICY STATEMENT
  - o Appropriate to the purpose of the organization in terms of what the core business of the organization is
  - o Appropriate to the nature and scale of the activities, products, and services of the organization

An example of a SHE Policy:

ACME SHE POLICY:

Management and Responsibility

We actively manage SHE as an integral part of our businesses, operations and practices.

We require regional and local management to comply with the applicable laws and the internal and external SHE requirements.

Security measures must be integrated into the SHE management system and must meet defined minimal requirements.

Control and Auditing

We operate only processes, apply technologies and develop and handle products which are assessed for their SHE risks.

All raw materials, intermediates and products are assessed for their regulatory compliance

We have emergency procedures in place and emergency response organizations at all our sites to control and to limit the impact of incidents and threats.

We periodically audit our operations as well as business and management practices at all our sites regarding SHE performance and compliance.

Security

Security measures must be integrated into the SHE management system and must meet defined minimal requirements. Site security is managed by the local site management and may involve third-party contracted organizations. It is of utmost importance, that the logistics security of mineral supply and product shipping and warehousing gets the required attention. Lonza engages in C-TPAT procedures to comply with security requirements.

Communication of Risks

We openly communicate and provide information on our SHE performance to all stakeholders. All communication channels in case of an emergency are defined and functional.

Another example of a SHE policy:

SAFETY, HEALTH AND ENVIRONMENT (SHE) POLICY OF ABC CHEMICALS

INTRODUCTION

We are a leading chemical distributor that adopts best practices for the safe distribution and disposal of chemical waste.

This is the safety, health, and environment (SHE) policy of the organisation.

Responsibility for SHE

Overall and final responsibility for health and safety is that of the board of directors. On a day-to-day basis responsibility lies with the Health and Safety Officer (HSO), who is responsible for the preparation and review of this policy. The HSO is also responsible for the development and implementation of organisational procedures to ensure a safe and healthy working practice that does not negatively impact the environment.

This involves:

- Ensuring current procedures are still effective and relevant to a changing business;
- Transparently passing on information relating to SHE to workers;
- Keeping up to date with legislative requirements that relate to SHE.

SHE Policy Statement

We recognise that we have a duty to secure the health, safety, and welfare of all our workers by ensuring that the working environment is as safe as is practicable without any threat to their health.

This will be done through:

Ensuring all our workers are competent to do their tasks and are given adequate training, information and supervision; maintenance of vehicles, plant and equipment; safe handling and use of substances.

Consultation with workers on matters affecting their health and safety

This duty extends to those people who visit our premises and to the public in general. These duties and responsibilities extend to all other workers in our plant. It is the duty of every worker to ensure that our working environment can be achieved and maintained by everyone working together, recognising that they are involved in the SHE process and have a duty to secure the well-being of others by observing all SHE regulations. To achieve this goal, good communication in relation to SHE matters is necessary. Behaviour which puts others at risk will result in disciplinary action being taken.

We will make every effort to see that our activities will not harm the environment and will do all in our power to minimise the environmental impact of our operations. To this end we will, whenever possible, attempt to recycle material rather than dispose of them. The remaining options, in order of priority, are: chemical treatment, with the intention of producing a less hazardous product; high temperature incineration; and landfill. We will reduce the amount of waste that goes to landfill on an ongoing basis and will comply with all prescribed legislative and regulatory requirements. We recognise that we a duty of care under the Environmental legislation for any waste that we consign and have a procedure for auditing any disposal site or transport contractor that are used for this purpose.

Procedure

We have written working procedures in place and our workers receive these in form of a training pack. Procedures are reviewed and updated bi-annually or when working conditions change significantly, whichever comes first.

Premises where hazardous substances are stored must comply with the site licence and must:

- Be always kept clean and tidy;
- Contain appropriate warning signs;
- Contain safety and emergency equipment including personal protective equipment, spillage kits, fire extinguished and first aid equipment to comply with SHE regulations.
- Provide adequate segregation of hazardous materials

Duties of workers

It is the duty of every worker while at work to take reasonable care for the health and safety of themselves and of other persons who may be affected by their acts or omissions at work. It is also the duty of every worker to co-operate with any duty or requirement imposed on us by any regulatory organisation. Our workers must not interfere with anything provided to safeguard their health and safety. Any health and safety concerns must be reported to the Health and Safety Officer or a Director of our organization.

Emergency procedures and emergency equipment

We have emergency procedures in place which are tested and reviewed periodically. All our workers are made aware of them as part of their training. Any incidents will be recorded and, if necessary, reported to the appropriate authority. We will ensure adequate provision of emergency equipment and will ensure that equipment will be tested or inspected periodically if applicable. Fire extinguishers are serviced annually by a third party. Workers must inform the organisation if the equipment is damaged or unusable. The Health and Safety Officer makes a periodic visual check of safety equipment in the transfer station.

Accidents and first aid

We have persons qualified in first aid. There are first aid kits in the main office, the transfer station and all our vehicles. The transfer station and vehicles also carry eyewash. The transfer station has an emergency shower. The Health and Safety Officer makes a periodic check of first aid stock. Any accidents and cases of work-related ill health are recorded in the accident book.

Work equipment

We will ensure adequate provision of personal protective equipment and will ensure that equipment will be tested or inspected periodically if applicable. The Health and Safety officer makes a periodic check of respirator masks and cartridges in the transfer station. A record is kept of PPE issued to our workers. Workers must inform us when equipment is damaged or unusable. Our directors will be jointly responsible for identifying all plant and equipment needing maintenance. Our directors will be responsible for ensuring effective maintenance procedures and the suitability of new equipment purchased.

Risk assessments and method statements

We realise that we have a responsibility to carry out risk assessments for our activities including the use of equipment, buildings, and hazardous substances. These assessments will be carried out by the Health and Safety Officer, in consultation with other workers. The findings will be presented to the Directors and made known to all affected workers. Our Directors will be responsible for ensuring that actions identified in the assessments are implemented. We have written method statements for our activities and these can be provided to any customer who may request them.

Training

The future success of our business is directly dependent on the people who make up the business. No matter what position an individual hold within our organization, he or she will have an effect, for good or for bad, on our performance in the marketplace. It is essential that everyone has the skills necessary to carry out their tasks safely and effectively. To facilitate this, all our workers will receive job-specific training and a written record is kept, such that it may be referred to in future. It is vital that we ensure every member of the team is equipped with the right skills and knowledge.

Welfare facilities

We will ensure that adequate welfare facilities are available at all our sites applicable to the level of staff on the sites.

Contractors and visitors

All contractors and visitors will be issued with the site rules to make them aware of the emergency procedures and the existence of hazards in the immediate environment.

Review of SHE policy

This policy and the organisational procedures to which it refers will be reviewed every two years or when working conditions change, whichever occurs first, and will be distributed to all workers as part of induction and training.

The Directors

21 May 2020

• COMPILING A SHE PROCEDURE

It is necessary to note the following:

- While written documented OHS procedures are not a requirement of the Act doing so is still one of the best ways of transferring information and illustrating due diligence to the provisions of the Act.
- On the other hand, the mining industry has certain mandatory codes of practice (COP) which need to be documented according to a specific set of criteria such as, for example:

Examples of Mandatory Codes of Practices:

- Ventilation (Prevention of flammable gasses)
- Self-propelled mobile machines
- General occupational health program
- Occupational health program for noise
- Occupational health program for airborne pollutants
- Minimum standards of fitness

- Thermal stress
- Eradication of silicosis program
- Asbestos management program
- Code of practice for pregnant and lactating women
- Material safety data sheets for items used on site
- Combat rockfall and rock burst accidents n metalliferous mines and mines other than coal.

Examples of Mandatory Procedures:

1. Drug use and alcohol abuse
2. Incident reporting and investigation
3. Risk assessment methodology
4. Daily vehicle and machinery inspections
5. Earth leakage, electrical installation and portable electrical equipment inspections
6. Emergency preparedness
7. Vehicle management
8. Internal audit process and responsibility
9. Non- conformities and legal non-compliance
10. Handling and storage of hazardous chemical substances
11. The refusal of a worker to do work due to health and safety issues
12. All safe work procedures

EXAMPLE OF A SHE PROCEDURE FOR LADDERS

Doc. No.	Form Title	Custodian	Location	Retention	Destruction
	Ladder inspection register				
	Ladder corrective action sheet				

	RESPONSIBILITY
a. Portable Ladders a. All ladders used must comply with the provisions of General Safety Regulations 13 A. b. A competent person must be appointed to inspect all portable ladders at least once per month and record the findings in the appropriate register. c. No custom or "homemade: ladders may be used unless it complies with the requirements of the Occupational Health and Safety Act 85 of 1993 and Regulations". d. All ladders must be constructed of sound material and be suitable for the purpose of which it is intended. e. The rungs must be properly set into the stiles and not fastened with nails, screws or spikes. f. No ladder longer than 5 meters which may require to be leaned against a structure or object may be used and all ladders must be lashed, held in place or otherwise secured while in use. g. The reach of a ladder will not be extended by fastening together two or more ladders. h. Suitable sheaths or receptacles in which hand tools may be kept must be provided and special precautions must be taken to prevent articles from falling when work is being done from a ladder. i. Wooden ladders may be treated with oil or wood preservation or be painted with clear varnish but not be painted or covered in any manner which disguise any cracks or inherent weaknesses. j. All ladders which are defective, or damage must be immediately removed from service for repair or be discarded and are not to be removed from safe storage for working purposes. Records of such actions should be kept and damage incident reports completed. **b. Staircases, Platforms and Stairways** • Staircases with more than four steps are to be provided with a single handrail.	

- Staircases with more than eight steps are to be provided with a handrail on either side.
- Toe-boards are to be fitted where needed, to prevent objects falling.
- When storing on top of any elevated platform/ roof or container, handrails, toe-boards and fixed ladders will be introduced.
- Stairways, landings, and platforms are to be kept clear and maintained in a safe condition.
- The surfaces of all staircases are to be non-slip.
- All work areas or platforms of more than 1500mm above floor level are to be equipped with a handrail which:
  a. Must be at least 900mm and not more than 1000mm above floor/platform level.
  b. To be equipped with a middle rail
  c. The top rail will be properly fixed to withstand an impact pressure of 100kg from any direction
  d. Vertical supports are to be provided at intervals of not more than 2500mm
  e. Toe-boards, 150mm high are to be provided.

c. Walkways

a. Elevated walkways are to be sufficiently wide to allow two persons to comfortably pass each other and are to be provided with a handrail between 900mm and 1000mm in height from the surface of the walkway and provided with toe-boards at least 250mm high.
b. Elevated walkways are to be used for access only and not used as work platforms.
c. Walkways are to be kept clear of materials, tools, and equipment at all times so as not to prevent emergency evacuations if it becomes necessary.
d. Walkways are to be wide enough to facilitate the quick and safe egress in the event of an emergency.

d. Fixed Ladders (where applicable)

a. All fixed ladders are to be identified.
b. Fixed ladders are to be inspected once a year.
c. All fixed ladders of which the height exceeds 5 meters are to be provided with a cage.
d. The cage is to extend, from a point, not more than 2.5 meters from the bottom, to a height of at least 900mm above the top level served by the ladder.
e. The cage is to provide firm support along the entire length for the back of a person and no part of the cage shall be mo0re than 700mm distant from the plane of the rungs
f. The rungs are to be at least 1500mm in distance from the structure to which the ladder is secured.
g. A landing of at least 600mm x 800mm is to be provided, where possible, at 900mm intervals. The stiles are to be at least 400mm apart.
h. All ladders are to be numbered in a standardised method 1 metre from the bottom of end of the stiles in black lettering.
i. The bottom rung of the cage of any fixed ladder must be closed off to prevent unauthorised access.

e. Ramps (where applicable)

- Every ramp shall be constructed
  - In accordance with accepted technical standards.
  - With a safety factor of at least 2 with respect to the load it is expected to carry. Provided that the design makes sufficient provision for the load if it is used for vehicles.
  - To have an incline of not more than 34 degrees from the horizontal.
  - Every ramp of which the inclination renders additional foothold, and in every case where the incline is more than 14 degrees, it shall be provided with stairs.
  - Are placed at suitable intervals that extend the full width of the ramp.
  - Stairs may be interrupted over the width not exceeding 230mm to facilitate movement of barrows.

• Every ramp that is higher than 2m shall be provided: • With handrails, at both sides, which are at least 900mm and not more than 1000mm in height. • With toe-boards that are at least 150mm high and so affixed that no open spaces exist between the toe-board and the ramp.	

• ADHERENCE TO SHE POLICIES AND PROCEDURES

One of the first steps to ensure compliance with SHE policies and procedures in the organization is to identify, for example, all the work activities, products produced, services delivered, and the geographic location of the organization. All applicable and relevant legal, regulatory, and other requirements associated with these activities, products and services need to be identified. These include what legislation, regulations, standards, and codes of practice are applicable to the organisation. The policies and procedures of the organization are developed from and aligned with these. The details of how these are going to be implemented is in direct relation to the legal requirements.

The SHE policy is a high-level statement by senior management declaring the intentions of the organization with regards to how it intends to manage its SHE system. It is a binding legal document or statement of intent which can be used against the organization in a court of law should it be found not to be complying with its own directives.

SHE procedures are documents that set out the ways in which the policy is operationalized in the organization. For example, the organization develops a procedure that details how it will control and operate with ladders in the organization. The procedure is developed based on the legal requirements and is an aid to inform all parties concerned as to how to use ladders. Therefore, not adhering to the SHE procedures of the organization is the same as not adhering to the law.

It is better practice to exceed the provisions of the legislative and regulatory requirements which in reality are generally minimum standards. It is preferable for the organization to not be restricted to a culture of mere compliance with legislation, regulations, standards, and codes of practice but rather to be driven by principles of continuous improvement and better practice.

• COMMUNICATION OF SHE POLICIES TO WORKERS

Many organizations have some of the most comprehensively developed SHE policies and procedures. However, if these are not effectively communicated to all workers in the organization the policies and procedures are worthless. Therefore, they must be communicated because:

- The SHE policy is the corporate standard of the organization and sets the targets or objectives that all procedures must achieve.
- The policies and procedures detail how activities, products and services are to be performed or produced in a safe, healthy, and environmentally compliant way.
- Under the Act and MHS Act employers have a duty to inform their workers of:
  - The hazards associated with their work activities.
  - The safest way to execute these work activities
  - The proper care, maintenance, use and limitations of PPE
  - How to protect themselves from exposures to hazards and harm

Workers are responsible to perform work activities and tasks as they are directed to by the organization and by the law and are best able to do so if they are properly informed of the requirements and expectations of the organization and the requirements of the relevant laws and regulations.

Duty to inform

Without derogating from any specific duty imposed on an employer by the OHS Act, every employer shall:

- As far as reasonably practicable, cause every worker to be made conversant with the hazards to his health and safety attached to any work which he has to perform, any article or substance which he has to produce, process, use, handle, store or transport and any plant or machinery which he is required or permitted to use, as well as with the precautionary measures which should be taken and observed with respect to those hazards;
- Inform the health and safety representatives concerned beforehand of inspections, investigations or formal inquires of which he has been notified by an inspector, and of any application for exemption made by him in terms of section 40; and
- Inform a health and safety representative as soon as reasonably practicable of the occurrence of an incident in the workplace or section if the workplace for which such representative has been designated.

Legislation frequently details the requirements for workers to be informed of relevant information such as, for example:

(a) Hazardous Chemical Substance Regulations
(b) The OHS Act
(c) MHS Act
(d) Construction Regulations

Workers can be informed about the content of the SHE policies and procedures of the organization through

- Mandatory SHE Inductions
- Formal competency-based training
- Informal on-the-job training
- Verbally through instructions given by supervisors
- Signage

One of the functions of management and supervisors includes the allocation of resources so that the entire workforce can work safely with no threat to their health. Management must ensure, inter alia,

- Appointment of health and safety representatives to serve on the health and safety committee
- Appointment of responsible workers and supervisors who are willing and able to work responsibly
- Allocation of a budget for resources such as PPE, and functional and safe equipment.
- Supply of materials needed to work safely
- Provision of plant and equipment that are safe and in good working order.

INFORMATION AND TRAINING THE EMPLOYERS MUST PROVIDE FOR THEIR WORKERS

• OHS ACT SECTIONS 8 AND 13

In terms of these sections, the employer shall inform his workers of the hazards relating to the tasks they are required to perform.

The following is a summary of training an employer is expected to provide to workers:

• INDUCTION TRAINING

- Health and Safety Policy of the organization.
- Duties of the Employer.
- Duties of the Workers.
- Colour Coding used on Site.
- Personal Protective Equipment.
- The importance of machine guarding.
- Health and Safety Representatives and Committee Structures.
- Scope of Authority.
- Hazardous chemicals on site.
- Emergency Evacuation Procedure.
- First-Aiders and Procedure on Site.
- Lockout/Isolation Procedure.

• WORK PROCEDURE TRAINING

Work procedure training on the following is recommended, namely:

- Hazards and risks attached to work to be performed.
- Step by step Safe Work Procedures for each task to be performed.
- Precautionary measures to be followed.
- Personal Protective Equipment to be used for each task.
- Emergency measures in case of exposure to hazards

• SPECIFIC TRAINING

Specific training is necessary for tasks or requirements in the workplace and can include:

- First-Aid Training.
- Health and Safety Representative Training.
- Forklift Operator Training.
- Use of Lifting Equipment Training.
- Scaffolding Inspection Training.
- Fire/Emergency Team Training.

• SHE OBJECTIVES AND TARGETS

The organization should have a policy statement in its SHE programme that reflects its commitment to effective SHE management in the workplace. An example of such a statement could be:

SHE Policy

The organisation .... (name of business) wants its workplace to be a healthy and safe environment. To achieve this objective, our organisation will establish and maintain a SHE programme designed to prevent injuries and disease and reduce negative impacts on the environment.

We are responsible for providing our workers with adequate instruction in SHE and for addressing unsafe situations in a timely, effective manner. All workers and service contractors are required to work safely and to know and follow our organisation's guidelines for safe work procedures.

Signed: ______________________________

Date: ________________________________

• RESPONSIBILITIES OF EMPLOYERS

These employer responsibilities include:

- Establishing the SHE programme in the organization
- Conducting an annual review of the SHE programme in each year
- Training supervisors in SHE
- Providing a safe and healthy work environment for the entire workforce

• RESPONSIBILITIES OF SUPERVISORS

The responsibilities of supervisors include:

- Ensuring that all new workers in the plant or site even if they are existing workers undergo SHE orientation or induction
- Providing ongoing on-the-job SHE training to workers
- Taking part in SHE inspections and investigations whenever these are conducted
- Reporting any safety, health, or environment hazards to senior management
- Correcting unsafe acts and conditions as soon as they manifest themselves

• RESPONSIBILITIES OF WORKERS

To effectively manage SHE is the organization every worker has responsibilities as well. The responsibilities of workers include:

- Learning and implementing all safe work procedures or operations (SWPs or SOPs)
- Correcting and avoiding exposure to hazards or reporting them immediately to their supervisors
- Participating in SHE inspections and investigations where appropriate and as required
- Using personal protective equipment where required after the likely exposure to hazards has been mitigated
- Helping to create a safe and healthy workplace by recommending ways to improve the SHE programme in the organization

• REVIEW QUESTIONS

1. What are the expected outcomes of a SHE policy?
2. What should be included in a SHE policy?

3. Why are SWPs or SOPs necessary?
4. What are the responsibilities of the various parties in terms of SHE objectives and targets?
5. Why should a SHE policy statement be concise and short?
6. Compile a SHE policy statement for your organization which is short and can fit on a credit card so that all the workers in your organization can learn and memorize it.

• REFERENCES

Agumba, Justus. and Haupt, Theodore. 2009. Management commitment and employee involvement metrics for improving H&S performance at project level of Small and Medium construction enterprises in the South African construction industry: An exploratory study. Proceedings of 6th CIDB Postgraduate Conference on Construction Industry Development, Sustainable Construction through Quality and Innovation Beyond 2010, 6-8 September 2009, Johannesburg, ISBN 0-86886758-6

Agumba, Justus. and Haupt, Theodore. (2011): Critical indicators for measuring health and safety performance improvement of Small and Medium construction Enterprises in South Africa, The 6th International Conference on Construction in the 21st Century (CITC-VI), "Construction Challenges in the New Decade," Kuala Lumpur, Malaysia, July 5-7

Agumba, Justus., Pretorius, Jan. Harm. and Haupt, Theodore. Conrad. 2013. "Health and safety management practices in small and medium enterprises in the South African construction industry" Acta Structilia, 20 (1), 66-88, ISSN 10 23 0564

Agumba, Justus. and Haupt, Theodore. 2015. Validating and Identifying Health and Safety Performance Improvement Indicators: Experience of using Delphi Technique, Journal of Economics and Behavioral Studies, 7 (3), 14-22, ISSN 2220-6140

Agumba, Justus. and Haupt, Theodore. 2018. The influence of health and safety practices on health and safety performance outcomes in small and medium enterprise projects in the South African construction industry. Journal of the South African Institute of Civil Engineering, 60(3): 61-72

Dabup, Na'ankwat. (2012). *Health, Safety and Environmental Implications in Nigeria's Oil and Gas Industry*. PhD. Nelson Mandela Metropolitan University.

Department of Labour (1993). Occupational Health and Safety Act No: 85 of 1993.

ISO (2015). *ISO14001 ENVIRONMENTAL MANAGEMENT SYSTEMS*. [online] Iso.org. Available at: https://www.iso.org/files/live/sites/isoorg/files/archive/pdf/en/pub100329.pdf.

OHSAS 18001:2007 (2007). *OHSAS 18001:2007 OCCUPATIONAL HEALTH AND SAFETY ASSESSMENT SERIES*. [online] Producao.ufrgs.br. Available at: http://www.producao.ufrgs.br/arquivos/disciplinas/103_ohsas_18001_2007_ing.pdf.

Manuele, Fred. (2013). *On the practice of safety, fourth edition*. Hoboken, N.J.: John Wiley & Sons.

Ridinger, Richard. (2012). *Policy on Environment Health and Safety*. [online] Lonza.com. Available at: https://www.lonza.com/~/media/Files/Global%20Citizenship/Pdfs/2012_01_EHS_EN.ashx?la=en.

Ridley, John. (1990). Employer's obligations for safety. *Safety At Work*, pp.306-317.

CHAPTER 5

COMMUNICATION

INTRODUCTION

Communication is a two-way process between a receiver and a sender that involves the sending and receiving of symbols, signs or signals that could be words, pictures, things, or actions. The accuracy and consistency of the words and symbols used to communicate information is important to achieve the desired mutual understanding. The success and effectiveness of communication involves the correct use of language that includes words, pictures, and body language, and how these are received and interpreted. It is speaking and listening, clarifying, and explaining, writing, and reading, behaving and observing behaviour, reinforcement and feedback. Its goal is to achieve understanding. South Africa with its 11 official languages besides colloquial slang make this goal more challenging. These differences in language must be accommodated in the forms of SHE-related communication.

Communication has been defined as simply: "Communication is what we do to give and get understanding." Communication has also been described as an action that provides information that is relevant and meaningful to persons receiving the information. This information might not have the same meaning to different people and may not produce the same outcomes that are typically manifested in behaviour and actions.

Effective communication involves cognition, a process of transforming and contextualizing sensory information to enable understanding, storage, recovery, and use. It therefore follows that when information stands on its own, it is not relevant to previous information and cannot be assimilated, cognition does not occur. Language used in the workplace or on the factory floor is different from that used in an office environment as workers create their own work-related vernacular to communicate among themselves and that vernacular should be taken cognizance of.

Further, information that is shared must account for personal capability to process the information itself as well as the volume of it. A phased approach over time is advocated using different communication methods that include reiteration and repetition to avoid boredom, monotony, and apathy. Correct interpretation of information received is dependent on personal experiences, previous knowledge, schooling, training, attitudes, and emotions.

Increasingly today social media platforms such as WhatsApp chat groups, Facebook groups, Instagram, and similar evolving platforms and Apps are proving to be popular and effective forms of communication and should be considered as part of organizational SHE communication strategies.

• 11 PRINCIPLES OF EFFECTIVE COMMUNICATION

1. Motivation to accomplish results tends to increase as people are informed about matters affecting those results.
2. Principle of line loss: The effectiveness of communication tends to vary inversely with its extension.
3. Principle of emotional appeal: Appeals to emotion are communicated more readily than appeals to reason.
4. Principal of application: The more a communication is applied or put to work, the better the message is understood and remembered.
5. Concrete: the message has to be concrete and supported with data that backs it up.
6. Coherent: is logical. It is a well-planned, logical and sequential manner of communication.
7. Clarity: Being clear about the goal or purpose.
8. Commitment: A well-committed message will leave a greater impact.
9. Consistency: The choice of words matters when communicating at work. Words should not leave the worker confused. Employers need to communicate messages in the least possible words, and have the consistency of tone, voice, and content so that they can save time.
10. Completeness: Each message must have a logical conclusion. People shouldn't be left wondering if there is more to come. Communication must be complete to ensure the need to be informed and act.
11. Courteous: Communication must be conducted in an honest, respectful, open, and polite manner. It should be conducted with proper care and it will be perfectly effective and important. Offensive words can put off people.

- GROUP MEETINGS

Effective group meetings serve several basic management purposes in the organization:

- ✓ They are often the only way that management can be sure that everyone in the workforce has received a critical message in the timeliest way.
- ✓ They create a co-operative climate through participation and group interaction.
- ✓ They help give everyone in the workforce the same attention and exposure to information.
- ✓ They give the person conducting a group meeting practice in improving communication and human relations skills.
- ✓ They help build the image of the person conducting the meeting to be viewed as a leader in the minds of other team members.

- PERSONAL COMMUNICATION

Personal communication normally occurs in five key areas of day-to-day responsibilities at work, namely:

- ✓ Individual job or work activity and workplace orientation or induction.
- ✓ Job/task/work activity instruction.
- ✓ Planned personal contacts.
- ✓ Key point tips whenever needed.
- ✓ Job performance coaching which is a more recent type of targeted personal communication.

Individual job orientation

Numerous SHE studies show that new workers are almost twice as likely to have an accident as experienced workers. This phenomenon also applies to workers who are being transferred between work locations, plants and sites. Organizations with formal orientation or induction for all new hires and transferees have average workers' compensation "modification rates" of around 25% lower than organizations that do not have any formal orientation programs for new hires and transferees.

Workers who are new to the job and to the work environment are especially at risk. Good and effective supervisors use proper orientation to help new or transferred workers get safely through that critical period without harm.

Orientation should take needs and feelings of workers into consideration. Therefore, good orientation or induction requires empathy, the ability to be in the shoes of the workers, and to see things from their viewpoint. It should not be a one-sided presentation of organization facts and figures. It should be a two-way process of reaching mutual understanding, of welcoming the new workers or transferees into the organization, helping them to become familiar with the work environment and laying the basis for the desired SHE knowledge, skills, and attitudes.

It is necessary for every supervisor to be conscious that first impressions tend to be lasting impressions, and that no one ever gets a second chance to make a good first impression. The orientation or induction process provides a unique opportunity to start building a solid supervisor-worker relationship, to let the behaviour of supervisors show their care and their concern for the new workers or transferees. It allows supervisors to determine what kind of mental image they want workers to have of them and then work to establish that very image.

Job or Task instruction

Proper job or task or work activity instruction is about how to get workers to do jobs or tasks correctly, quickly, conscientiously, and safely. Effective instruction is a systematic substitute for trial-and-error learning and a reliable replacement for hit-or-miss or random instruction. Its two basic goals are:

- ✓ To help motivate the worker to do the task or work activity properly and safely.
- ✓ To ensure the worker knows how to do the job properly and safely.

Instruction technique

The best technique involves four easy to remember steps which are, namely Motivate, Tell and Show, Test, and Check.

a) Motivate

- ✓ Put the worker at ease.
- ✓ Find out what the worker knows about the task or work activity.
- ✓ Position the worker properly.
- ✓ Stimulate the interest of the worker.
- ✓ Emphasize the importance of the task or work activity and the performance of the worker.

b) Tell and show
- ✓ Demonstrate the operation.
- ✓ Use a step-by-step approach.
- ✓ Emphasize key points.
- ✓ Give clear and complete Instructions.

c) Test
- ✓ Have the worker tell and show to determine whether the worker has learnt the task or work activity.
- ✓ Have the worker explain the key points.
- ✓ Ask questions and correct or prevent errors.
- ✓ Continue until the worker knows what they should know.

d) Check
- ✓ Tell the worker who to go to for help.
- ✓ Allow the worker to work on his or her own.
- ✓ Follow up often; answer questions; review key points.
- ✓ Reinforce positive parts of performance.

Planned Personal Contacts

Whether or not one has planned group contacts such as, for example, SHE meetings, toolbox talks, or tailgate sessions, one should have planned individual contacts with each worker. Individual contact is an excellent supplement to group meetings and is especially critical when conditions do not permit getting people together for group meetings.

Many personal contacts are spontaneous. They are not planned, but just happen in the normal course of events. These informal contacts provide hundreds of chances to give key point tips and reinforce desired attitudes and behaviours. However, these chance contacts should be supplemented with planned personal contact. Furthermore, by keeping an open line of communication, one is better positioned to receive valuable feedback from workers without having to pursue it. Worker contact and ongoing communication is one task that can demand a lot of time, but the rewards are huge.

Planned individual contacts give opportunities for:

- ✓ Personalizing critical aspects of SHE, quality, productivity and cost control for each worker.
- ✓ Building better SHE awareness and attitudes.
- ✓ Showing each worker personal concern for proper practices and conditions.
- ✓ Improving supervisor-worker relationships.

Planned individual contacts enable the best use of the time invested by direct contact with individual workers. Using SHE contacts as an example; a planned personal contact is an intentional get-together of a supervisor and a worker to discuss a critical SHE topic related to the work of the worker. This allows one to cover a topic most suitable to the individual and tailoring the discussion to that person and his or her own work ... a specific topic for a specific person doing a specific job in a specific situation.

These contacts should be frequent enough to influence the attitudes, knowledge and/or skills of the worker. The specific number will vary with the number of workers the supervisor has, the size of the area they work in, and similar factors. Effort should be made for contact with each person at least every week. A minimum standard should be at least one planned contact with each worker each month. The use of any or all the available means of communication should be encouraged including social media platforms that include, for example, Skype, Hangout, Twitter, Instagram and Zoom.

Five key steps in conducting planned personal contact are, namely:

(1) pick a critical topic
(2) prepare the contact
(3) make the contact
(4) record the contact and
(5) follow up.

Key point tips

"Key points" are those vital bits of information that make or break the job; special tricks of the trade that make the task more efficient; the feel or knack that is the mark of a professional; critical points of quality, productivity, cost control or health and safety.

"Tip" means just what it does in everyday life; a piece of information given to be helpful; a small gift; a hint or suggestion. Key point tipping, then, is the organized process of giving workers helpful hints, suggestions, reminders, or tips about key quality, production, cost or health and safety points in their work.

The best tips are short tips. And they should always be given as reminders rather than formal instruction. Here are some examples:

- ✓ **Health and safety tip** — "Skosana will you ensure that you always keep the guard of that circular saw in place as it would be tragic if you lost a finger if you did not"
- ✓ **Production tip** — "Koos, you can increase the amount of bricks that you lay if you had stacks of bricks laid out along the length of wall you are building"
- ✓ **Cost tip** — "Piet will you go around the site every day and pick up all the nails that are lying around? Not only will the site be safer, but those nails can be reused. "
- ✓ **Quality tip** — "Siya you know that you always have time to do something over but never enough time to do it right the first time. Let's change that today!"

Job performance coaching

Job performance coaching is the day-to-day actions taken to help workers perform as best as they can. Effective coaching is "leadership in action" and involves motivating, communicating, and developing people. It is "management control in action", built on the activities of work identification, standards, measurement, evaluation, correction, and commendation. Effective job coaching is an excellent means for building better relationships between supervisors or managers and their workforce.

Each worker in the organization must know what is expected of them, how they are doing and what they should do to improve. They look to the supervisor or manager for guidance. Supervisors or managers must provide that guidance and to lead workers to a winning performance in an efficient, safe, healthy, and productive work environment

Job performance coaching is based on the simple, basic principle that every worker has a right to know, namely

- ✓ What their job is.
- ✓ The job performance metrics.
- ✓ How they are doing.
- ✓ Specific steps to improvement.

When workers do not know these, the consequences are almost always unpleasant such as, for example:

- ✓ When workers do not really know what their jobs are, the results are confusion, wheel spinning, and losses such as waste, damage and injuries.
- ✓ When workers do not know their performance metrics or goals, the results are guesswork, wrong priorities, misdirected energies, or hit-and-miss efforts.
- ✓ When workers do not know how they are doing, the results are low motivation, poor morale, and measurement by mind reading.
- ✓ When workers do not know specifically how to improve, the results are glittering generalities, procrastination, and status quo.

• COMMUNICATION STRUCTURE

The awareness that communication is both a core skill and a central problem for the SHE manager is not a new one. Most workers do not like being ineffective: being effective is easier and more rewarding that being ineffective. The reason why good intentions are simply not sufficient can be understood if consideration is given about what SHE communication is, and how SHE communication is thought about within the organization. Examples include the following:

- ✓ SHE communication is too commonplace.
- ✓ SHE communication is too massive.
- ✓ SHE communication is too fast.
- ✓ SHE communication is too intangible and nebulous.
- ✓ SHE communication is a gift.
- ✓ SHE communication is an art.

The importance of effective communication in the workplace cannot be overemphasized. It is important in problem solving, conflict resolution, for positive working and personal relationships, and in reducing the stresses associated with interpersonal interactions. Here are a few extra final tips extracted from https://www.nifc.gov/hrsp/tools/keys_to_effective_communication.pdf:

- ✓ Open the door to two-way conversation. Encourage questions. Ask questions.
- ✓ Consider the communications strengths and weaknesses of the person you are engaging with and communicate in the manner that is best accepted by them.
- ✓ Paraphrase what you heard back to the speaker, to ensure you have a common understanding.
- ✓ Do not be thrown off course by words that affect you emotionally. Continue to listen even when the urge is to start debate.
- ✓ Communicate to be understood. Many people communicate to impress – not express. Use short words that communicate clearly and concretely; present one idea, at the most two ideas, in one sentence. Avoid jargon. Use strong verbs. Avoid passive voice as much as possible.
- ✓ Be open to feedback.
- ✓ Be an active listener.

Active Listening Tips

- ✓ Listen with purpose. Ask yourself "What worthwhile idea is being expressed?" ... "What is being said that I can use?"
- ✓ Judge content not delivery. Look beyond the person's delivery and concentrate on what is being said.
- ✓ Keep your emotions in check. Avoid becoming over stimulated by what the person says. Avoid allowing your own biases or values to detract from the person's message.
- ✓ Listen for the main idea or central themes of the message.
- ✓ Be flexible. Find a variety of ways to remember what you hear.
- ✓ Find a variety of note keeping techniques to help you remember.
- ✓ Work at listening. Give your full attention to the speaker. Face the speaker. Use facial expressions that indicate you are following what the speaker is saying.
- ✓ Resist distractions. Concentrate on what the speaker is saying. Stay focused on the situation at hand.
- ✓ Keep an open mind. Avoid developing blind spots regarding cherished convictions. When you hear "red-flag" words keep your emotions in check.
- ✓ Capitalize on thought speed. Most of us talk about 120 words a minute. Our thinking speed is about 500 words a minute. Therefore, there is a lot of time to spare while a person is speaking to us. Don't let your mind wander and then dart back to the conversation. Use the spare time to advantage by thinking about what is being said; try to anticipate the point; mentally summarize the point so far; mentally question any supporting points; look for nonverbal clues to the meaning.

Non-Verbal Communication

About 65% of the message is sent non-verbally. Become sensitive to non-verbal messages. Look for such things as, for example, body position and movement, gestures, facial expressions, eye contact, silence, and use of space and time.

- Eye Contact. If someone is looked in the eye, they pay more attention to what is being said.
- Posture. Good posture is a sign of confidence and creates a sense of trust in the skills and abilities of the one communicating.
- Gestures. Body movements and gestures should be used that are necessary to make the point. Excessive motions are distracting to the listener.
- Expression. Expressions sometimes say more the words being spoken. Smiling and showing interest while speaking are effective.
- Voice. Speak with a firm and assertive quality. This tone implies directness and honesty.

Avoid Communication Stoppers

Examples of behaviours and phrases that often stop a person from communicating and which should be avoided if effective communication is to happen, namely:

Ordering	"Don't talk like that."
Warning	"If you do that, you'll be sorry."
Moralizing	"You ought/should...."
Advising	"I suggest that you..."
Reason with	"Let's look at the facts."
Diagnosing	"You feel that way because..."
Judging	"You are wrong about that."
Name calling	"You are acting like...."
Distracting	"Let's talk about something else."
Interrupting	"But what about..."

• REVIEW QUESTIONS

1. Why is effective communication so important?
2. How would you improve your communication with workers in your workplace?
3. Why is it important to be an active listener?
4. What would be some of the pitfalls to avoid that negatively affect communication at work?
5. How would you use various ICT platforms to communicate within your organization in the event of a national disaster such as the recent COVID-19 pandemic?

• REFERENCES:

Anderson, David. S. and Miller, Richard. E. 2017. *Health and Safety Communication: A Practical Guide Forward.* Routledge (Taylor and Francis Group). ISBN 978-1-138-64744-2

CCOHS (n.d.). *Worker Orientation Checklist.* [online] Ccohs.ca. Available at: https://www.ccohs.ca/oshanswers/hsprograms/orientation.html.

Haupt, Theodore. Conrad., Munshi, Meenakshi., and Smallwood, John. 2004. Combating HIV and AIDS in South African Construction through Effective Communication, *Acta Structilia*, 11 (1 and 2), 26-43, ISSN 1023-0564

Haupt, Theodore. Conrad., Munshi, Meenakshi., and Smallwood, John. 2004. The Role of Communication in Combating HIV and AIDS in Construction," *Revista Ingenieria De Construccion*, 19 (2), 111-120, ISSN 0716-2952

NIFC. (n.d.). Keys to effective communication. https://www.nifc.gov/hrsp/tools/keys_to_effective_communication.pdf

Nkolimwa, Dominic., Jani, Dev., and Dominic, Theresia. 2020. Management practices on occupational health and safety in the Tanzanian's small scale mining firms: does compliance cost matter? *Business Management Review,* 22(2). 16 pages

OSHA (n.d.). *Education and Training | Occupational Safety and Health Administration.* [online] Osha.gov. Available at: https://www.osha.gov/shpguidelines/education-training.html.

Pfeiffer (1998). *The Pfeiffer Library.* [online] Home.snu.edu. Available at: http://home.snu.edu/~jsmith/library/body/v25.pdf.

University of Wisconsin--Milwaukee (n.d.). *Principles of Communication.* [online] Www4.uwm.edu. Available at: https://www4.uwm.edu/cuts/bench/commun.htm#princ.

Vigorosa, Lucia., Caffaro, Federica., and Cavallo, Eugenio. 2020. Occupational safety and visual communication: User-centred design of safety training material for migrant farmworkers in Italy. *Safety Science.* Vol. 121. 562-572

CHAPTER 6

IMPORTANCE OF RISK MANAGEMENT

INITIAL THOUGHTS

When one thinks about risk and risk management the following come to mind:

- '…the entire history of the human species is a chronology of exposure to misfortune and adversity and of errors to deal with these **risks**...'
- Where action can be deemed to have consequences, then it is the degree of uncertainty in those circumstances that can be **risk**
- Every time a choice is purposely made, **risks** are played off against each other
- '**Risk** is the hidden danger lurking beneath everything that we do…'
- No single **risk-free** aspect to our lives
  - Can be apparent
  - Can be obscure

Risk is pervasive and can either be considered consciously or most of the time sub-consciously. Risk can be perceived as situations in which a decision is made, the consequences of which depend on the outcomes of future events having known probabilities. One can also suffer from risk homeostasis in terms of which

- The reduction or elimination of risk will increase other risks or likely acceptance of other risks in return or
- The more certain risks are mitigated the more blind one becomes and willing to accept of other risks

For example, if one wears a seatbelt to prevent flying through the windscreen in the event of an accident, one could conceivably be willing to drive faster as a result which in itself would be another risk.

• FUNDAMENTAL PRINCIPLES OF RISK MANAGEMENT

Risk management is a management function with the objective of protecting people, assets, environment, and income by avoiding or minimising the potential for loss from pure risks, and the provision of funds to recover losses.

The primary functions of management are planning, organising, leading and control.

• PLANNING

The purpose of every plan is to facilitate the attainment of the objectives, mission, and goal of an organization. Plan your work and work your plan! Effective SHE management is an objective which any organisation must achieve. Therefore, the organisation must plan to reach its SHE goals. Some of the contributions of planning include:

- ✓ Giving direction to an organisation.
- ✓ Promoting co-operation between departments and workers in an organisation.
- ✓ Compelling managers to look towards the future for modern technology.
- ✓ Emphasizing interdependence of various functions to achieve common objectives.
- ✓ Encouraging proactive management of change.
- ✓ Increasing worker efficiency.
- ✓ Improving flexibility of people and processes.
- ✓ Reducing costs.
- ✓ Reducing risks in operational and maintenance activities.
- ✓ Achieving effective environmental controls.

Steps in the planning process

Planning always starts with the formulation of the goal or objective which an organisation wants to achieve. Typical questions that should inform the process include: -

- ✓ What is the objective of the plan?
- ✓ What would it mean to the organisation?
- ✓ What results should be achieved and be avoided?
- ✓ What resources are needed?
- ✓ What restrictions need to be considered?
- ✓ How is commitment fostered?
- ✓ How is buy-in realized?

Senior or top management need to emphasize clearly that they attach the same importance and are committed the risk prevention and control programme as to any other organisational objectives. Only then will it be possible to have an effective SHE programme.

- ORGANISING

In order to execute the SHE plan, management needs to organise resources and activities by assigning responsibility and authority to various persons in the organization. This assignment is done in various ways that include:

- ✓ Determining the organisational structure.
- ✓ Clarifying responsibilities of each person.
- ✓ Establishing accountability.
- ✓ Establishing clear channels of communication.
- ✓ Allocating organization resources meaningfully.
- ✓ Promoting cooperation when workers work together to produce a whole that is greater than the sum of the parts.
- ✓ Dividing the total workload into activities.

The importance given to SHE management is reflected in the position of SHE staff in the overall organizational management structure. Equally important is the quality and qualifications of SHE staff and how they are perceived by the workforce. They need to earn recognition and the respect of entire the workforce to be functionally effective.

Organisational relationships must be created to clarify to the workforce who will assign their tasks to them and whom they must report to. Responsibility and authority must be carefully delegated. Authority to do the job means that they have the authority to deploy resources to perform the task. Workers often find that their recommendations are rejected. It is therefore important for management to decide what type and level of authority will be granted to a particular manager which authority could, for example, be in the form of:

- ✓ Advisory capacity;
- ✓ Compulsory advice;
- ✓ Concurrent authority; or
- ✓ Command authority.

The following are examples of each type of function and authority, namely:

Advisory: -	Line managers can choose whether they want to consult the human resources department on the academic qualification required for a SHE officer.
Compulsory: -	The manager will be obliged to consult the human resources department on academic qualifications for SHE officers.
Concurrent: -	The manager and the line manager must agree on qualifications for SHE officers.
Command: -	The SHE managers will give orders on SHE matters that other departments will have to carry out.

To reduce levels of conflict, which may arise between line and other managers, activities should be properly integrated. Responsibility and authority should be clearly demarcated.

Management considers the plan

↓

Analyse primary task in detail

↓

Subdivide primary task into sub-tasks

↓

Combine tasks into departments

↓

Implement co-ordination mechanisms

↓

Evaluate results

The Organizing Process Cycle

- LEADERSHIP

Leadership is one of the most controversial subjects in management. Leadership involves taking the lead in the formulation of plans, influencing people, giving orders, motivating, and handling people, managing conflict and communicating with subordinates. Leadership should happen in such a way that workers willingly pursue the objectives and goals of the organisation. Through leadership members and resources are activated to get things into motion and keep them in motion.

- CONTROL

When individuals have control over certain activities, they can influence the course of events. Control is the process whereby management ensures that the activities of the organisation fit in with objectives and planned activities. Until actual performance is compared with the standards required, management will not know whether activities have been carried out according to plan. Nor will they be able to identify weaknesses.

Control in the SHE system is a very important phase. It is a continuous process, which is interlinked with planning. Performance in terms of SHE must be regularly measured and compared against standards. The knowledge, experience, information, and facts acquired during the control process are the most important inputs in the next cycle of planning.

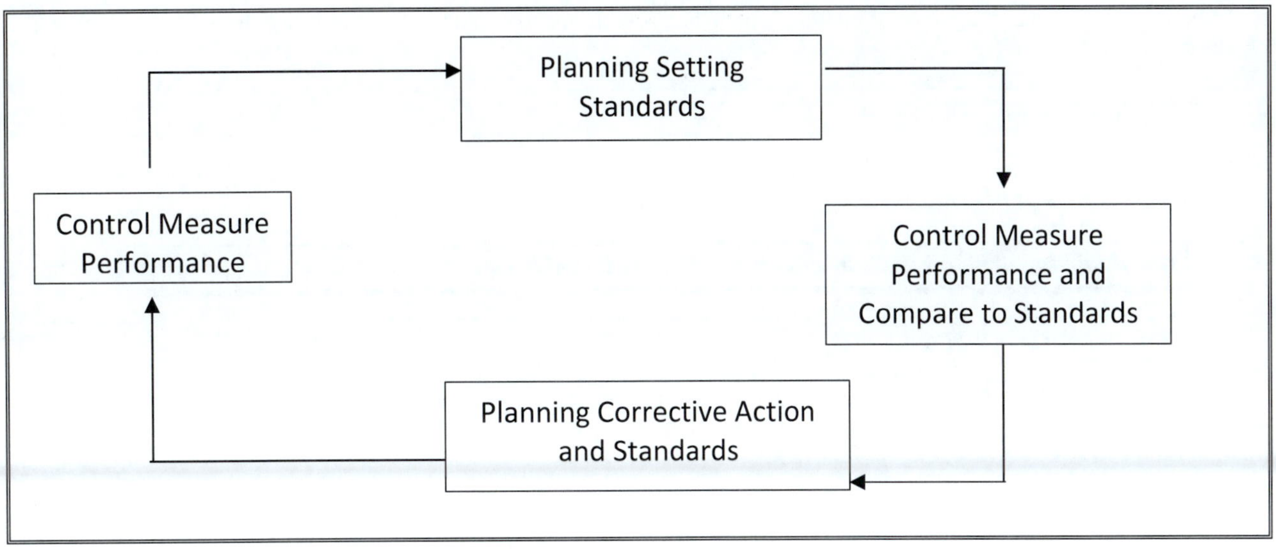

The Link Between Planning and Control

OBJECTIVE OF RISK MANAGEMENT

The objective of risk management includes

- Recognition that risks are difficult to eradicate
- Avoid downside risks (risk avoidance)
- Exploit opportunities
- Make decisions against set of predetermined set of objectives, rules and priorities based on relevant knowledge, data, and information

Put simply, it is the

Protection of people, assets, income, and environment

DEFINITION OF PURE RISKS

A pure risk is a risk, which results in loss, damage, disruption, or injury. Pure risk can usually be insured. Risk management is primarily concerned with pure risk. Examples are fire, theft, floods, incidents, and diseases.

PROVISION OF FUNDS

Risk financing is the provision of funds to cover losses that occur.

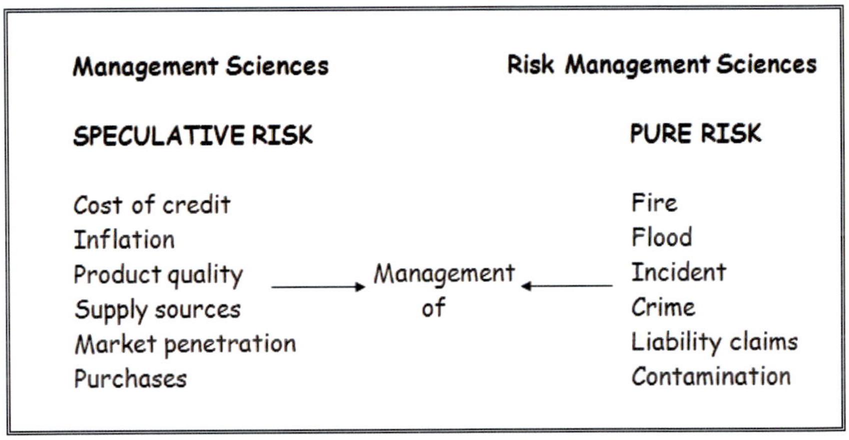

Risk Management Versus Management Sciences

Risk management is a management function, which focuses on specific areas of the business to achieve corporate objectives.

- DEFINITION OF RISK

Risk is the presence of uncertainty and it is measured as the deviation from the expected outcome of a given situation or event. Two aspects of risks are considered when defining a risk, namely frequency and severity.

A definition used in law is: "The possibility of an undesirable change".

- DEFINITION OF FREQUENCY

Frequency is the number of times the event occurs over a given period. One could ask the question: how often will or did it happen? What are the chances or probability of it happening?

- DEFINITION OF SEVERITY

Severity represents the cost of the damage, consequential loss or injury. Here one could ask the question; how serious or severe is it? What is the extent of the impact or consequences?

- CLASSIFICATION OF RISK

There are many ways to classify risk. For example, universal categories are shown in the figure.

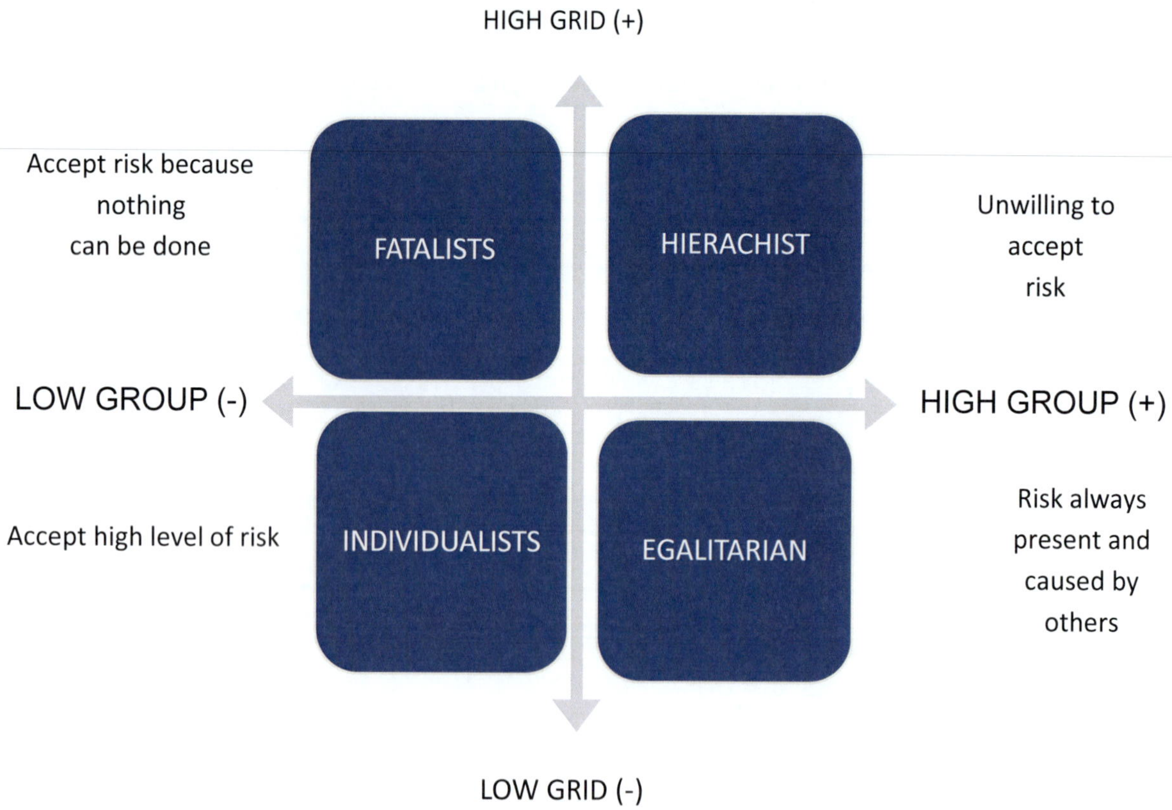

Risks may also be classified either as pure risks (insurable risks) or speculative risks.

Pure risk examples are loss or damage resulting from:

- ✓ Fire
- ✓ Environmental impairment
- ✓ Occupational incidents
- ✓ Occupational diseases
- ✓ Natural and national disasters
- ✓ Crime

- ✓ Business interruption
- ✓ Product liability
- ✓ Contract liability
- ✓ Business information

Speculative risk examples are uncertainty relating to:

- ✓ New products
- ✓ New production processes
- ✓ Changing of supply sources
- ✓ New trade restrictions
- ✓ New market conditions.

Finally, risks may also be classified as:

- Known risks
  - Minor variations in productivity
- Known unknown risks
  - Occurrence is predictable
- Unknown unknown risks

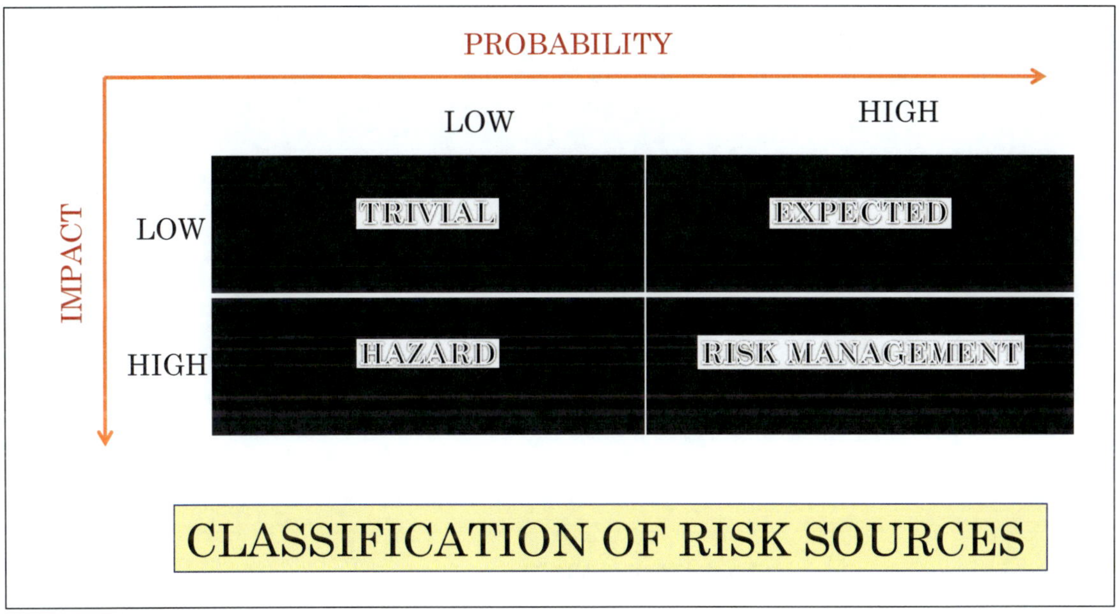

• BENEFITS OF RISK MANAGEMENT

Some of the benefits of risk management include:

- Clarification of project issues from start
  - Promotes uninterrupted progression of activities
  - Promotes communication
- Decisions supported by thorough analysis
- Instils confidence
  - Project
  - Third parties
  - Project team
- Continual monitoring of project definition and structure
- Clearer understanding of specific project risks
- Build-up of historical data
- Promotes project success

- CAUSES OF RISK MANAGEMENT FAILURES

The failure of risk management in organizations can be caused by, for example:

- Human error
- Irrationality
- Cultural misunderstandings
- Failed communications
- Systemic failure
- Homeostatic need

- RISK VERSUS UNCERTAINTY

The concepts of risk and uncertainty while often used interchangeably can cause confusion.

- **Risk** – When decision is expressed in terms of range of possible outcomes and when known probabilities can be attached to outcomes
- **Uncertainty** – When there is more than one possible outcome of course of action but probability of each outcome is not known

- DISTRIBUTION OF RISK

Generally, risk is not distributed evenly across any project or business undertaking. Typically, the risk exposure is greatest during the early stages of the project or business undertaking and decreases over time. The following figure illustrates this trend. There also tends to be a spike as the project or installation draws to a close.

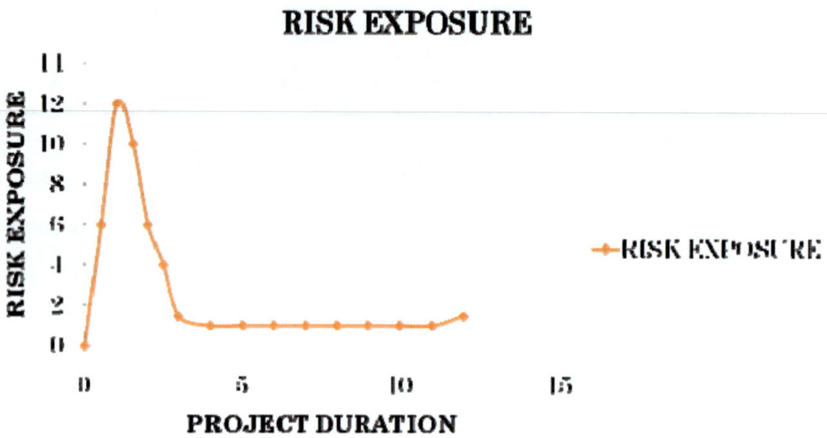

RISK RELATED CONCEPTS

- HAZARD

A hazard is a condition, activity, object, or substance that creates, introduces, or increases the frequency or severity of a loss-producing event. Hazards are anything that may cause a loss-producing event.

- PERIL

A peril describes the risk that is or could be involved. Perils are the source of the loss. (Flammable liquid as a hazard could be the cause of a fire (peril) which constitutes a risk of explosion.)

- PROBABILITY

Probability is an indication of the likelihood that a certain event could occur in the long run. This should be distinguished from risk. Risk is a concept of variability and reflects the uncertainty of an event rather than its underlying probability.

- RISK MANAGEMENT PROCESSES

Several processes are involved in the effective management of risk. These include the following, namely:

• RISK IDENTIFICATION

Risk identification aims at identifying potential situations that might cause losses. These can be due to, or be enhanced by, natural phenomena or by contravening laws.

• HAZARD IDENTIFICATION, RISK ASSESSMENT AND CONTROLS

Each organization should have in place procedures for ongoing hazard identification, risk assessment, and necessary controls. Hazard identification and risk assessment need to be proactive instead of reactive methodologies. Risks need to be identified, prioritized, and documented together with appropriate controls and interventions. The hierarchy of risk reduction needs to be always implemented, namely:

- Elimination;
- Substitution;
- Engineering controls;
- Signage, warnings and/or administrative controls; and
- Personal protective equipment.

• RISK IDENTIFICATION METHODS

There are multiple approaches that can be followed to identify risk. It is important for each organization to decide which approach or method to use depending on the nature and scale of its operations. The aim is to identify, analyse and control hazardous situations before these situations turn into negative incidents. The earlier the SHE personnel become involved, the greater the chances are of identifying risks. This intervention/precaution reduces the probability of occurrence and the severity of the consequences.

Macro-identification

This involves identifying risks, which could arise from major sources and create significant losses and referred to as major perils.

Micro-identification

This is the identification of sub-peril risks and hazards within situations and these should be related to one another.

• RISK IDENTIFICATION TECHNIQUES

Physical inspection

Physical inspection of the workplace can help to identify hazards, which might pose a threat to health, safety, and the environment.

Interviews

These should take place during the inspection to determine levels of health, safety and environmental awareness. People at all levels in the organisation should be interviewed.

Documentary information

The study of documentary information includes, for example, statutory records, process descriptions, logbooks, monthly production reports, standards and codes of practice, organisational charts, accounting data, flow charts, and information from contracts. Other scientifically proven techniques are hazard and operation studies, failure mode and effect analysis, fault tree analysis and hazard indices.

The following figure illustrates the use of a cause-and-effect diagram. In this case there was a need to change the location of activities which resulted in likely changes in other aspects of the activity. The diagram illustrates the cause for each change and its probable effect.

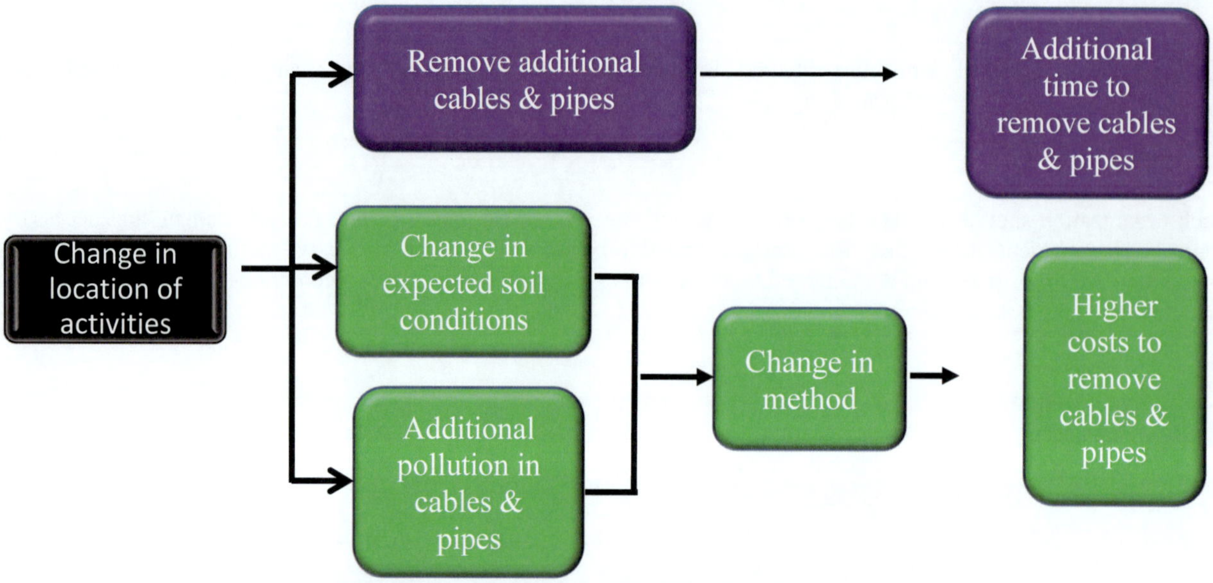

CAUSE AND EFFECT DIAGRAM

Another option could be to use a fault tree analysis as illustrated in the following diagram where discrepancies were found in the records being kept in the stores for cables and pipes. Several areas of likely fault or defect in the system were identified in this way.

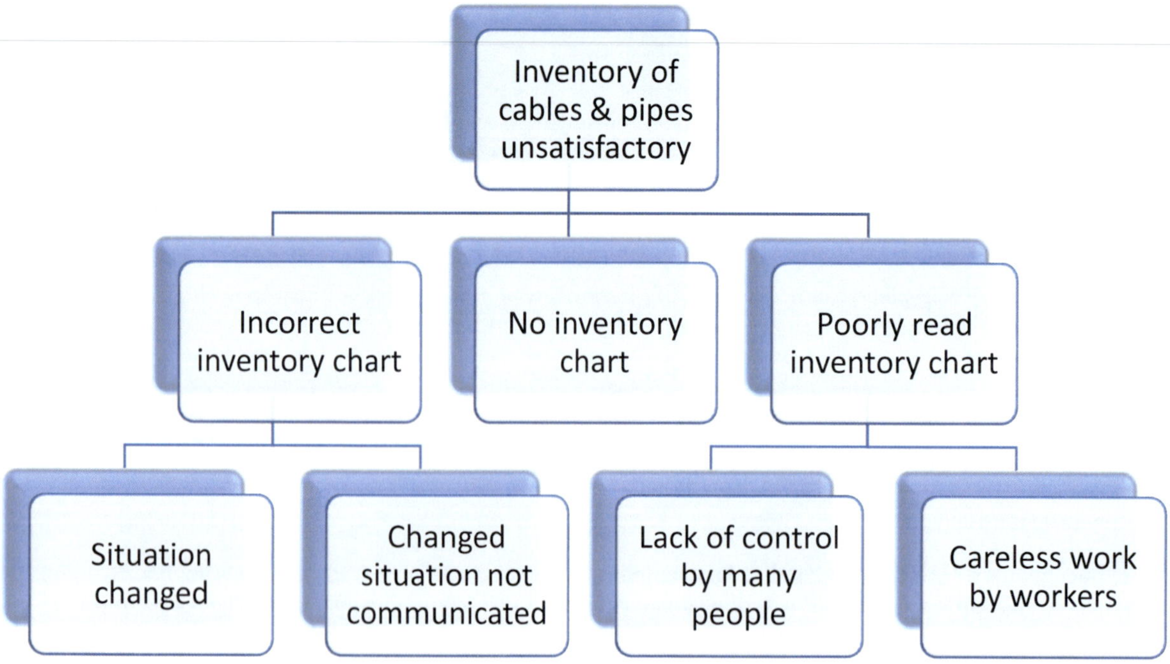

- HAZARD IDENTIFICATION AND RISK ASSESSMENT (HIRA)

The HIRA should be conducted to ensure the reduction or elimination of risk to those who will be involved in any stage of a process, project, or installation or during the maintenance equipment in the plant. The process does not need to be precise, but should deal with known hazards, issues or risks where possible, or reasonable. A competent person is required to conduct the HIRA, and an SHE specialist could be used to assist in this process. However, the HIRA process must commence during the initiation phase, and be addressed during all the other phases. For example, in the case of a major installation in the plant, the design related aspects could include, *inter alia*, concept design in addition to detailed design, while evolving details, schedule, method of fixing, and specifications.

a. Steps to be followed

Steps may vary slightly depending on the type of risk assessment used. It is advisable to assemble a risk assessment team given the subjective nature of the process. However, the risks need to be determined from the initial concept or decision to proceed, followed by the drawings where required such as in the case of a major installation, and the work divided into the various activities followed. Thought must be given to the method of procurement, schedule, installation processes to be followed, and products used to limit exposure.

Material safety data sheets (MSDSs) should be obtained from suppliers to determine short or long-term health risks to workers, where after it is possible to determine if alternative products should rather be used. For example, have solvent-based paints been specified for use in a closed environment where ventilation is likely not to be optimum? In such a case, consideration needs to be given as to whether a water-based paint could be used, as PPE may only be used as a last resort, and where no other safer or healthier option is available.

Previous knowledge is useful, where the work to be done is similar. However, it must be remembered that no two processes, projects, installations, or maintenance operations are alike. For example, the placing of storm water culverts in position invariably entail the same procedure. However, the environmental factors such as ground (need to shore), surrounding traffic (national road or farm area), may alter the risks significantly. Installation of roof sheeting will depend on the pitch of the roof, and the size of the sheets. Workers need to be protected from falling, and attachments for fall arrest lines may be required.

Since the HIRA in cases is required to be part of a SHE specification and plan it must be recorded. Furthermore, the Occupational Health and Safety Inspectorate may require the responsible persons or department in the organization to indicate endeavours made to identify hazards and mitigate risks particularly if an accident should occur, resulting in either fatalities or other injuries.

A quantitative process is recommended, and risks should be graded from low to extreme. It must be remembered that the medium to extreme risk areas must be addressed. The HIRA needs to be made available to all parties involved and such HIRAs should be kept on file. Should it not be possible to eliminate or mitigate the hazard, the requisite insert should be made in the relevant documentation, and where appropriate, and financial implications should be shown.

b. Risk quantification

To make an informed decision with respect to deciding, among other, for example whether a design, detail, method of fixing, or specification is appropriate, the risk needs to be quantified. Based upon the principle of reasonably practicable, which includes cost and other goals, a decision can be made to amend or leave the design, or other relevant aspects unchanged. Furthermore, a precise estimate is not required, as doing so would be too time consuming, and often there is a lack of data.

Loss frequency is a direct measure of the frequency at which a particular loss occurs over a given period. All incidents must be reported even if no loss occurred. An increase in minor injuries should be taken as an indication of a deteriorating situation. This is an important database for risk evaluation. Investigations in many organisations over a long period of time have indicated the relationship.

It is important to bear in mind that the extent of a loss is not limited to the cost to replace or repair. The immediate or obvious loss or damage may be only "the tip of the iceberg". It is accepted in many organisations to adopt and follow the "Pareto Principle". In simple terms this rule can be defined as follows:

- ✓ 80% of the losses will normally be associated with 20% of the exposure, or
- ✓ As the extent of the loss doubles, the claims frequency will halve.

Other techniques such as statistical estimates, asset values and impact studies are also available to determine possible loss extent.

The following figure illustrates how the risk can be determined by plotting the likelihood of the risk occurring (probability/frequency) on the vertical axis and the likely severity (consequences / impact) on the horizontal axis. The scores relative to probability and consequences / impact are as follows:

- Likely severity (consequences / impact) - three categories: high, medium, and low:
  - High – fatality, major injury or illness causing long–term disability;
  - Medium – injury or illness causing short–term disability, and
  - Low – other injury or illness

- Likelihood of occurrence (probability) - three categories: high, medium, and low:
  - High – certain or near certain to occur;
  - Medium – reasonably likely to occur, and
  - Low – very seldom or never occurs.

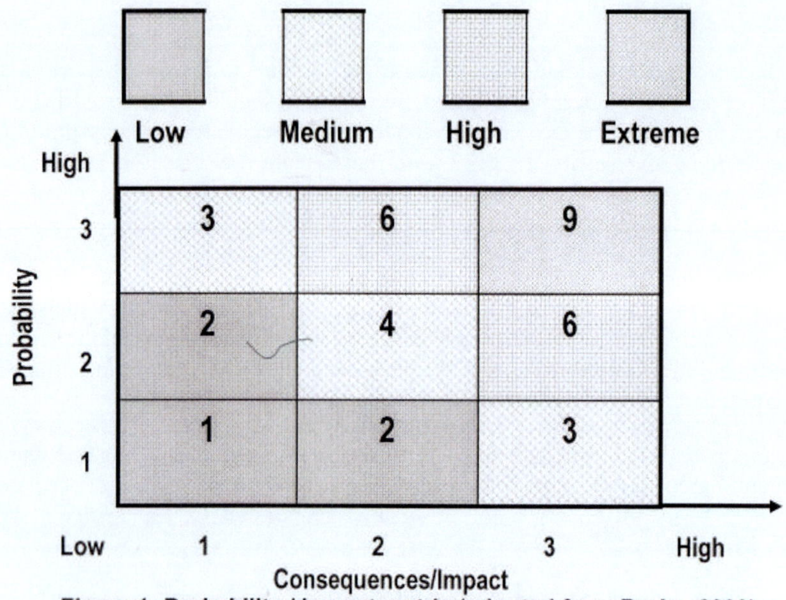

Figure 1: Probability / Impact matrix (adapted from Burke, 2003)

A score of '9' would require elimination of the cause of the problem such as, for example, selection of a flat roof in lieu of a steep pitched roof structure. A '6' would require elimination, or at least substitution such as, for example, water in lieu of a solvent-based spray on paint finish. A '4' would require elimination, substitution, or at least containment such as, for example, screening of areas in which stripping of asbestos-containing materials will be required, or the application of a decorative spray-on solvent-based paint finish. A '3' would require elimination, substitution, containment, or at least an engineering intervention, which often would also require work in accordance with a safe work or operating procedure (SWP/SOP), which would also entail the wearing of personal protective equipment (PPE) such as, for example, a balustrade wall to a flat roof. A '2' would require work in accordance with a SWP/SOP, and entail the wearing of PPE such as, for example, construction of the balustrade wall in concrete, should that be the solution relative to the '3' score. A '1' would require no intervention. The HIRA needs to be communicated and reviewed together with all identified controls. The final step in the HIRA is to evaluate and prioritize the risks as all risks, their loss frequency and consequences are not the same.

• RISK EVALUATION OR QUANTIFICATION

The aim of risk evaluation is to measure the magnitude of one risk relative to another. The aim is also to establish the impact (financially and otherwise) that the risk would have on the organisation. There are multiple approaches to evaluation and quantification of risks. The method used is not important. What is important are the decisions that flow from the process. For example, one could

1. Assign points to the most significant risks

Risk	1	2	3	4	5	6	7	8	9	10	
											Total
A	10	5	0	10	0	0	0	25	30	20	100
B	5	10	30	0	0	15	10	0	30	0	100
C	10	25	15	20	0	0	5	0	5	20	100
D	5	20	17	15	33	10	0	0	0	0	100
E	5	18	20	7	20	30	0	0	0	0	100
Total	**35**	**78**	**82**	**52**	**53**	**55**	**15**	**25**	**65**	**40**	**500**
Rank	**8**	**2**	**1**	**6**	**5**	**4**	**10**	**9**	**3**	**7**	

ASSIGNING POINTS METHOD

2. Assess probability or loss frequency and consequence separately using numbers where **Degree of risk = probability x consequence**

Class	**Corresponding probability**	**Description**
1	0-5%	Unlikely
2	5-25%	Possible
3	25-50%	Likely
4	50-100%	Nearly certain

Probability Classes

	Time	**Money**	**Quality**	**Information**	**Organization**
Class	**Delay**	**Costs (millions)**	**Decrease in lifespan (years)**	**Damage done to req'd level of documents**	**Reduction in support**
1	0-3 mths	R0-1	0-1	Limited	Full
2	>3<6 mths	R1-2	1-3	Unsatisfactory accuracy or detail	For most part
3	>6<12 mths	R2-5	3-5	Insufficient accuracy or detail	Limited
4	> 1 year	>R5	>5	No documents	None

Consequence classes

The risks are now prioritized using the formula and the probability and consequence classes as shown in the example in the following table.

Description	P	C	Risk (PxC)	Rank
More design modifications than planned	4	1	4	3
More land needed	3	4	12	1
Additional soundproofing facilities	4	2	8	2
Removing unexpected obstacles	2	1	2	4

An example of prioritization of risks

A probability-consequence matrix is compiled as shown in the following table.

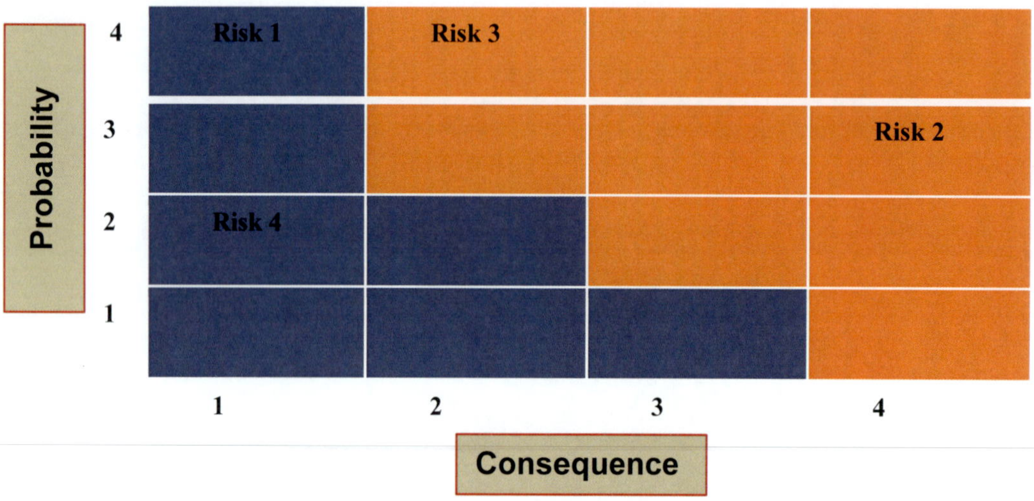

Probability-Consequence Matrix

3. One could also use a combination of both these methods.

It is also important to note that generally there are two types of personalities that will determine the level of risk acceptance and strategy to be adopted. These are risk takers and risk avoiders.

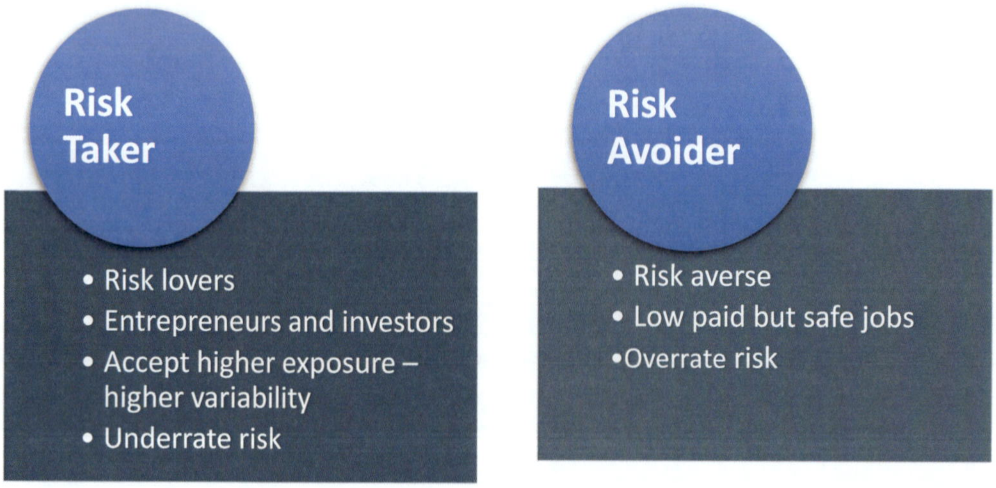

Risk takers vs risk avoiders

Typically risk takers will adopt a more lenient approach to hazard identification and eventual risk assessment and quantification. On the other hand, risk avoiders will take a much more conservative approach. For this reason, it is preferable where possible for hazard identification, risk assessment and quantification to be done collectively by more than one person to ensure a balanced approach to the process. This process is shown in the following figure. Note that the various steps in the risk control process are collectively iterative.

• RISK CONTROL

Risk control is defined as the provision of appropriate levels and standards of protection for people, assets, income, and environment to avoid or minimise the pure risks, which have been identified and assessed in an organisation.

Many organizations consider risk control as the basis for effective risk management. Risk control has a significant impact on the organisation's stability and continuity of operations. Risk control is preferred to the term loss control. Loss control suggests measures that only deal with incidents as, and after they have occurred. The more important preventative and proactive aspects are included in the term risk control.

It is important to see risk control in the context of achieving corporate objectives. Anticipated benefits should exceed losses, which could occur. Risk control techniques prevent losses from occurring or reduce the effect of losses once they have occurred. Most pure risks respond favourably to control measures and/or financing. Risk reduction techniques serve to prevent losses from occurring or reduce the effects of a loss once it has occurred.

A risk control programme needs to be managed properly. This requires that performance be measured against prescribed standards. The following system gives guidance as to what should be included in such a programme.

• RISK CONTROL MEASURES: GENERAL PRINCIPLES

For risk control measures to be effective they have to be:

- Compatible with other organisational objectives
- Appropriate and relevant
- Relative to other measures
- Applied at the appropriate stage.
- Flexible.
- Capable of gradual response – for those that have a gradual onset.
- Integrate with other forms and levels of protection.

• RISK CONTROL TECHNIQUES AND OPTIONS

There are four main techniques to control risk, they are:

- Risk avoidance - not to take on a risk in the first place.
- Risk elimination - eliminating an existing risk.

- Risk reduction - reducing the frequency and/or the severity of a risk.
- Risk transfer - transferring the risk to someone else or to another place where it poses a smaller risk for the organisation.

These techniques are shown in the following figure.

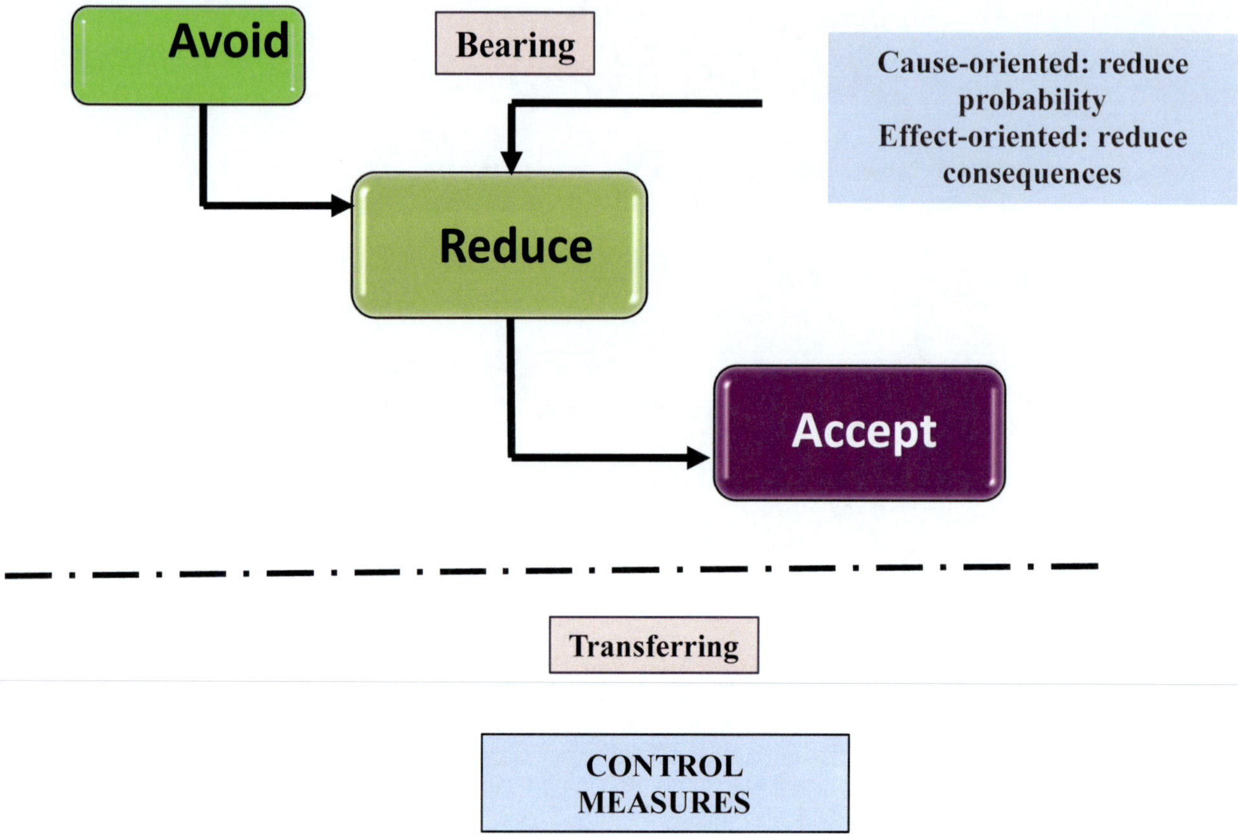

The risk will be borne if it is avoided, reduced, or accepted. If the decision is made not to bear the risk, then it needs to be transferred to another party or agency which could be in the form of insurance or contracting out.

Within these techniques the following options are available: -

- Physical.
- Procedural.
- Education and training.
- Legal

Any, or a combination of these options, may be used to achieve the technique decided on. To explain how they may be used in practice, the following examples are given.

- PRACTICAL EXAMPLES OF RISK CONTROL

Using the physical options:

Avoidance: -	Not to erect an operation within the 100-year flood line of a river.
Elimination: -	Stand-by equipment to eliminate production loss due to breakdown of a vital piece of equipment.
Risk reduction: -	(Frequency). A security fence would reduce the incidence of illegal entry.
Risk reduction: -	(Severity). A high-pressure release valve would prevent a vessel from over pressurising. This would limit damage to the vessel.
Risk transfer: -	Removal of hazardous substances by contractor.

Using the procedural options:

Avoidance: -	Prohibit the use of certain chemicals on the premises.
Elimination: -	Safe work procedures, eliminate unsafe acts.
Risk reduction: -	(Frequency). Buying procedure to ensure compatibility.
Risk reduction: -	(Severity). Evacuation procedure
Risk transfer: -	Procedure to have hazardous waste removed by a contractor.

The option of education and training

Risk avoidance: -	On the job training.
Risk elimination: -	First-aid training to eliminate the severity of injuries by applying proper treatment.
Risk reduction: -	(Frequency). Fire prevention programmes to reduce frequencies of fire.
Risk reduction: -	(Severity). Fire drill to reduce severity of a fire.
Risk transfer: -	Supervision.

Using the legal option

Risk avoidance: -	Parking on own risk notice.
Risk elimination: -	Following manufacturers' requirements.
Risk reduction: -	Comply with legal requirements.
Risk transfer: -	Insurance contract, disclaimer notice or inspection by an accredited authority.

• RISK EXPOSURE PREVENTION PRINCIPLES

The following principles will assist in preventing exposure to risks and experiencing the consequences of such exposure:

- Avoid the risks
- Evaluate the risks that cannot be avoided
- Combat the risks at source
- Adapt to technical progress
- Replace dangerous by non-dangerous
- Adapt work to worker design of workplace, choice of equipment and methods
- Develop coherent prevention policy – technology, work organization, working conditions, social relationships, for example
- Give collective protective measures priority over individual protective measures
- Give appropriate instructions to workers

• RISK CONTROL AREAS

Each organisation, whether financial, service, or manufacturing, is exposed to pure risks. These organisations will all have their own peculiar risks and risk control measures.

Suppliers: -	Failure to supply, supply incorrect material, late delivery.
Service: -	Failure of service, inferior service.
Electronic data processing: -	Description of information flow, breakdown, fraud
Plant and equipment: -	Breakdown, theft, malicious damage.
Stock: -	Pilferage, destruction, contamination.
Transport: -	Accidents, hi-jacking, theft.
Customer: -	Loss of customers.
Labour force: -	Riots, strikes, work stoppages, injuries, illness.
Liabilities: -	Injury to third parties, inferior products, professional advice, environmental damage.
Vital records: -	Intelligence or information services, list of clients.

• STATUTORY RISK CONTROL REQUIREMENTS

Many risk control measures are prescribed by law. Acts that apply include:

- ✓ Occupational Health and Safety Act No 85 Of 1993.
- ✓ Mine Health and Safety Act No. 29 Of 1996.
- ✓ Compensation for Occupational Injuries and Diseases Act No 130 Of 1993.
- ✓ Hazardous Substances Act No 15 Of 1973.

- ✓ National Building Regulations and Building Standards Act No 103 Of 1973.
- ✓ Explosives Act No 26 Of 1956.

There are many standard codes of practice, such as, for example, SABS 0157-1979 for quality control systems and Good Manufacturing Practices (GMP) for the pharmaceutical industry.

• INCENTIVES TO IMPLEMENT A RISK MANAGEMENT PROGRAMME.

In any industry it makes good sense to limit the liabilities of the organization as much as possible. Making profits is fine but having cost factors in terms of claims can be detrimental to profitability and sustainability of the organisation. Management therefore has economical, legal and social responsibilities. The bottom line is that organizations will have to do more with less. They will have to be more efficient in their use of capital and human resources. A systematic risk management approach identifies leverage points within the system. Initiating action at those areas that will provide the greatest leverage and highest potential for success are addressed. In order to develop leveraged action, the fundamental issues must be worked with and not the symptoms. By following the risk management process, attention will focus on the real issues to ensure that the organisation will meet its economic, environmental, social, and moral obligations.

• APPLYING RISK MANAGEMENT PRINCIPLES

To manage pure risks, the performance of the risk control activities should be measurable. How does one measure cost savings on losses that have not yet occurred, and which could or could not have been prevented by risk control measures? Frequency of incidents is not regarded as a reliable guide to SHE performance. There is no clear correlation between work conditions, the severity of the incident and the frequency of the incident. Incidents may occur by chance factors alone. This could cause deviations from the previous or expected norm. The most common method used is to measure performance in terms of the achievement of previously agreed and adopted objectives.

• RISK CONTROL COSTS

Risk control costs should not be confused with the total costs of risks. Risk control costs refer only to costs of activities aimed at preventing or reducing the frequency and severity of losses from pure risks.

The total cost of risk is a much broader concept which comprises the following components:

- Cost of insurance premiums.
- Cost of uninsured losses (retained losses).
- Cost of measures to prevent or reduce losses (risk control costs).
- Administrative cost of risk management.

• ASPECTS TO CONSIDER

To decide whether to install a risk control measure or not, consideration should be given to the following costs, namely:

Property loss or damage	Tangible Intangible
Consequential loss	Business interruption Supply interruption Loss of profits on finished goods Reduced rental income Interest on borrowed funds Increased insurance Hired labour/machinery Loss of market share Increased advertising costs
Cost of paying for liability claims	Product liability Professional liability Environmental impairment liability Employers' liability Claims arising from industrial relations issues Retained liability losses

Cost associated with injury or death of individuals (persons)	Statutory insurance Personal accident insurance Group life insurance Retained costs
Pure financial or economic loss	Any costs that are not included in the previous categories

• THE COST- BENEFIT APPROACH

The most obvious way of increasing profitability is by increasing revenue, reducing costs or both. More often than not competitive pressure prevents an organisation from simply raising its revenue by increasing sales or prices. The only sure way to maximise profits is by minimising costs.

Management in this light often views the cost of risk control. If the projected benefit exceeds the cost of the risk control measure by more than say 18%, the "investment" is regarded in a favourable light. In the risk management context, the term "income" or "benefit" refers to effective income (the amount of money saved by the implementation of risk control measures). The cost-benefit equation looks like this:

$$\text{Income} = \frac{\textit{Benefit}}{\text{Cost of risk control}}$$

$$\text{\% return on investment} = \frac{\text{Income}}{\text{Capital invested}} \times 100$$

Effective income equals expected loss before risk control measures are subtracted, and minus expected loss after risk control.

Effective income = [expected loss (before risk control measures)] — [expected loss (after risk control)].

This will serve to rank the various risk control recommendations in order of importance. This will assist management in their decision-making on risk control expenditure. The process is called "decision-making under uncertainty" It involves the uncertainty of whether a loss will occur, and if it does occur, what the consequences will be. To decide whether to implement measures to prevent or reduce losses, management will have to consider the risk tendency of the leaders in the industry.

• RISK CONTROL STANDARDS

Risk control measures should be applied in the form of standards, which are to be met. Standards are normally grouped according to risk control disciplines, such as, for example, fire standards, security standards, occupational health and safety standards.

Standards are based on the following criteria. -

- ✓ Legal requirements — both common law and statutory requirements.
- ✓ South African Bureau of Standards (SABS) specifications and Codes of Practice (CoP's).
- ✓ Accepted risk management practices.
- ✓ The collective advice of risk control consultants, brokers and insurers.
- ✓ The experience and knowledge in the organization itself.

• RISK CONTROL DISCIPLINES

Risk control activities call for a range of disciplines, which include:

- ✓ Fire protection.
- ✓ Security.
- ✓ Occupational health and safety.
- ✓ Natural perils risk control
- ✓ Motor vehicle risk control
- ✓ Transit risk control
- ✓ Emergency planning and crisis management.
- ✓ Legal liability risk control
- ✓ Business interruption risk control
- ✓ Environmental protection.
- ✓ Engineering risk control

• RISK FINANCING

Risk financing measures are used to arrange financial resources, which would compensate an organisation for those losses that it may suffer from pure risks. If properly assessed and managed, it could be a contributor to the profitability of the organization. There are two main options in risk financing, namely self-funding and insurance. Self-funding takes place when losses are financed from internal resources. Self-funding could be funded on a "pure funded" basis or on a day-to-day basis as an operating expense.

• SELF FUNDING

Insurance will compensate for the insured losses suffered because of an insured event. Self-funding is a means of self-finance to pay for losses that might occur. There are three different funding methods or options. The ultimate choice will depend on the requirements of the organisation in question. These options are:

- **Unfunded option**

The unfunded option implies the deliberate self-funding of the cost of loss. In the event of loss, the cost is met as a current expense from existing resources. This method is intrinsically the least expensive since losses are covered without recourse to an insurer or any other party. The size of the loss that can be funded by this method will depend on the ability of the organisation.

- **Funded option**

The funded option requires that the cost of a loss be covered by deliberate acceptance. Unlike unfunded options, losses are not charged to current expenses but to a reserve fund. This fund is established in the accounts with the express purpose to provide for a loss. There must be a self-insurance reserve, which is systematically credited with the amounts necessary to meet the cost of projected claims.

- **Captive insurance company**

A captive insurance company is an insurance company that is owned by a parent company or companies. The captive insurance company writes the insurance business, in sum or in part, of its parent company. Captive insurance is the logical development and extension of the self-insurance concept. The organisation uses an owned or controlled insurance company to manage its self-insurance reserve. The decision to use a captive is normally related to obtaining insurance cover for risks not normally insurable. It could also be used to provide consistent insurance protection for multinational companies, which are obliged to insure their interests in the country in which they reside.

Selecting the option that is likely to be less costly to the organisation is the subject of a financial analysis. The financial analysis compares the economic effectiveness of each option. In the final analysis, the mix of options selected must not expose the organisation to any unbudgeted or unexpected financial strain.

• TRANSFER (INSURANCE)

Insurance is the process whereby the responsibility for financing losses is transferred to a professional financial risk carrier. Note that the risk is not transferred but only the financial consequences of the risk.

• INSURABLE RISKS

Insurance is only applicable to certain pure risks. Ideally the risk has to qualify as such in terms of a number of criteria: -

- Large number of independent exposures offering similar risks for insurance. Thereby spreading the "few" losses anticipated among the "many" more fortunate.
- Large enough statistical base to offer a viable premium.
- Losses anticipated should be fully definable in terms of time, place, cause and amount.
- The conditions directly responsible for or leading up to the loss should be accidental. The intention is not to insure anyone against intentional loss.

Many insurers will insure risks that do not qualify in terms of these criteria. One such example is the risk of death. Although it exposes many independent people, death is ultimately inevitable and not always accidental.

• TYPES OF INSURANCE

The insurance industry may be divided into two distinct sections, namely life and short-term insurance.

Life insurance entails the following: -

- Provision of financial security.
- Ensuring payment of outstanding loans.
- Making investments in retirement plans.
- Possible inclusion of personal accident and disability cover, medical expenses, and illness cover.

Short term insurance is divided into the following major classes: -

- **Fire and natural disasters**: – Insurance against assets or against goods for which s/he is responsible, such as, for example, the risk of fire, flood, storm, tempest, earthquake.
- **Crime: -** Insurance against assets, money, or goods for which the insured is responsible. Examples are theft, burglary, robbery, fidelity, and fraud.
- **Accidental damage (all risks): –** Insurance against the risk of accidental damage to or loss of insured items because of falling, impact, breakage etc.
- **Breakdown: -** Insurance against the accidental, mechanical, or electrical breakdown or malfunctioning of machinery and equipment. Such machinery could include large manufacturing machines or personal computers.
- **Transit: -** Insurance against the loss of or damage to goods while in transit, or during temporary storage.
- **Motor: -** Insurance against accidental damage to the insured's motor vehicle resulting from accident, theft, fire, storm etc. The legal liability that the insured may incur in terms of damage to third party property or third-party death or injury.
- **Liability: -** Insurance against the legal liability the insured may incur in terms of:
  - Damage to third party property.
  - Injury to or death of third parties.
  - Damage to the environment.
  - Consequences of defective products.
  - Malpractice or wrongful advice.
  - Libel and slander.
  - Product guarantees.
- **Consequential loss (Business Interruption/Loss of Profits Insurance): -** An insured may suffer a loss of profit because of any of the various insurance classes. This loss of profit especially fire and breakdown could be insured. The insurance would be provided for a given indemnity period. During this time the insured would be indemnified for loss of profits. This is a simplified explanation of a complex class of insurance.
- **Politically motivated damage: –** Because of the nature of this risk and the high potential cost of a single event, these risks are usually separately insured. The South African Special Risks Insurance Association (SASRIA) insures these risks. SASRIA had its origin due to the reluctance of the professional underwriters to accept these risks. The government was the motivating force in the establishment of this Association.
- **Personal incident: -** Insurance against death or injury of workers for which the employer elects to provide compensation. This class also includes insurance for which the employer may be legally liable. A case in which the worker is not protected in terms of the COID Act is an example. There are many other types of short-term insurance, which are just as important. A few major classes are: -
  - Marine hull and liability.
  - Aviation hull and liability.
  - Credit guarantee insurance.
  - Export credit insurance.
  - Construction all risks.
  - Erection all risks.
  - Advance profits.

All these classes are aimed at indemnifying the insured against loss sustained as a result of an insured peril. The insured will pay the insurer a premium for this indemnity. The insurer will indemnify the insured in terms of the conditions set out in the insurance policy. Such indemnity would take the form of replacing the asset or repairing the damage or alternatively a financial compensation. Normally at the option of the insurer, but strictly in accordance with the terms of the policy.

- REVIEW QUESTIONS

1. Given that there are two types of persons in terms of how they view risk, explain which type you are.
2. Why is it important to conduct a full HIRA which includes risk prioritization and quantification?
3. What is the most important aspect of a risk assessment process?
4. Discuss the various options when the deciding to transfer a risk.
5. Why would you consider introducing a risk management program in your organization?
6. Complete the matrix below using other examples for controlling risks in your organization.

Techniques Options	Avoidance	Elimination	Reduction	Transfer
Physical				
Procedural				
Education and Training				
Legal				

• REFERENCES

Anonymous (n.d.). Loose notes on SHE management

EU (2014). *The business case for safety and health at work: Cost-benefit analyses of interventions in small and medium-sized enterprises*. [online] Osha.europa.eu. Available at: https://osha.europa.eu/en/publications/business-case-safety-and-health-cost-benefit-analyses-interventions-small-and-medium.

Gauthier, Francois., Chinniah, Yuvin., Burlet-Vienney, Damien., Aucourt, B. and Larouche, Stephane. 2018. Risk assessment in safety of machinery: Impact of construction flaws in risk estimation parameters. *Safety Science,* 421-433

Haupt, Theodore. 2019. Notes of CEng8042 Risk Management course delivered at Bahir Dar University, Ethiopia, April 2019

HSE (2014). *Risk management: Cost Benefit Analysis (CBA) checklist*. [online] Hse.gov.uk. Available at: https://www.hse.gov.uk/risk/theory/alarpcheck.htm.

Hughes, B. and Marullo, J. (2009). *Understanding Risk Control -- Occupational Health & Safety*. [online] Occupational Health & Safety. Available at: https://ohsonline.com/Articles/2009/10/02/Understanding-Risk-Control.aspx.

ISO 31000:2018 (2018). *ISO 31000:2018(en) Risk management — Guidelines*. [online] Iso.org. Available at: https://www.iso.org/obp/ui/#iso:std:iso:31000:ed-2:v1:en.

Nunes, Isabel. (2016). *OSHwiki*. [online] Oshwiki.eu. Available at: https://oshwiki.eu/wiki/Occupational_safety_and_health_risk_assessment_methodologies.

OSHA (2020). *Risk Management Principles and Application*. [online] Oshatrain.org. Available at: https://www.oshatrain.org/notes/4bnotes02.html.

OSHA (n.d.). *Hazard Identification and Assessment | Occupational Safety and Health Administration*. [online] Osha.gov. Available at: https://www.osha.gov/shpguidelines/hazard-Identification.html.

Smith, Nigel. J., Merna, Tony., and Jobling, Paul. (2006): *Managing Risk in Construction Projects,* Blackwell Publishing, Oxford

CHAPTER 7

LEGAL CONCEPTS AND PRINCIPLES

INTRODUCTION

Laws are necessary because they guide society on what is the right and wrong so that that there can be peace and order. In the absence of laws life would be complete chaos. The law is defined as a rule of conduct or action prescribed or formally recognized as binding or enforced by a controlling authority such as a command or provision enacted by a legislature. It also is the discipline and profession concerned with the customs, practices, and rules of conduct of a community that are recognized as binding by the community. Enforcement of the body of rules is through a controlling authority. The law consists of "legal rules". Legal rules. deal with the conduct of individuals in society. Legal rules apply to all members of the community. It is important to explain what the nature and practical effect of legal rules are. Generally, the members of a community will conduct themselves according to certain accepted customs and rules.

- One would, for instance, find that certain linguistic rules are being adhered to, so that members of the community can communicate effectively with one another. For example, on construction sites workers use their own lingo to communicate with each other about various tools. Someone who does not work on construction sites will not necessarily be able to understand what is being referred to.
- Other rules are social ones, which enable members to get along with one another. They would, for example, congratulate other members on their birthdays, children are taught to always say "please" and "thank you" when they want or receive something. If they do not, they could be rude or unmannered.

"Legal rules" are similar in nature and regulate the rights and duties of each community member. On the other hand, non-legal rules. apply to select groups in the community. For example, the rules of a sporting club apply only to the members of the club. Members can decide to be part of the group and to abide by its rules, or they can opt out.

The main difference between legal rules and other rules of conduct in a community is that legal rules are binding upon members of the community.

This claim means that community members may be forced to comply with the rules and that non-compliance will have a certain prescribed detrimental effect on the person violating them. For example, should he destroy property belonging to another member of the community he may be punished by the state or be ordered to pay compensation to another member of the community that has suffered damages because of his conduct. Going over the speed limit or driving without a valid driver's licence will incur consequences in the form of a fine.

• LEGAL TERMINOLOGY

In order to have a better understanding of the legislative and regulatory framework which governs the SHE space in South Africa the understanding of a few legal terms and concepts is useful. While the list is not exhaustive, the following will be helpful:

- **Precedent** - A prior ruling or judgment on any case is known as a precedent.

- **Stare Decisis/Precedent** – (to stand by decided matters)
  A doctrine or policy of following rules or principles laid down in previous judicial decisions unless they contravene the ordinary principles of justice. It is a legal doctrine that obligates courts to follow historical cases when making a ruling on a similar case. Stare decisis ensures that cases with similar scenarios and facts are approached in the same way. Simply put, it binds courts to follow legal precedents set by previous decisions. Stare decisis dictates that courts look to precedents when overseeing an on-going case with similar circumstances.
  Lower courts are bound by the decisions of the higher court, and bound by decisions of the same divisional court such as, for example, KwaZulu-Natal, Gauteng, and Western Cape.

- **Appeal**
  An appeal is a request to a higher court to review a decision made by a lower court. The higher court can review the entire case, certain aspects of the case, or the sentence imposed by the lower court. The appellant, who may be the plaintiff or the defendant in the lower court case, must show the higher court that mistakes or errors were made during the previous trial, and that the case should be overturned, dismissed, or re-tried. The right to appeal must be granted by the court before the appeal process can commence.

- **Review**
  To re-examine judicially or administratively; a judicial reconsideration for purposes of correction of, for example, the examination of a case by an appellate court.

- **Locus Standi**
  This doctrine is defined in legal lexicons as the right of a person or entity to bring an action, to be heard in court, or to address the Court on a matter before it. It is the right of any person or entity to initiate proceedings in court. The right to be heard.

- **Liability**
  Legal responsibility for one's acts or omissions. It is frequently defined in various legal lexicons as the failure of a person or entity to meet that legal responsibility for his or her acts or omissions which leaves him or her open to a lawsuit for any resulting damages or a court order to perform such as, for example, in a breach of contract or violation of statute.

- **Onus or Burden of Proof**
  The duty of presenting a certain amount of evidence to meet the legal requirements for establishing the entitlement of the party in a case to the outcome sought. It is the obligation to offer credible evidence in a court of law in support of a contention or accusation. In a civil matter the legal requirement is for the plaintiff to present sufficient evidence to prove its case based on the preponderance of probability whereas in a criminal matter the legal requirement is for the State or prosecution to present sufficient evidence to prove its case beyond reasonable doubt.

- **Mandamus/Interdict (**do something/stop doing something)
  An extraordinary writ or order that is issued by a court that commands a natural or juristic individual to perform, or refrain from performing, a particular act, the performance or omission of which is required by law as an obligation. It is made without the benefit of full judicial process, or before a case has concluded. It may be issued by a court at any time that is appropriate, but it is usually issued in a case that has already begun.

- **Authorisations**
  To empower another with the legal right to perform an action. Examples include permits, licenses or permissions.

- **Audi Alteram Partem** (hear the other side (of the argument)/ the right to hear and be heard)
  It has been quoted as being one of the most cherished and sacrosanct principles of law. It embodies the concept that no person should be condemned, punished, or have any property or legal right compromised by a court of law without having heard that person. The notion that every individual whose life, liberty, or property are in legal jeopardy, has the right to confront the evidence against him or her in a fair hearing. Therefore, no case or judgment can be decided without listening to the point of another party.

- **Bona Fide** (In good faith)
  Suggests that the person acts in good faith and is honest; genuine, actual; authentic; acting without the intention of defrauding.
  The law requires all persons in their transactions to act with good faith and a contract where the parties have not acted bona fide is void at the pleasure of the innocent or aggrieved party.

- **Mala Fide** (In bad faith)
  A person acts in bad faith through the fraudulent deception of another person; the intentional or malicious refusal to perform some duty or contractual obligation. A person acts in bad faith by committing an Intentional dishonest act by not fulfilling legal or contractual obligations, misleading another, entering into an agreement without the intention or means to fulfil it, or violating basic standards of honesty in dealing with others.

- **Delict**
  In civil law, a delict is an intentional or negligent act which gives rise to a legal obligation between parties even though there has been no contract between them. The act by which one person, by fraud or malignity, causes some damage to some other. In its most enlarged sense, this term includes all kinds of crimes and

misdemeanours, and even the injury which has been caused by another, either voluntarily or accidentally without evil intention. Delicts may be either public or private. The public are those that affect the whole community by their hurtful consequences while the private is that which is directly injurious to a private individual. Because of its civil nature the Plaintiff must prove on preponderance of probabilities that they have suffered damage. So, in a civil action, if the plaintiff proves that it is more likely than not that the defendant was responsible for the injuries caused or loss suffered, the plaintiff wins.

The following are key elements of delict:

- Conduct - The plaintiff must prove that there was a voluntary/human/positive act or omission by the defendant that caused harm or damage to the plaintiff
- Wrongfulness
- Culpability
- Causation
- Damage or harm

MHS Act 29 of 1996 as amended - *86. Negligent act or omission - (1) Any person who, by a negligent act or by a negligent omission, causes serious injury or serious illness to a person at a mine, commits an offence.*

OHS Act 85 0f 1993 as amended - *37. (1) Whenever a worker does or omits to do any act which it would be an offence in terms of this Act for the employer of such worker or a user to do or omit to do, then, unless it is proved that-*

(a) in doing or omitting to do that act the worker was acting without the connivance or permission of the employer or any such user;

(b) it was not under any condition or in any, circumstance within the scope of the authority of the worker to do or omit to do an act, whether lawful or unlawful, of the character of the act or omission charged; and

(c) all reasonable steps were taken by the employer or any such user to the employer or any such user himself shall be presumed to have done or omitted to do that act; and shall be liable to be convicted and sentenced in respect thereof; and the fact that he issued instructions forbidding any act or omission of the kind in question shall not, in itself, be accepted as sufficient proof that he took all reasonable steps to prevent the act or omission.

- COMMON LAW

Common law is a body of unwritten laws based on legal precedents established by the courts. Common law influences the decision-making process in unusual cases where the outcome cannot be determined based on existing statutes or written rules of law. Common law, also known as case law or precedent, is law that has been developed by judges through court decisions rather than through legislative statutes or the normal course of proclaiming laws. Common law relies on detailed records of similar situations and statutes because there is no official legal code that can apply to a case at hand. Common law is based on the principle that it is unfair to treat similar facts differently on different occasions; common law therefore binds future decisions. For example, one person has been given a life sentence for rape, everyone else should receive the same sentence.

Common law is often referred to as "unwritten law" because it cannot be found in our legislation

Common law consists of those legal rules which have been in existence for generations, and which have through the ages been recognised by people as binding upon them. They are undocumented and unwritten.

Legal cases involving common law are interpreted in light of the common law tradition. They leave a number of things unsaid because they are already understood from the point of view of pre-existing case law and custom, that is, what happens now is based on court cases in the past. The common law applicable in different countries differs to a certain extent from country to country.

South African common law is known as Roman Dutch law, because it is that which applied in Rome during the time of the Roman Empire. It had been refined by the Romans before it eventually became applicable in the Netherlands and, more particularly, the Province of Holland. It had developed further in Holland before being introduced at the Cape by Jan van Riebeeck around 1652 and being made applicable here.

Despite the fact that much of South African contemporary law consists of legislation, important parts of it are still not regulated and are still part of our common law. Most of the well-known crimes such as murder, culpable homicide, assault, rape, robbery, theft, housebreaking and extortion are not defined in legislation and are still part of South African common law.

In order to determine what the common law provides with regard to a specific matter, one has to consult the reported judgments of the courts in which the courts explain what the common law provides with respect to that matter. In order to determine which reported judgments to consult in a particular instance, one may consult the textbooks by lawyers dealing with that particular topic or case law records.

• BASIC PRINCIPLES OF COMMON LAW

Statutes are understood to always be interpreted considering the common law tradition, and so may leave a number of things unsaid because they are already understood from the point of view of pre-existing case law and custom.

By contrast to the statutory codifications of common law, some laws are purely statutory, and may create a new course of action beyond the common law. An example is wrongful death, which allows certain persons, usually a spouse, child or estate, to sue for damages on behalf of the deceased. Where the claim is grounded in common law, then all damages traditionally recognised historically for that tort may be sued for, whether there is mention of those damages in the current statutory law.

For instance, a person who sustains bodily injury through the negligence of another may sue for medical costs, pain, suffering, loss of earnings or earning capacity, mental and/or emotional distress, loss of quality of life, disfigurement, and more. These damages need not be set out in statute as they already exist in the tradition of common law.

Common law is based on the 'reasonable person' principle. It is not easy to define a 'reasonable person' and make it applicable in general. In fact, it is not possible at all. Each set of circumstances will determine its own reasonable person. In other words, what a reasonable person would have done under the specific circumstances.

The following questions need to be answered if the person is to be considered 'reasonable':

- Could the person in question have foreseen that the incident will occur?
- Would the person who has foreseen the incident have implemented the necessary corrective measures?

• STATUTORY LAW

Statutory law or statute law is written law passed by a body of legislature. This is as opposed to oral or customary law or regulatory law promulgated by the executive or common law of the judiciary. Statutes may originate with national parliament, provincial legislatures, or local municipalities. In those instances, where common law did not provide for certain situations the State promulgated statutes to deal with those situations. The majority of crimes in South African criminal law have been created by statute.

In order to establish exactly what South African law provides on a specific topic, it is always the safest to ascertain whether there is any statute which deals with that topic. If no such statute exists, common law would provide a solution to the problem.

• STATUTORY SOURCES OF SOUTH AFRICAN LAW

(a) statutes adopted by the Parliament of the Republic (Acts of Parliament). Parliament is the supreme legislative authority;
(b) statutes of the various former Republics and colonies before they were included in the Union of South Africa in so far as they had not been repealed or amended;
(c) ordinances, proclamations, and regulations of the various provincial authorities (Regulations);
(d) city councils, municipalities, divisional councils and village management boards are subordinate legislative bodies authorised to pass by-laws and regulations. The by-laws and regulations such as traffic regulations are adopted with the consent of the Provincial Government or Parliament of the province;
(e) proclamations and regulations which have been promulgated by the President, Ministers and Premiers of provinces.

- LEGAL HIERARCHY

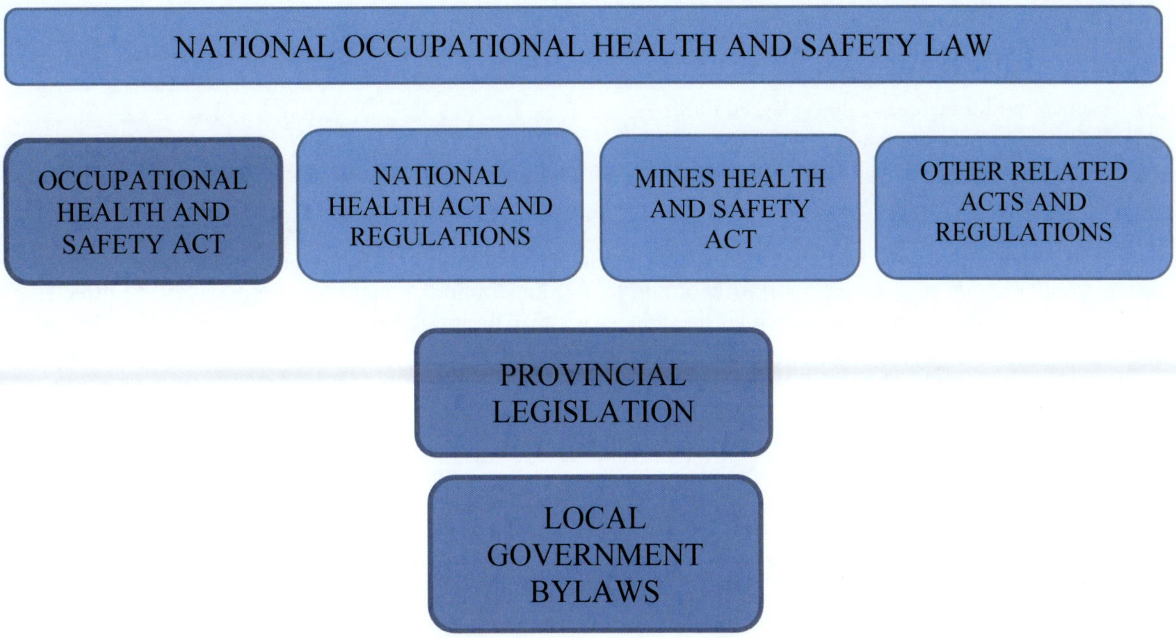

The purpose of statutory law is to satisfy the changing needs and demands of the country. Statutory laws are acts, ordinances and by-laws introduced by bodies with legislation powers.

- THE SUPREME LAW

A constitution contains those sets of laws that establish a state; an array of laws that *constitutes* the state, in the sense that the state is established, exists, and operates within the parameters of those rules. The origin of the laws that deal with SHE in South Africa is the Constitution of the Republic of South Africa Act No. 108 of 1996 as amended. The Constitution entrenches and protects the basic human rights of all people in South Africa.

The Constitution is the supreme law in South Africa. Therefore, any law or conduct inconsistent with it is invalid and the obligation imposed by it must be fulfilled. It supersedes all other rules contained in statutes, common law and custom. Further, any rule inconsistent with a constitutional rule is an invalid rule. Any conduct that contradicts the Constitution, including failing to fulfil an obligation imposed by the Constitution is also invalid. Chapter 2 of the Constitution, which is commonly referred to as the Bill of Rights, guarantees certain rights to various persons in South Africa which includes the right to an environment which is not detrimental to the health or wellbeing of individuals:

- Equality
- Human Dignity
- Life
- Freedom and security of person
- Privacy
- Freedom of religion
- Freedom of expression
- Freedom of association
- Residence
- Property
- Environment
- Housing
- Health Care
- Children
- Education Culture
- Access to information

The Constitution entitles private parties to institute action to enforce their environmental right against any other private party or corporation. A law or conduct can be unconstitutional because it violates a right in the Bill of Rights.

Constitution of the Republic of South Africa Act No. 108 of 1996 as amended - Section 24: "Everyone has the right":

- To an environment that is not harmful to their health or well-being
- To have the environment protected, for the benefit of present and future generations, through reasonable legislative and other measures that:
  a) Prevent pollution and ecology degradation
  b) Promote conservation and
  c) Secure ecology sustainable development and use of natural resources while promoting justifiable economic and social development

Therefore, in the absence of any SHE legislative and regulatory framework in South Africa the provisions in the Constitution would provide such a framework. According to the Constitution, a business enterprise will be affected by any legislation promulgated and by any challenges that may be made to its applications for permits. No Act of Parliament or Regulations or Municipal Bylaws can contradict the Constitution - otherwise it can be challenged in the Highest Court of Law in South Africa namely The Constitutional Court.

South African law is divided into different sections. Criminal law, the law of criminal procedure and the law of evidence are examples of such divisions or sections of the law.

• CRIMINAL LAW

Criminal law is a body of rules and statutes that defines conduct prohibited by the State because it threatens and harms public safety and welfare and that establishes punishment to be imposed for the commission of such acts. In particular, it defines criminal offences, regulates the apprehension, charging, and trial of suspected persons, and fixes penalties and modes of treatment applicable to convicted offenders. Conduct which is punishable by the State is known as a crime. Criminal law therefore deals with crimes and punishment of the guilty parties.

• THE REQUIREMENTS FOR A CRIME

A crime is unlawful human conduct, performed with a blameworthy state of mind by a criminally responsible person and which is punishable by the State.

If all the crimes in South African law are considered carefully, the basic requirements for all of them are similar. The requirements are the following:

No	Requirement	Description
1	Legality	Conduct constitutes an offence (a crime) only if the law states that it is an offence and is punishable by the State. An act therefore constitutes an offence only if it is punishable by the State. The requirement for a crime is known as the "requirement of legality" or the "principle of legality".
2	An Act	It is a requirement of every crime that a human being has performed an act. This does not necessarily mean that one can be guilty of a crime only if one has actually done something. One can also be guilty of a crime if something, which the law requires one to do, was not done (e.g., non-submission of an income-tax return). This "failure to act" is called "an omission". For the purposes of our criminal law, an omission is also regarded as an act.
3	Unlawfulness	It is a requirement of every crime that the conduct (act) must be unlawful. Conduct will be unlawful if it contravenes a statutory provision or a common law rule or if it conflicts with the general public's idea of what is right or wrong
4	Fault	It is a requirement of most crimes that the unlawful conduct must have been accompanied by a blameworthy state of mind. This "blameworthy state of mind" is called "Fault" or "Mens Rea" and is a state of mind for which a person may be blamed.

• LEGALITY

Conduct constitutes an offence or a crime only if the law states that it is an offence and is punishable by the State. An act therefore constitutes an offence only if it is punishable by the State. This requirement for a crime is known as the "requirement of legality" or the "principle of legality".

- REGULATIONS

An Act of Parliament informs of WHAT to do and not to do in terms of health, safety and environment matters. The details of HOW to do and not to do are prescribed in regulations, which form part of the Act.

Regulations are made by the Minister of Labour and form part of the related Act. The role of regulations is to prescribe standards and procedures with a specific sector of trade or industry in mind to address health and safety issues unique to that particular sector. Examples of regulations pertaining to the OHS Act are:

- Asbestos Regulations 2001.
- Certificate of Competency for Mechanical Engineering.
- Construction Regulations 2014.
- Diving Regulations.
- Driven Machinery regulations.
- Electrical Installation Regulations.
- Electrical Machinery Regulations.
- Environmental regulation for workplaces.
- Explosives Regulations.
- Facilities Regulations.
- General Administrative Regulations.
- General Safety Regulations.
- Hazardous Biological Agents.
- Hazardous Chemical Substances.
- Lift Escalator and Passenger Conveyor regulations.
- Major Hazard Installation.
- Noise Induced Hearing Loss.
- Vessels under pressure regulations.

- CODES OF PRACTICE

Legislation – The regulations and related codes of practice are an excellent guide to identifying and controlling hazards. A comprehensive knowledge of those relevant to the operations of an organization is imperative if exposure to hazards is to be effectively managed.

Codes of practice are not legislation. Codes provide advice on how to comply with specific parts of the regulations. For example, recently, the new Ergonomics Regulations, promulgated in December 2019 are another aspect to be incorporated into the employer's already existing SHE programmes. It aims to eliminate, or reduce, ergonomic hazards in the early stages of designing systems and equipment, rather than trying to find a solution once there is a final product. Codes of practice while not law in themselves need to be developed taking into cognisance the provisions and requirements of these regulations.

Unfortunately, industries have potential hazards to SHE which could result in accidents and occupational diseases the consequences of which could be that workers can get injured, disabled or even killed. It has been asserted that no matter the level of awareness of the SHE factors related to work-related tasks and activities, incidents or accidents will never be eliminated completely. This assertion needs to be challenged because it is based on a false premise that incidents or accidents are inherently part of the job. Rather, jobs may be hazardous and exposure to the hazards can be proactively mitigated.

Human factors as well as environmental factors play a major contributory role in the causes of accidents. It has been claimed that human beings are by nature lazy and will always look for a short cut to complete a task or assignment and will often "get away with it", but then run out of luck, and someone will get hurt or killed. This claim raises many questions and should be challenged.

Further, it has also been asserted that …

accidents don't just happen –
they are caused by unsafe acts
and/or
unsafe conditions.

The reality in fact based on multiple research and analytical studies is that all accidents are failures of management and the management system and are preventable.

• STANDARDS

A "standard" means any provision occurring in a specification, compulsory specification, code of practice or standard method as defined in section I of the Standards Act, 1993 (Act No. 29 of 1993).

Standards can also be defined as the "minimum requirement" against which performance is measured.

The South African Bureau of Standards (SABS) is described as a South African statutory body established in terms of the Standards Act, 1945 (Act No. 24 of 1945) and operates according to the Standards Act, 2008 (Act No. 29 of 2008) as the national institution that promotes and maintains the standardisation and quality of commodities and the rendering of services.

In the past the SABS as the national standardisation authority, was responsible for maintaining South Africa's database of national standards, as well as developing new standards and revising, amending, or withdrawing existing standards as required. However, the role of the SABS has changed. Specifications are now being done by SANS which is a separate, independent body that reports to Department of Trade and Industry (dti).

The SABS is a certification body that is accredited by the South African National Specifications (SANS). It is therefore incorrect to refer to the SABS as a standard, as it is a testing and certification body who is allowed to sample and test products and certify a producer's product to a specific SANS standard, through their SANS accreditation. SANS, on the other hand, refers to a standard that specifies the performance requirements of a specific product. SANS standards are not the property of SABS. most of the specifications are done by SANS and have either been developed nationally, or are ISO adopted standards. The thinking behind this is to bring the SANS specifications in line with International specifications as far as possible. Some confusion might still exist with the public, seeing that SANS specifications are still obtainable from the SABS in Pretoria and SANS uses the same building as SABS.

Most of the standards used in industry in South Africa were originally set by the South African Bureau of Standards (SABS) such as for example, "SABS 0177: Part II" means the South African Bureau of Standards' Code of Practice entitled Fire Resistance Test for Building Elements, SABS 0177: Part II – 1981. This code of practice will dictate the minimum requirements in terms of fire resistance to which building elements must conform.

These standards become part of regulations which in their turn form part of the Act.

The O.H.S. Act 85 of 1993 (Occupational Health and Safety) governs health and safety in the workplace

CONTRAVENING THE LAW

• LEGAL LIABILITY

In various instances employers rely on insurance against liability for damages. This cannot be done for all liabilities faced by the employer. The consequences of exposure to a hazard could lead to either a civil or criminal action instituted against the employer or workers. The South African legal system comprises of three basic forms of liability which have a bearing on the management of SHE, namely:

- Civil,
- Criminal, and
- Vicarious liability.

• CIVIL LAW AND CIVIL LIABILITY

No law exists that clearly stipulates when one becomes civilly liable and each case would have to be judged on its own merits. The South African civil law provides the basic grounds on which a person can recover damages or losses suffered through the wrongful actions of another and this is dealt with by the Law of Delict. If one intentionally or even just mistakenly injure someone or damages the property of someone, he/she could end up being responsible for paying for the other person's loss. This is known as civil liability.

Regardless of the activity one is involved in, the law requires an individual to act or behave toward other individuals in a certain, definable way, not causing them any injury or harm. The way the individual must act or behave is called a standard of care. The standard of care is how the reasonable person with similar training and experience would act under similar circumstances.

The duties of care principle rests upon foreseeability. Foreseeability is part of the reasonable person test; which test is applied by our Courts to establish negligence.

The reasonable person test implies the following:

(a) Would a reasonable person, placed in the position of the Defendant, have foreseen the possibility of causing harm or loss?
(b) Would a reasonable person have taken steps to guard against the harm or loss from occurring?
(c) Did the Defendant take such steps?

An additional test by the South African courts is the "Reasonable Expert" test. This test, although similar to the thought process of the reasonable person test, should be seen as a more subjective test due to the knowledge and skills that may be expected from a person in a professional capacity.

If the negligence of someone (defendant) causes loss or injury to another person (third party or plaintiff), this person could sue for compensation in a Civil Court. In these cases, the State is generally not involved.

A civil action is a lawsuit filed by a private person not the government against another private person. Usually, these lawsuits seek monetary damages for injury or loss that the party suing (the plaintiff) alleges the party sued (the defendant) caused. A defendant who loses in a civil action does not face the risk of prison or fines. A classic civil lawsuit would be a lawsuit by a homeowner against his neighbour, seeking damages (money) for the ruination of his car due to the falling tree of the neighbour.

The onus normally lies with the Plaintiff to prove on a balance of probabilities that the Defendant was negligent and caused loss or injury to him for which he must be compensated. The case is heard before a Judge or Magistrate, depending on the damages suffered and amount claimed. Compensation is awarded by the Court to the Plaintiff if he has proved his case.

• CRIMINAL LIABILITY

One must differentiate between:

(a) Common Law Crimes	**(b) Statutory Law Crimes**
Assault	Discrimination **(Constitution)**
Culpable Homicide	Speeding **(Traffic Act)**
Murder	Tax Evasion **(Income Tax Act)**
Rape	Unhealthy Work Conditions **(OHS Act)**
Robbery	Unsafe Acts **(OHS Act)**
Theft	Unfair Dismissal **(Basic Conditions of Employment Act)**
Criminal Procedure Act	Unhealthy Work Conditions **(MHS Act)**

Prosecution will always be by the State. In cases of non-compliance with legislation, or a crime committed by an individual or business, the State prosecutes the offender, called the Accused.

The onus lies with the State to prove beyond a reasonable doubt that the Accused is guilty. The case is heard before a Judge or Magistrate, depending on the severity of the crime. The Accused is either acquitted or convicted and sentenced.

Again, the sentence depends on the crime, and can range from a reprimand to a lifetime jail sentence and/or fine. Sometimes a criminal trial is avoided by allowing the Accused to pay an admission of guilt fine as outlined in the summons.

An admission of guilt remains a conviction and can have serious repercussions in later civil action. Legal advice should therefore be sought.

• VICARIOUS LIABILITY

At times, the law imposes responsibility for civil wrongs on people or entities other than those actually engaging in the conduct that led to injury or damage. Vicarious liability relates either to criminal or civil liability and refers mainly to one entity responsible for the faults or contraventions of another.

Examples of vicarious liability:

A parent could be held accountable for the loss caused through the negligence of his child, or an employer can be held accountable for a contravention by his workers such as being held vicariously liable for the sexual harassment of one worker by another worker if the employer knew about the harassment and failed to address it effectively.

The concept of vicarious liability evolved from the following ancient doctrines:

King Hammurabi's Code of Laws "The Careless Supervisor":

The supervisor failed to check that the workers under his control were performing their job functions without carelessness. As a result, one worker lost his hand, and because the supervisor was guilty, one of his hands was amputated.

Lex Aquila – "Killing a Slave":

A slave owner was entitled to do anything with his slave, except kill him. In other words, the owner had to ensure that his slave was able to work without danger of death. If he failed to do so, he was guilty of a crime.

Actiones Noxales – "Sins of the Slave to be visited upon the Master":

The master had to use his control over the slave to ensure the slave behaved like a reasonable person. If the slave committed a crime, the master had failed to exercise control. This failure of control was as serious as the crime itself and the master would be responsible for the crime as if he had committed it himself.

This concept has not changed much and the supervisor today would generally be held liable for the wrongdoing of the workers under his supervision. Therefore, if a worker violates any provision of a statute or commits a criminal common law offence, the supervisor could be held criminally liable. And if a third party suffered loss or damage due to the worker's negligence, he could lodge a civil claim against the supervisor. In both cases, the supervisor is liable as if he himself had committed the wrongdoing.

Vicarious liability will however, in the majority of cases, only be invoked if a master and servant relationship in reality exists.

• STRICT LIABILITY

Strict liability is where a person could be held liable without any fault or negligence on his or her side. This is, as already mentioned, a liability that is very seldom imposed on a person. A typical example of this kind of liability is where civil claims have been laid against Ford Motor Company for accidents in which the drivers of Ford Kuga have died or were seriously injured. Although the accidents were due to faulty Goodyear tyres and not because of any fault from Ford Motor Company's side per se, they are liable in terms of strict liability.

The argument was that the product in its entirety was purchased and not the individual parts.

• REASONABLY PRACTICABLE

The notion of reasonably practicable is in fact the same as the reasonable person principle found in common law, but with a few specific guidelines the employer must consider, which are-

- The scope and the severity of the hazard. A hazard could be regarded as exposure to anything, which can injure a person or affect the health of such person. The scope means the number of persons exposed to the hazard and the severity relates to the degree to which such will be affected.
- The means and methods to eliminate, mitigate or to reduce the effect of the hazard. This is exactly where the employer must think like a reasonable person and implement the most appropriate measure, which might differ from situation to situation.
- The suitability of corrective measure. Again, the suitability might differ from employer to employer. The suitability will be measured against what the reasonable employer under the specific circumstances would have done. That might include aspects like culture, level of education, attitude, honesty, respect, willingness to learn, and cooperativeness of workers.
- The cost implication of the corrective measures. Many still use the "cost" versus "benefit" argument. This argument is not a valid one because who is able to determine the value of a life. That does not mean that employers do not have the right to say that no money is available to implement certain measures. In that case they either must implement alternative measures which they can afford or stop the worker from being exposed to that hazard.
- It is important to point out that there are TWO competent persons referred to in the Act namely:
    - The dictionary meaning of a competent person, and
    - A competent person as defined in various regulations.
- Although various dictionaries define the word "competent" in a number of different ways, common among these definitions are the words "able, skilful, properly, qualified, proper and suitable". All persons involved in one or another activity, or who render a service, must be competent to do so and therefore employers must

ensure that such persons carry out certain activities. This is only relevant in those instances where the term "competent person" is not defined in regulation 1 of the Regulations in question.

• EXAMPLES OF DISTINCTION BETWEEN CIVIL AND CRIMINAL LEGAL PROCEEDINGS

Civil

Simphiwe Ndlovu was a crane driver operating inside a sugar processing factory lifting large square metal drums filled with syrup. The drums were lifted by chains running from the grab of the crane to each corner of the drum. Before lifting started it was critical to ensure that the grab was centred on the chains. On this occasion Simphiwe started the lift without allowing time for the loaders to check that the grab was centred. The load swung and pinned the arm of another worker Pumi Molo against a truck. Pumi Molo sued the organisation for damages

Criminal

A fire broke out in a plant owned by Smart Tanks where metal tanks were being manufactured and five workers were killed as a result.

The fire was particularly intense because the atmosphere inside the working area in the plant had become oxygen-enriched. The oxygen that had escaped from a hose left by another worker employed by a sub-contractor to whom some of the work had been outsourced. Smart Tanks had distributed a book of rules for the safe use of oxygen equipment to their own workers. However, this book was not distributed to the workers of the sub-contractor except when they asked for it.

• REVIEW QUESTIONS

1. Had there not been any SHE legislation in South Africa, how would workers still be protected?
2. What is the difference between common and statutory law?
3. Who is responsible for the burden of proof in a civil case and in a criminal case?
4. Explain the legal precept of stare decisis.
5. What are the differences between the various types of liability?

• REFERENCES

Anonymous. Nd. Loose notes on SHE management

Bamford, Margaret. (2013). *Work and Health.* New York, NY: Springer.

Bosman, Ab. (n.d.). *Manufacturers Liability and the Contingency Fees Act.* [online] Saimh.co.za. Available at: http://www.saimh.co.za/beltcon/beltcon10/paper1015.html.

Department of Agriculture, Forestry and Fisheries (1998). National Forests Act No. 84 of 1998.

Department of Environmental Affairs (1965). Atmospheric Pollution Prevention Act No. 45 of 1965.

Department of Environmental Affairs (1989). Environmental Conservation Act No 73 of 1989.

Department of Environmental Affairs (1998). National Environmental Management Act No: 107 of 1998.

Department of Labour (1997). Basic Conditions of Employment Act 75 Of 1997. Department of Labour.

Department of Labour (1998). Employment equity act 55 of 1998.

Department of Labour (1995). Labour Relations Act 66 Of 1995. Department of Labour.

Department of Labour (1973). Occupational Diseases in Mines and Works Act 78 of 1973.

Department of Labour (1993). Compensation for Occupational Injuries and Diseases Act no of 103 of 1993.

Department of Labour (1993). Occupational Health and Safety Act No: 85 of 1993.

Department of Mineral Resources (1996). Mine Health and Safety Act No: 29 of 1996.

Department of Safety and Security (2003). Explosives Act No. 15 of 2003.

Department of Water Affairs (1998). National Water Act No. 36 of 1998.

Department of Water and Sanitation (1997). Water Services Act No 108 of 1997.

Gibson, Clive. and Flood, Patric. (2015). *Everyone's Guide to Labour Law.* Cape Town: Penguin Random House South Africa.

Olivier, Belinda. (n.d.). Corporate Governance. [online] Financialmarketsjournal.co.za. Available at: http://financialmarketsjournal.co.za/oldsite/8thedition/printedarticles/corporategovernance.htm.

Penelope.uchicago.edu. (n.d.). *LacusCurtius • Roman Law — Noxalis Actio (Smith's Dictionary, 1875).* [online] Available at: http://penelope.uchicago.edu/Thayer/E/Roman/Texts/secondary/SMIGRA*/Noxalis_Actio.html.

Republic of South Africa. 1996. Constitution of the Republic of South Africa No. 108 of 1996.

Roman Law II - Lex Aquilia Notes (n.d.). *Roman Law Lex Aquilia.* [online] Available at: https://www.studocu.com/en/document/anglia-ruskin-university/law/summaries/roman-law-ii-lex-aquilia-notes/1788700/view.

Shaw, G. (2010). *RM Model for the Tourism Industry | Risk | Tourism.* [online] Scribd. Available at: https://www.scribd.com/document/188917955/RM-Model-for-the-Tourism-Industry.

The laws of Hammurabi 1792 BC to c. 1750 BC (1903). *The Laws of Hammurabi, King of Babylonia - Wikisource, the free online library*. [online] En.wikisource.org. Available at: https://en.wikisource.org/wiki/The_Laws_of_Hammurabi,_King_of_Babylonia.

Union of South Africa (U of SA) 1956. Water Act, No. 54, in U. of SA, Statutes of the Union of South Africa 1956. Part II Nos. 47–73. Cape Times Ltd, under supervision of the Government Printer, Parow, pp. 1046–1305.

Van der Luit-Drummond, John. (2013). The common law: A student's tale. [online] JvdLD. Available at: https://jvdld.com/2013/02/07/the-common-law-a-students-tale/.

Work Health and Safety Codes of Practice 2011 (2011). *Work Health and Safety Codes of Practice 2011*. [online] Legislation.gov.au. Available at: https://www.legislation.gov.au/Details/F2011L02804.

CHAPTER 8

STATUTORY REQUIREMENTS

OCCUPATIONAL HEALTH AND SAFETY ACT 85, 1993 AS AMENDED

• THE ACT AND REGULATIONS

The Occupation Health and Safety Act (Act 85 of 1993) as amended (the Act) consists of 50 sections approved by the South African Parliament. The purpose of the Act is to provide for the health and safety of persons at work or in connection with the use of plant and machinery. It also provides for the protection of persons other than persons at work from hazards arising out of or in connection with the activities at work. The Minister of Labour incorporates various regulations, on specific topics, into the Act from time to time.

Whereas the Constitution protects the rights of all citizens and those in the country in terms of health and safety, the Occupational Health and Safety Act 85, 1993, as amended (the Act), and other relevant legislation prescribe the different responsibilities and liabilities of both the employer and the worker.

The Act replaced the old Machinery and Occupational Safety Act in 1993, which in turn replaced the older Factories Act of 1941 in 1983. As technology develops it is necessary for legislators to regularly revise and adapt legislation to keep abreast of technological developments and related issues. As new serious health threats arise which threaten the entire country or a particular region on a large scale, existing legislation, and regulations may need to be reviewed, revised and adapted to cope with those threats.

• THE PURPOSE OF THE ACT

The purpose of the Act is:

- To provide for the health and safety of persons at work and for the health and safety of persons in connection with the use of plant and machinery.
- The protection of persons other than persons at work against hazards to health and safety arising out of or in connection with the activities of persons at work.
- To establish an advisory council for occupational health and safety; and to provide for matters connected therewith.

• THE ACT

The Occupational Health and Safety Act 85 of 1993 as amended requires every employer to bring about and maintain, as far as reasonable practicable, a workplace that is safe and without risk to the health of the workers.

Therefore, the employer must take the following into account, namely:

Ensuring that the workplace is free of hazardous substances, such as, for example, benzene, chlorine, and microorganisms, articles, equipment, and processes that may cause injury, damage or disease. The employer needs to ensure proper and regular sanitisation of all workplaces and surfaces. Where this is not possible, the employer must inform workers of these dangers, how to work safely, and provide personal protective equipment for a safe workplace.

However, it is not the sole responsibility of the employer to adhere to the Act. Workers also have responsibilities that they must carry out regarding certain regulations within this legislation.

In short this means that the worker and the employer share responsibility in a partnership regarding SHE in the workplace.

• OBJECTIVES OF THE ACT

The purpose of the Act as stated earlier is to provide for the health and safety of persons at work or in connection with the use of plant and machinery. It further provides for the protection of persons other than persons at work from hazards arising out of or in connection with the activities at work.

The Act supports the constitutional right of every worker to a safe and healthy workplace. The duty for creating and maintaining a healthy and safe workplace falls on every worker in the workplace, to the degree they have the authority and ability to do so. Whether they are the Chief Executive Officer or the newest worker hired, everyone has in the partnership between the employer and workers a personal and shared responsibility for working together co-operatively to prevent occupational injuries, disease, illnesses and fatalities.

• RESPONSIBILITIES

The Act clearly defines the responsibilities of employers and workers

• EMPLOYER RESPONSIBILITIES

• CHIEF EXECUTIVE OFFICER (CEO)

In terms of Section 16 of the Act:

- The CEO must ensure the proper execution of the duties of the employer in terms of the Act.
- The CEO may assign duties to any person under his control, who shall act subject to the control and directions of the CEO.
- The head of department of any department of State shall be deemed to be the chief executive officer of that department.

• GENERAL DUTIES OF EMPLOYER: SECTION 8

The employer must provide and maintain all the equipment that is necessary to do the work, and all the systems according to which work must be done, in a condition that will not affect the health and safety of workers.

Before personal protective equipment may be used, the employer must first try to reduce any danger that may affect the health and safety of his workers. Only when this is not practical, should the employer resort to personal protective equipment. Therefore, the use of personal protective equipment should always be a measure of last resort to, inter alia, eliminate giving the worker a false sense of security that they are safe and without any threat to their health and wellbeing while the hazard is in reality still present.

The employer must take measures to protect the health and safety of workers against hazards that may result from the production, processing, use, handling, storage or transporting of articles or substances, including all workplaces and surfaces. In other words, anything that workers may come into contact with at work.

To ensure that these duties are complied with, the employer must do the following, namely:

1. Identify potential hazards through conducting a HIRA, which may be present while work is carried out, something is being produced, processed, used, stored or transported, and any equipment is being used.
2. Establish the precautionary measures that are necessary to protect his or her workers against the identified hazards and provide the means to improve the means to implement these precautionary measures.
3. Provide the necessary information, instructions, training and supervision while keeping the extent of the competence of workers in mind. In other words, what they may do and may not do.
4. Not permit any worker to carry on with any task unless the necessary precautionary measures have been taken.
5. Take steps to ensure that every person under his or her control complies with the requirement of the Act.
6. Enforce the necessary control measures in the interest of health and safety.
7. Ensure that the work being done and the equipment used, is under general supervision of a worker who has been trained to understand the hazards associated with the work. Such a worker must ensure that the precautionary measures are implemented and maintained.

• ARTICLES

Manufacturers, designers, importers, sellers and suppliers must ensure that the following are adhered to regarding articles:

- Their articles are safe and without risk to health and comply with all prescribed requirements.
- When a structure or an article is installed on any premises, it must be done in such a way that neither an unsafe situation nor a health risk is created.

- SUBSTANCES

Manufacturers, designers, importers, sellers, and suppliers must ensure that the following is adhered to regarding substances that they supply or make available:

✓ Such substances are safe and without risk to the health and safety of persons when properly used.
✓ Information is available on the following:
  - Use of the substance at work
  - Health and safety risk associated with the substance
  - Conditions that are necessary to ensure that the substance will be safe and without risk to health when properly used.
  - Procedures in case of an accident are in place.

If a person to whom an article or substance has been sold or supplied, undertakes in writing to take specified steps to ensure that the article or substance will meet all the prescribed requirements, and will be safe and without risk to health, the duties of the importer, designer, seller, supplier or manufacturer will subsequently shift to the person who undertakes to take such steps.

- DUTY TO INFORM WORKERS: SECTION 13

The Employer must see to it that every worker is informed and clearly understands the health and safety hazards in the workplace or any work he or she is to do, anything being produced, processed, used, stored, handled or transported, and any equipment or machinery being used. The Employer must then provide information about precautionary measures against any likely exposure to these hazards.

The Employer must inform Health and Safety Representatives when an inspector notifies him or her of inspections and investigations to be conducted at the premises. The Employer must also inform Health and Safety Representatives of any application for exemption made, or of any exemption granted to him or her in terms of the Act. Exemption means being exempted from certain provisions of the Act, regulations, notices or instructions issued under the Act.

The Employer must, as soon as possible, inform the Health and Safety representatives of the occurrence of an incident in the workplace. An incident is an event that occurs in the workplace where a person is killed, injured or becomes ill. It also includes the spillage of a hazardous chemical substance, for example, where a tank leaks formaldehyde which is a chemical product used in industry due to a faulty valve, or where machinery run out of control, without necessarily killing or injuring anyone.

- EMPLOYER LIABILITIES
- OFFENCES, PENALTIES AND SPECIAL ORDERS OF COURT

In terms of Section 38(2) of the Act:

"Any employer who does or omits to do an act, thereby causing any person to be injured at a workplace, or, in the case of a person employed by him, to be injured at any place in the course of his employment, or any user who does or omits to do an act in connection with the use of plant or machinery, thereby causing any person to be injured, shall be guilty of an offence if that employer or user, as the case may be, would in respect of that act or omission have been guilty of the offence of culpable homicide had that act or omission caused the death of the said person, irrespective of whether or not the injury could have led to the death of such person, and on conviction be liable to a fine not exceeding R100,000 or to imprisonment for a period not exceeding two years or to both such fine and such imprisonment."

- WORKER RESPONSIBILITIES
- GENERAL DUTIES OF WORKERS AT WORK (SECTION 14)

In terms of Section 14 of the Act:

It is the duty of all workers in the workplace to ensure that they comply with the following:

- Take care of their own health and safety, as well as that of fellow workers in the workplace, who may be affected by their actions or negligence to act which includes horseplay at work. Many people have been injured and even killed due to horseplay in the workplace.
- Where the Act imposes a duty or requirements on workers, to fully cooperate with the employer.
- Give information to inspectors from the Department of Labour if they should require it.

- Carry out any lawful instruction which the employer or authorized person prescribes with regard to health and safety.
- Comply with the rules and procedures that the employer gives them.
- Wear the prescribed safety clothing or the prescribed safety equipment where required.
- Report unsafe or unhealthy conditions to the employer or Health and Safety Representative as soon as possible
- If they are involved in an accident or incident that may influence their health, report it to the employer or authorized person or the Health and Safety Representative as soon as possible, but not later than the end of their shift.
- Comply strictly with all COVID-19 protocols in the workplace to prevent spread of infections.

• DUTY NOT TO INTERFERE WITH, DAMAGE OR MISUSE THINGS (SECTION 15)

In terms of Section 15 of the Act:

No worker or other person shall intentionally or recklessly interfere with, damage or misuse anything which is provided in the interest of health or safety.

• THE RIGHTS OF WORKERS

The Act has extended the rights of workers to include the following, namely

- The right to information
- The right to participate in inspections
- The right to comment on legislation and make representations
- The right not to be victimized
- The right to appeal

The right to information

Workers must have access to:

- The Act and associated Regulations under the Act.
- Health and safety rules and procedures of the workplace and organization.
- Health and safety standards of the workplace that must be kept by the employer.

Workers may request the employer to inform them about the following:

- Health and safety hazards in the workplace.
- The precautionary measures which must be taken.
- The procedure that must be followed if they are exposed to substances hazardous to their health.

Workers may request that their private medical practitioner investigate their medical and exposure records. If the worker is a Health and Safety Representative, he or she may investigate and comment in writing on exposure assessments and monitoring reports.

The right to comment on legislation and make representations

Workers may comment or make representations on any regulation or safety standard published under the Act.

The right not to be victimized

Employers may not dismiss any worker from their service, reduce the salary of any worker or reduce the service conditions of any worker because:

- The worker supplied information, which is required in terms of the Act, to someone who is charged with the administration of the Act.
- The worker complied with a lawful notice such as, for example, a prohibition or a contravention notice.
- The worker did something in terms of the Act that should have been done.
- The worker did not do something in terms of the Act that is prohibited.
- The worker has given evidence before the Labour Court or a court of law on matters regarding health and safety.

The right to appeal

Workers may appeal against the decision of an inspector by making a written submission to the Chief Inspector.

Duty not to interfere with or misuse of objects

No one may interfere with or misuse any object that has been provided in the interest of health and safety. A worker may, for example, not remove a safety guard from a machine and use the machine or allow another worker to use it without such a guard.

CONSEQUENCES OF NON-COMPLIANCE

• NEAR-MISSES OF INCIDENTS

Not complying with the Act may result in accidents and workers and other people possibly getting hurt in unsafe working conditions, which could result in prosecution or legal action.

A near-miss is an accident that has simply not yet resulted in an observable or visible physical consequence. All near misses should be investigated in the same way that accidents are. If treated properly, the collection, analysis and investigation of near misses will prevent the injury and damage outcomes of those events if remedial action is identified and applied. Targets set to reduce lost-time accidents will be achieved due to an underreporting culture. Dealing with near misses is injury prevention and not accident prevention. There are more 'no injury' accidents than those that result in injury. There is still the problem of overcoming the reluctance or inertia of workers who may never report a near-miss. Reasons for underreporting include:

- The fear of blame
- The apparent triviality of the event measured by its outcome
- Embarrassment
- Suspicion that nothing will be done
- Perceived enormity of the task of filling in forms
- Apathy
- Lack of a workable system
- Personal fault

It is important that all near-misses are reported and fully investigated. They may be catastrophic the next time.

• COST OF INCIDENTS

All unwanted incidents and accidents cost money. While the total costs of accidents are difficult to determine, there are always hidden expenses in the form of indirect costs that must be identified and considered. Costs must be measured in terms of the effect on the individual and the community, as well as the financial loss.

It has been stated that if the true costs of injuries were well defined, management would be in a better position to make informed decisions concerning health and safety. Rather than addressing health and safety solely from an altruistic point of view, employers should also consider health and safety from a purely economic perspective.

• SOCIAL IMPACT

All elements of society such as workers, families, employers, the economy, and resources have probably been affected somehow by the occurrence of an accident at work. The economic impacts of incidents and accidents are, for example, decrease in family income, decrease in the standard of living, negative effects on education and schooling expenses, increase in debts and difficulty to pay bills, insurance policies and mortgage bonds. Victims and their dependents could suffer great pain, discomfort, hardship, sorrow, and even psychological effects for many years, following a disabling occupational disease or injury. The loss of earning power of someone who has contracted asbestosis or lost an arm will affect him or her in an unimaginable way.

- FINANCIAL IMPACT

The iceberg depicts the situation quite clearly, namely that the portion below the surface is far greater than the exposed peak. This applies to the cost of a loss-producing incident. The accumulative effects of all the incidents have a dramatic effect on the profits or bottom line of an organization, as well as the national economy.

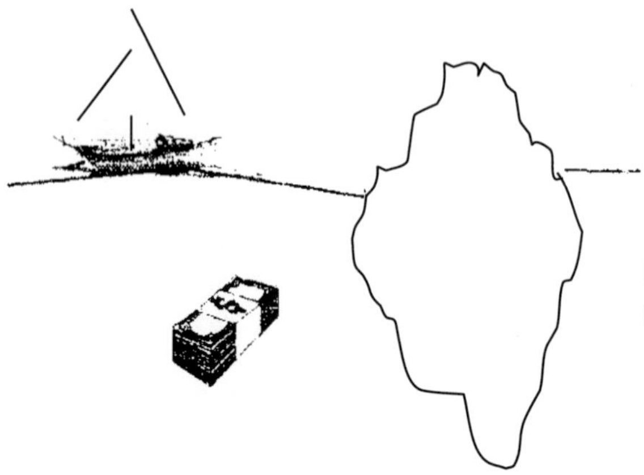

The costs of occupational injuries and illnesses can be divided into the following broad categories, namely:

- direct costs that include the costs of hospital, physician, and allied health services, rehabilitation, nursing home care, home health care, medical equipment, burial costs, insurance administrative costs for medical claims, payments for mental health treatment, police, fire, emergency transport, coroner services and property damage;
- indirect costs that refer to victim productivity losses that include wage losses and household production losses, administrative costs that include the cost of administering workers' compensation wage replacement programs and sick leave, disruption of work schedules, administrative time for investigations and reports, training of replacement personnel, wages paid to the injured workers and others for time not worked, clean up and repair, adverse publicity, third-party liability claims, and equipment damage;
- costs of health and safety programs that include salaries for health and safety, medical and clinical personnel, health and safety meetings, inspections of tools and equipment, orientation or induction sessions, workplace inspections, personal protective equipment, health programs, and miscellaneous supplies and equipment; and
- quality of life costs that refer to value attributed to the pain and suffering that victims and their families experience because of the injury, fatality or illness. Loss of quality-of-life costs can be six times greater than the workplace disruption costs.

The indirect costs of incidents and accidents are greater than their direct costs. Estimates of indirect costs as a proportion of direct costs have ranged from ratios that are 1:1 to more than 20:1, depending on the specific industry and methodology utilised. Given that indirect costs of accidents always exceed the direct costs; it makes business sense to give greater attention to the indirect costs of worker injuries. Even where indirect and direct costs are similar, indirect costs balloon exponentially when liability litigation is instituted for a worker's injury. The following example illustrates how much an injury could potentially affect the output or bottom-line of an organisation trying to achieve a three percent profit margin. If an organisation sustained a ZAR50,000 loss due to injury, illness or damage and still tries to make a three percent profit, the organisation theoretically must increase sales of services by an additional ZAR1,667,000 The Health and Safety Executive in the United Kingdom suggest the following Accident Cost Calculator to determine and record the cost of incidents and accidents at work:

Category	Example	Time spent	Cost (£)
Dealing with incident (immediate action)	• First-aid treatment • Taking injured person to hospital/home • Making the area safe • Putting out fires • Immediate staff downtime • Other		
Getting back to business	• Assessing/rescheduling work activities • Recovering work/production (including staff costs) • Cleaning up site and disposal of waste, equipment, products etc. • Bringing work up to standard such as product reworking time/cost • Repairing any damage/faults • Hiring or purchasing tools, equipment, plant, and services • Other		
Business costs	• Salary costs of injured person while off work • Salary costs of replacement workers • Lost work time including people waiting to resume work, delays, reduced productivity, and effects on the productivity of other workers • Overtime costs • Recruitment costs for new workers • Contract penalties • Cancelled and/or lost orders • Other		
Category	Example	Time spent	Cost (£)
Action to safeguard future business	• Reassuring customers • Providing alternative sources of supply for customers • Other		
Sanctions and penalties	• Compensation claim payments • Attorneys' fees and legal expenses • Staff time dealing with legal cases • Fines and costs imposed due to criminal proceedings • Increase in insurance premiums • Other		
Other	• Own examples		
Total Cost			

These are all important factors that must be considered in the estimation of the amount of monetary compensation. Each case will be different and the severity of the case will certainly have an effect on the costs.

• NON-COMPLIANCE: (SECTION 38)

The following is an extract from the Act as amended about the consequences of non-compliance with its requirements (Clause 38):

1. Acts or omissions by workers or mandataries

(1) Whenever a worker does or omits to do any act which it would be an offence in terms of this Act for the employer of such worker or a user to do or omit to do, then, unless it is proved that-

(a) in doing or omitting to do that act the worker was acting without the connivance or permission of the employer or any such user;

(b) it was not under any condition or in any circumstance within the scope of the authority of the worker to do or omit to do an act, whether lawful or unlawful, of the character of the act or omission charged; and

(c) all reasonable steps were taken by the employer or any such user to prevent any act or omission of the kind in question, the employer or any such user himself shall be presumed to have done or omitted to do that act, and shall be liable to be convicted and sentenced in respect hereof; and the fact that he issued instructions forbidding any act or omission of the kind in question shall not, in itself, be accepted as sufficient proof that he took all reasonable steps to prevent the act or omission.

(2) The provisions of subsection (1) shall mutatis mutandis apply in the case of a mandatory of any employer or user, except if the parties have agreed in writing to the arrangements and procedures between them to ensure compliance by the mandatory with the provisions of this Act.

(3) Whenever any worker or mandatory of any employer or user does or omits to do an act which it would be an offence in terms of this Act for the employer or any such user to do or omit to do, he shall be liable to be convicted and sentenced in respect thereof as if he were the employer or user.

(4) Whenever any worker or mandatory of the State commits or omits to do an act which would be an offence in terms of this Act, had he been the worker or mandatory of an employer other than the State and had such employer committed or omitted to do that act, he shall be liable to be convicted and sentenced in respect thereof as if he were such an employer.

(5) Any worker or mandatory referred to in subsection (3) may be so convicted and sentenced in addition to the employer or user.

(6) Whenever the worker or mandatory of an employer is convicted of an offence consisting of a contravention of section 23, the court shall, when making an order under section 38 (4), make such an order against the employer and not against such worker or mandatory.

2. Offences, penalties and special orders of court

(1) Any person who

(a) contravenes or fails to comply with a provision of section 7, 8, 9, 10 (1), (2) or (3), 12, 13, 14, 15, 16 (1) or (2), 17 (1), (2) or (5), 18 (3), 19 (1), 20 (2) or (4), 22, 23, 24 (1) or (2), 25, 26, 29 (3), 30 (2) or (6), 34 or 36;

(b) contravenes or fails to comply with a direction or notice under section 17 (6), 19 (4) or (7), 21 (1) or 30 (1) (a), (b) or (c) or (3), (4) or (6);

(c) contravenes or fails to comply with a condition of an exemption under section 40 (1);

(d) in any record, application, statement or other document referred to in this Act wilfully furnishes information or makes a statement which is false in any material respect;

(e) hinders or obstructs an inspector in the performance of his functions; refuses or fails to comply to the best of his ability with any requirement or request made by an inspector in the performance of his functions;

(f) deleted

(g) refuses or fails to answer to the best of his ability any question which an inspector in the performance of his functions has put to him;

(h) wilfully furnishes to an inspector information which is false or misleading;

(i) gives himself out as an inspector;

(j) having been subpoenaed under section 32 to appear before an inspector, without sufficient cause (the onus of proof whereof shall rest upon him) fails to attend on the day and at the place specified in the subpoena, or fails to remain in attendance until the inspector has excused him from further attendance;

(k) having been called under section 32, without sufficient cause (the onus of proof whereof shall rest upon him)-

(i) refuses to appear before the inspector;

(ii) refuses to be sworn or to make affirmation as a witness after he has been directed to do so;

(iii) refuses to answer, or fails to answer to the best of his knowledge and belief, any question put to him; or

(iv) refuses to comply with a requirement to produce a book, document or thing specified in the subpoena or which he has with him;

(l) tampers with or discourages, threatens, deceives or in any way unduly influences any person with regard to evidence to be given or with regard to a book, document or thing to be produced by such a person before an inspector under section 32;

(m) prejudices, influences or anticipates the proceedings or findings of an inquiry under section 32 or 33;

(n) tampers with or misuses any safety equipment installed or provided to any person by an employer or user;

(o) fails to use any safety equipment at a workplace or in the course of his employment or in connection with the use of plant or machinery, which was provided to him by an employer or such a user;

(p) wilfully or recklessly does anything at a workplace or in connection with the use of plant or machinery which threatens the health or safety of any person, shall be guilty of an offence and on conviction be liable to a fine not exceeding R50000 or to imprisonment for a period not exceeding one year or to both such fine and such imprisonment.

(2) Any employer who does or omits to do an act, thereby causing any person to be injured at a workplace, or, in the case of a person employed by him, to be injured at any place in the course of his employment, or any user

who does or omits to do an act in connection with the use of plant or machinery, thereby causing any person to be injured, shall be guilty of an offence if that employer or user, as the case may be, would in respect of that act or omission have been guilty of the offence of culpable homicide had that act or omission caused the death of the said person, irrespective of whether or not the injury could have led to the death of such person, and on conviction be liable to a fine not exceeding R100 000 or to imprisonment for a period not exceeding two years or to both such fine and such imprisonment.

(3) Whenever a person is convicted of an offence consisting of a failure to comply with a provision of this Act or of any direction or notice issued thereunder, the court convicting him may, in addition to any punishment imposed on him in respect of that offence, issue an order requiring him to comply with the said provision within a period determined by the court.

Whenever an employer is convicted of an offence consisting of a contravention of a provision of section 23, the court convicting him shall inquire into and determine the amount which contrary to the said provision was deducted from the remuneration of the worker concerned or recovered from him and shall then act with respect to the said amount mutatis mutandis in accordance with sections 28 and 29 of the Basic Conditions of Employment Act, 1983 (Act No. 3 of 1983), as if such amount is an amount underpaid within the meaning of those sections.

The primary reasons for evaluating, implementing, and maintaining SHE in the workplace include:

- The creation of a safe and healthy working environment. If this objective is achieved, then compliance with legislation is much easier
- To create a safe system of work, that ensures that hazards are identified, risks are assessed and controlled.
- To ensure that machinery is provided in a safe working order and maintained in a safe working order.

Unfortunately, some organizations use their external compliance audits as the gauge of success, but then in the same vein they try their best not to be transparent on audits. Therefore, the gauge of success is perverted. SHE practitioners need to ensure a safety and healthy working environment while still allowing the business to make a profit.

What are the consequences of not complying with legislation?

- Workers are injured or killed
- Persons can be exposed the harmful substances and be diagnosed with an occupational disease
- People can be maimed by machinery by possibly losing a limb such as an arm, leg or hand
- There could be fines or imprisonment for the CEO
- Businesses can be stopped or closed down which will impact severely on the financial resources of the business
- The local community and family system are severely impacted by the potential loss of a bread winner
- Imagine children growing up without a father who has been killed or permanently disabled at work? How that will affect the family?
- Criminal prosecution of the employer and individual.
- Civil action against the employer and individual.
- Increases in Compensation Commissioner premiums.
- Additional compensation to the injured.
- Organization closing down or being shut down, leading to loss of income for everyone

• CLOSING OF OPERATIONS

Below is an article released by Department of Labour in relation to not complying with OHS.

Source: https://www.sanews.gov.za/south-africa/labour-dept-shuts-down-company-issues-warnings

• LABOUR DEPARTMENT SHUTS DOWN COMPANY, ISSUES WARNINGS

Wednesday, March 7, 2012

Pretoria - The Department of Labour closed down a company for non-compliance with the Occupational Health and Safety Act and issued several warnings to other businesses during an inspection operation in Johannesburg on Tuesday.

The department issued 24 businesses with written warnings for contravening the Basic Conditions of Employment Act, including underpaying workers. A total of 24 illegal immigrants were arrested at the firms.

The department in conjunction with the police, immigration officials, traffic cops and the South African Revenue Services (SARS) conducted a joint-inspection aimed at clamping down on companies not complying with labour laws and illegal business operations.

Gauteng Department of Labour spokesperson, Mishack Magakwe, said the notices of non-compliance that were issued will be followed up on. "We will continue to enforce labour legislation," he said.

According to the department, in some instances, workers were working without any contracts, pay slips or attendance registers and were being underpaid.

"Some of the workers were not registered with the UIF and the compensation fund while the Occupational Health and Safety Act was found to be non-existent." Magakwe said, adding that employers were given 21 days to correct matters.

"They were issued with written undertakings to comply with or risk being dragged to court," he said.

The department said more joint inspections with SAPS, Home Affairs, SARS and Metro Police in the province could be expected.

• APPLICABLE LEGISLATION

Depending on the industry that is worked in, any of the following Acts as amended could also apply to the place of work, namely

- Mine Health and Safety Act, 1996 (Act No: 29 of 1996)
- National Environmental Management Act, 1998 (Act No: 107 of 1998)
- Environmental Conservation Act 1989 (Act No: 73 of 1989)
- Water Act 1956 (Act No: 54 of 1956)
- National Water Act 1998 (Act No: 36 of 1998)
- Water Services Act 1997 (Act No: 108 of 1997)
- Forestry Act 1998 (Act No: 84 of 1998)
- Atmospheric Pollution Prevention Act 1965 (Act No: 45 of 1965)
- Compensation for Occupational Injuries and Diseases Act (Act no: 103 of 1993)
- Occupational Diseases in Mines and Works Act 1973 (Act No:78 of 1973)
- Explosives Act 1956 (Act No: 26 of 1956)
- Municipal Bylaws
- Health, safety and environment agreements

• DEPARTMENT OF LABOUR

• CHIEF DIRECTORATE OF OCCUPATIONAL HEALTH AND SAFETY

The Chief Directorate of Occupational Health and Safety of the Department of Labour administers the Act. To ensure the health and safety of workers, provincial offices have been established in all 9 provinces in South Africa. To this end, Occupational Health and Safety inspectors from these provincial offices carry out inspections and investigations in the workplaces.

• INSPECTIONS

Inspections are usually planned based on accident statistics, the presence of hazardous substances, such as the use of benzene in laundries, or the use of dangerous machinery in the workplace. Unplanned inspections, on the other hand, usually arise from requests or complaints by workers, employers, or members of the public. These complaints or requests are treated confidentially.

• INSPECTORS

• DESIGNATION OF INSPECTORS BY MINISTER

In terms of Section 28 of the Act:

- The Minister may designate any person as an inspector to perform the functions assigned to an inspector by this Act.
- Each designated inspector must have a certificate signed by or on behalf of the Minister, stating that he has been designated as an inspector.
- Inspectors may be required to produce such a certificate on demand during performance of their functions under this Act.

• FUNCTIONS OF INSPECTORS

In terms of Section 29 of the Act:

An inspector may, for the purposes of this Act –

- Without previous notice, at all reasonable times enter any workplace.
- Question any person in the workplace on any matter to which this Act relates.
- Require from any person in possession or control of a book, record, or other document on or in the workplace, to produce to him such book, record, or other document.
- Examine any such book, record or other document or make a copy thereof or an extract therefrom.
- Require from such a person an explanation of any entry in such book, record or other document.
- Inspect any article, substance, plant, or machinery which is used in the workplace, or any work performed, or any condition prevalent in the workplace. He may also remove for examination or analysis any article, substance, plant or machinery or a part or sample thereof.
- Seize any such book, record, or other document or any such article, substance, plant or machinery or a part or sample thereof which in his opinion may serve as evidence at the trial of any person charged with an offence under this Act or the common law.
- The employer may make copies of documents before such seizure.
- Direct any employer, worker or user, including any former employer, worker, or user, to appear before him on any matter to which this Act relates.
- Perform any other function as may be prescribed.

An interpreter, a member of the South African Police or any other assistant may, when required by an inspector, accompany him when he performs his functions under this Act. For the purposes of this Act an inspector's assistant shall, while he acts under the instructions of an inspector, be deemed to be an inspector.

When an inspector enters any workplace the employer and workers performing any work must always provide the inspector with facilities to enable him and his assistant (if any) to perform effectively and safely his or their functions under this Act.

When an inspector removes or seizes any article, substance, plant, machinery, book, record, or other document he must issue a receipt to the owner or person in control thereof.

• SPECIAL POWERS OF INSPECTORS

In terms of Section 30 of the Act:

If an Inspector finds dangerous or adverse conditions at the workplace, he or she may set requirements to the employer in the following ways and serve the following documents on the employer, namely:

- Prohibition notice
- Contravention notice
- Improvement notice
- Other powers

Prohibition notice

In the case of threatening danger an inspector may prohibit a particular action, process, or the use of a machine or equipment, by means of a prohibition notice. No person may disregard the contents of such a notice and compliance must take place with immediate effect.

Contravention notice

If a provision of a regulation is contravened, the inspector may serve a contravention notice on the workers or the employer. A contravention of the Act can result in immediate prosecution, but in the case of a contravention of a regulation, the employer may be given the opportunity to correct the contravention within a limit specified in the notice, which is usually 60 days.

Improvement notice

Where the health and safety measures which the employer has instituted do not satisfactorily protect the health and safety of the workers, the inspector may require the employer to bring about more effective measures. An improvement notice, which prescribes the corrective measures, is then served on the employer.

Non-compliance notice

This notice directs an employer to appear at an enquiry to investigate alleged non-compliance with the requirements of the Act.

Other powers

To enable the inspector to carry out his or her duties, he or she may enter any workplace or premises where machinery or hazardous substances are being used and question or serve a summons on persons to appear before him or her. The inspector may request that any documents be submitted to him or her, investigate and make copies of the documents, and demand an explanation about any entries in such documents. The inspector may also inspect any condition or article and take samples of it and seize any article that may serve as evidence.

Should an employer not be happy with the actions of the inspector, he may appeal to the Chief Inspector in writing within 60 days.

Note: The powers of inspectors are not absolute. Any person, who disagrees with any decision taken by the inspector, may appeal against that decision in writing to the Chief Inspector Occupational Health and Safety at Department of Labour.

• HEALTH AND SAFETY REPRESENTATIVES (SECTION 17)

In terms of Section 17 of the Act:

- Employers, who employ more than 20 workers at any workplace, must appoint health and safety representatives in writing for such workplace, or for different sections thereof.
- An employer and the worker representatives or workers, where there are no such representatives, must consult in good faith regarding the election, period of office and subsequent designation of health and safety representatives. If such consultation fails, the matter shall be referred for arbitration to a person mutually agreed upon, whose decision shall be final.
- Only full-time workers, who are acquainted with conditions and activities at that workplace, will be eligible for designation as health and safety representatives for that workplace.
- The number of health and safety representatives for a workplace or section thereof shall in the case of shops and offices be at least one health and safety representative for every 100 workers or part thereof, and in the case of all other workplaces at least one health and safety representative for every 50 workers or part thereof: Workers working another workplace than that where they ordinarily report for duty, shall be deemed to be working at the workplace where they report for duty.

- Inspectors of the Department of Labour (DoL) may direct employers in writing to increase the number of health and safety representatives if they feel that, under certain circumstances, the prescribed number is not adequate.
- All activities in connection with the activities and training of health and safety representatives shall be performed during ordinary working hours, and any time reasonably spent by any worker in this regard shall for all purposes be deemed to be time spent by him in the carrying out of his duties as a worker.

It is strongly recommended that despite the legislative requirements which are minimum requirements in any case organizations should always consider appointing at least one health and safety representative in every workplace or site even if the number of workers is less than 20.

• FUNCTIONS OF HEALTH AND SAFETY REPRESENTATIVES

In terms of Section 18 of the Act:

A health and safety representative may perform the following functions in respect of the workplace or section of the workplace for which he has been designated, namely-

- Review the effectiveness of health and safety measures in the workplace.
- Identify potential hazards and major incidents at the workplace.
- Examine the causes of incidents at the workplace.
- Investigate complaints by any worker relating to that worker's health or safety at work.
- Make representations to the employer or a health and safety committee on matters arising from the above.
- Make representations to the employer on general matters affecting the health or safety of the workers at the workplace.
- Inspect the workplace with a view to the health and safety of workers, at times agreed upon with the employer, who may be present during the inspection. The employer must be given reasonable notice of such inspections.
- Participate in consultations with inspectors at the workplace and accompany inspectors on inspections of the workplace.
- Receive information from inspectors as contemplated in section 36.
- Attend meetings of the health and safety committee of which he is a member, in connection with any of the above functions.

• HEALTH AND SAFETY COMMITTEES

In terms of Section 19 of the Act:

- Employers must establish one or more health and safety committees where two or more health and safety representatives have been designated, and, at every meeting of such a committee, consult with the committee on measures to ensure the health and safety of his workers at work.
- Nominees on a health and safety committee shall be designated in writing by the employer for a specific period. Health and safety representatives must be members of the committee for the period of their designation.
- Health and safety committee meetings must be held at least once every three months, at a time and place determined by the committee. Under certain circumstances a DoL inspector may in writing direct a safety committee to hold a meeting.
- A health and safety committee may appoint advisory members with particular knowledge of health or safety matters. These advisory members will have no voting rights.
- Inspectors of the DoL may direct employers in writing to increase the number of safety committees if they feel that, under certain circumstances, the prescribed number is not adequate.

As in the case of health and safety representatives, it is strongly recommended that despite the legislative requirements which are minimum requirements in any case organizations should always consider having at least one health and safety committee in the organization even if the number of health and safety representatives is less than 2.

• FUNCTIONS OF HEALTH AND SAFETY COMMITTEES

In terms of Section 20 of the Act:

- A health and safety committee may make recommendations to the employer or an inspector regarding any matter affecting the health or safety of persons at the workplace, discuss and report to an inspector on any incident causing injury or death at the workplace.
- A health and safety committee must keep record of recommendations made to an employer and of any report made to an inspector.

Employers must take the prescribed steps to ensure that a health and safety committee complies with the provisions of the Act.

• PURPOSE OF HEALTH AND SAFETY COMMITTEES

Members meet to initiate, promote, maintain and review measures of ensuring the health and safety of workers.

• ESTABLISHMENT OF HEALTH AND SAFETY COMMITTEES

At least one committee must be established when two or more representatives are designated. It is recommended that there always be a health and safety committee irrespective of the number of health and safety representatives especially considering the role that the committee plays in the overall health and safety effort.

• MEMBERSHIP OF A HEALTH AND SAFETY COMMITTEE

The employer determines the number of Health and Safety committee members based on the following:

- If only one committee has been established for a workplace, all the representatives must be members of that committee.
- If two or more have been established for a workplace, each representative must be a member of at least one of those committees.
- Therefore, every representative must be a member of a committee.
- The employer may also nominate other persons to represent him or her on a committee but such nominees may not be more than the number of representatives designated on that committee.
- If, however, an inspector is of the opinion that the number of committees in the workplace is inadequate, he or she may determine the establishment of additional committees

• FREQUENCY OF HEALTH AND SAFETY COMMITTEE MEETINGS

- They meet whenever it is necessary, but at least once every three months.
- The committee determines the time and place.
- However, if 10% or more of the workers put a request for a meeting to the inspector, the inspector may order that such a meeting be held at a time and place which he or she determines.

Again, it is recommended that the employer use his discretion as part of the health and safety improvement effort in the organization and arrange meetings irrespective of whether the number of workers requesting a meeting is less than 10% of all the workers in the organization.

• PROCEDURE OF THE MEETINGS

The members of the committee elect the chairperson and determine, for example, his or her period of office, and meeting procedures.

• CONSULTATION WITH EXPERTS FOR ADVICE

Committees may co-opt persons as advisory members for their knowledge and expertise on Health and Safety matters. However, an advisory member does not have any voting powers.

The committees only deal with health and safety matters in the workplace or sections thereof, for which such committees have been established.

Generally, health and safety committees have the following functions:

- **Make recommendations:** A committee must make recommendations about the health and safety of workers to the employer. Where these recommendations do not lead to solving the matter, the committee may make recommendation to an inspector.

- **Discuss Incidents:** A committee must discuss any incident that leads to the injury, illness, or death of any worker and may report about it in writing to an inspector.
- **Record Keeping**: A committee must keep record of every recommendation to the employer and every report to an inspector.
- **Other Functions:** Committee members must perform any other functions required of them by regulation.

• DEDUCTIONS

An employer may not make any deduction from the salary of a worker about anything he or she is required to do in the interest of health and safety in terms of the Act.

• REPORT TO THE CHIEF INSPECTOR REGARDING OCCUPATIONAL DISEASES

If a medical practitioner examines or treats a worker for a disease that he or she suspects arose from that worker's employment, the medical practitioner must report the matter to the employer of the worker, to the Chief inspector and notify the patient.

• CO-OPERATION WITH THE INSPECTOR

Given the pivotal role of inspectors of the DoL in ensuring health and safety compliance in organizations it is important to co-operate with them as far as is reasonably practicable.

Compliance with directions, subpoenas, requests or commands

Employers and workers must comply with the directions, subpoenas, requests or orders of inspectors. In addition, no one may prevent anyone else from complying.

Answering questions

The questions of inspectors should be answered, but no one is obliged to answer a question by which he or she might incriminate him or herself. To incriminate oneself means that one is suggesting that one is responsible for a contravention.

Investigations

When the inspector so requires, he or she must be provided with the necessary means and be given the assistance he or she may need to hold an investigation. The inspector may also request that individuals attend investigations that may assist the inspector with the investigation. No one may insult the inspector or deliberately interrupt the investigation.

Prosecutions

When the worker does something which in terms of the Occupation Health and Safety Act is regarded as an offence, the employer is responsible for that offence, and he or she could be found guilty and sentenced for it, unless the employer can prove that:

- He or she did not give his or her consent
- He or she did take all reasonable steps to prevent it from happening
- The worker did not act within the scope of his or her competence, in other words, that the worker did something, which he or she knew he or she should not have done

The Act or regulations can be purchased from the Government printer in Gazette form or bound form from various publishers.

WORK PROCEDURES

No specific section or regulation in the Act requires the employer to have written safe work procedures for every task performed by a worker. What the Act does require is for the employer to ensure that every worker is informed of the hazards and risks associated with the tasks the worker is to perform.

Section 37 of the Act makes the employer liable for the acts or omissions of workers, unless the employer can prove the following:

(a) The worker acted without the permission of the employer;
(b) It was not within the scope of authority of the worker to do the act; and
(c) The employer took all reasonable steps to prevent the act in question.

NB: The fact that the employer issued instructions forbidding the act in question shall in itself not be regarded as proof that all reasonable steps were taken.

12.3Detailed work procedures would assist with the following:

(a) Documented proof of hazards identified.
(b) Documented proof of risks identified.
(c) Documented proof of identified precautionary measures.
(d) Documented proof of worker's scope of authority.
(e) Immediate availability of training material.
(f) Proof that reasonable measures were taken to ensure that workers followed the correct work procedures.
(g) Assistance with the worker/job specifications.
(h) Provide the standard for planned task observations.

• LEGISLATIVE PROCEDURES

Every organisation has a health and safety risk to both staff and customers. To minimise the risk of injury to workers and customers, workers need to acknowledge and practice certain health and safety procedures. The following is an example:

	Action
1	Maintain in good condition insulated stands, trestles, mats or other such protective equipment as may be necessary to prevent accidents for use of persons working in close proximity to electrical equipment.
2	Workers must take reasonable care for their own health and safety.
3	Workers must take steps as may be reasonably practicable to eliminate or mitigate any hazard or potential hazard to their own health and safety, as well as that of other workers and customers.
4	Workers must take precautionary measures which are prescribed.
5	Workers must ensure that relevant signage is visible.
6	Affix a prominent notice or sign in a conspicuous place at the workplace, indicating where the first aid box or boxes are kept as well as the name of the person in charge of such first aid box or boxes.
7	Where more than 10 workers are employed at a workplace, take steps to ensure that for every group of up to 50, a minimum of one qualified first aid representative is available at that workplace.

The health and safety of workers is regulated by the Act and associated Regulations. All organisations need to subscribe to legislative procedures in order to ensure the health and safety of their workers and customers. The table contains some suggestions for what employers and workers need to do.

• COMPANY PROCEDURES

Depending on the type of organisation, procedures are implemented to suit the specific needs of the organisation. There are however general expectations about company procedures concerning safety. For every 50 workers in the hotel, there should be one Health and Safety Officer. This person attends Health and Safety Committee meetings and is responsible for ensuring that health and safety hazards are dealt with promptly. Note the earlier recommendation for appointing health and safety representatives irrespective of the number of workers in the organization. Examples of company procedures are shown in the following table.

	Action
1	Do not run in the workplace.
2	Wear shoes with non-slip soles
3	Clean up spills immediately
4	Put up signs to mark wet areas
5	Be familiar with the location of the First Aid Box
6	Make sure that workers know the names of the First Aid and Safety Officers in their unit / department and how to contact them.
7	Have knowledge of the relevant emergency authorities and how to contact them.
8	Display emergency and caution signs where they are easily accessible and observed.
9	Effectively display fire exits.

• RESPONSIBILITIES OF MANAGEMENT

It is commonly accepted that management has three main areas of responsibility:

• FINANCIAL

- To maximise profits and minimise losses to sustain longevity of the business.
- Prevention of accidents and incidents which can be a major loss factor in most businesses.

• LEGAL

To ensure that the organization complies with applicable legislation such as the Occupational Health and Safety Act, No 85 of 1993 and the Compensation for Occupational Injuries and Diseases Act, No 130 of 1993. (OHS Act and COID Act).

• MORAL

To ensure that workers are safe and healthy at work, so they are able to return home in as good a condition as they left.

• FUNCTIONS OF MANAGEMENT

- Line managers represent the CEO and therefore managing health and safety is not just something they do - it forms one of their core or main duties.
- Under the Act, every employer and line manager must, as far as possible, ensure the safety and health of all workers.
- They must ensure that safety and health is managed and that all workplace processes and activities are included in the health and safety management program of the organization.
- Their leadership, on-going commitment and involvement to health and safety can play an important role in positively motivating and encouraging workers to improve health and safety in the workplace. Rules are therefore necessary and compliance to these rules is very important. Line managers who do not enforce these rules and discipline persons who break these rules are not doing their work.

Some see management as a continuous cycle of planning, organising, leading and controlling the efforts of others in order to achieve specific goals. Management is a process aimed at the realisation of objectives or goals through the utilisation of the efforts of others.

- The goal of most organisations is to manufacture or supply goods or services in such a way that a reasonable profit can be earned.
- Accidents/Incidents are costly and can therefore have an adverse effect on the profitability of any organisation. To make a profit, management must exercise control in such a way that accidents/incidents are kept to a minimum.

The employer must make sure that the workplace is safe and healthy and must not allow any worker to do work which is potentially dangerous.

• INFORMATION AND TRAINING ABOUT DANGERS

The management of the organization must

- Inform workers of the dangers in the workplace.

- Ensure that there are warnings and notices on all dangerous machinery.
- Train workers on how to use dangerous machinery or substances safely.

• REDUCING DANGERS IN THE WORKPLACE

The management of the organization must

- Ensure that equipment is properly maintained.
- Provide protective clothing and equipment where necessary.
- Reduce any dangers to a minimum before issuing protective clothing.
- Make sure that a supervisor oversees operations and enforces safety requirements.
- Set out precautionary measures to prevent dangers.

• HEALTH AND SAFETY REPRESENTATIVES

Management must

- Recognize the selection of health and safety representatives (1 for every 20 workers as per the Act). Although there should be at least one irrespective of whether there are 20 workers employed or not.
- Explain the responsibilities of the health and safety representatives.
- Create a health and safety committee. This committee must meet at least every three months and deals with all safety and health issues that affect workers.

• PROCEDURES FOR DEALING WITH ACCIDENTS

Management must

- Ensure that workers can escape from danger if necessary.
- Create an environment that is conducive to conducting a threat-free investigation with the aim of avoiding the recurrence of the accident and not to lay blame on anyone involved.

• REPORTING ACCIDENTS OR INCIDENTS

Management must

- Keep a report of all accidents and safety or health incidents in the workplace.
- Report certain types of accident or incident to the health and safety representative and to the Department of Labour.

• THE CHIEF EXECUTIVE OFFICER (CEO)

The Chief Executive Officer (CEO) of an organisation represents the employer and is as such personally responsible for each workers' health and safety as well as to ensure that all the employer duties stipulated by the Act are met. Therefore, he / she can be held personally liable should this not happen.

In practice, the CEO will assign (section 16.2) specific functions to various line managers and appoint specific officials in the organisation to assist with this duty.

The following still applies:

- Delegation never means that the final responsibility is delegated. The CEO stays responsible and liable for all the regulations as stipulated in the Act.

Officials to whom specific health and safety duties have been assigned will perform these duties under the control of the CEO. Moving beyond duties and responsibilities is part of that unwritten function.

Health and safety in the workplace should go beyond compliance with legal provisions and actions aimed primarily at avoiding prosecution. It should instead aim at better practice. It is also a moral issue that reflects on the ethical values of the organisation and has economic and cost implications for both the organisation as well as the worker. It is a workplace issue that should be dealt with through constructive participation and joint problem solving between the relevant role players in the organisation. Health and safety should not be a priority of the organization but an organizational value and part of a positive organizational health and safety culture.

• LEGAL RESPONSIBILITY OF SUPERVISORS

Most supervisors and foremen are of the opinion that their most critical area of responsibility lies with ensuring that production targets are met. While it is true that no business can operate without ensuring that quality products and services are timorously made available to their customers, the Act does not concern itself with production output of the employer, but with the health and safety of workers and others that may be affected by the activities of the employer. Ensuring adherence to these standards cannot only be placed on management shoulders. Further assistance is required, and this is where the role of the supervisors and foremen becomes important in the organizational health and safety effort. Research has shown that the involvement of supervisors and foremen in the health and safety effort of the organization because of their relationship with their workers is critical and produces positive outcomes.

• REQUIREMENTS FROM THE SUPERVISORY STAFF (SECTION 8)

In terms of Section 8 (e) and (i), supervisors and foremen are required to provide such information, instructions, training and supervision as may be necessary to ensure, as far as is reasonably practicable, the health and safety at work of its workers; and to ensure that work is performed and that plant and machinery is used under the general supervision of a person trained to understand the hazards associated with it and who has the authority to ensure that precautionary measures taken by the employer are implemented.

REGULATIONS OF THE ACT

Under the Act:

(a) The Minister may make regulations (Section 43) as to any matter which in terms of this Act shall or may be prescribed;

(b) which in the opinion of the Minister are necessary or expedient in the interest of the health and safety of persons at work or the health and safety of persons in connection with the use of plant or machinery, or the protection of persons other than persons at work against risks to health and safety arising from or connected with the activities of persons at work, including regulations as to-

  i. the planning, layout, construction, use, alteration, repair, maintenance or demolition of buildings;
  ii. the design, manufacture, construction, installation, operation, use, handling, alteration, repair, maintenance or conveyance of plant, machinery or health and safety equipment;
  iii. the training, safety equipment or facilities to be provided by employers or users, the persons to whom and the circumstances in which they are to be provided and the application thereof;
  iv. the health or safety measures to be taken by employers or users;
  v. the occupational hygiene measures to be taken by employers or users;
  vi. any matter regarding the biological monitoring or medical surveillance of workers;
  vii. the production, processing, use, handling, storage or transport of, and the exposure of workers and other persons to, hazardous articles, substances or organisms or potentially hazardous articles, substances or organisms, including specific limits, thresholds or indices of or for such exposure;
  viii. the performance of work in hazardous or potentially hazardous conditions or circumstances,
  ix. the emergency equipment and medicine to be held available by employers and users, the places where such equipment and medicine are to be held, the requirements with which such equipment and medicine shall comply, the inspection of such equipment and medicine, the application of first-aid and the qualifications which persons applying first-aid shall possess;
  x. the compilation by employers of health and safety directives in respect of a workplace, the matters to be dealt with in such directives and the manner in which such directives shall be brought to the attention of workers and other persons at such a workplace;
  xi. the registration of persons performing hazardous work or using or handling plant or machinery, the qualifications which such persons shall possess and the fees payable to the State in respect of such registration;
  xii. the accreditation, functions, duties and activities of approved inspection authorities;
  xiii. the consultations between an employer and workers on matters of health and safety;
  xiv. subject to section 36, the provision of information by an employer or user to workers or the public on any matter to which this Act relates;
  xv. the conditions under which any employer is prohibited from permitting any person to partake of food or to smoke on or in any premises where a specified activity is carried out;
  xvi. the conditions under which the manufacture of explosives and activities incidental thereto may take place;

(c) as to the preventive and protective measures for major hazard installations with a view to the protection of workers and the public against the risk of major incidents;

(d) as to the registration of premises where workers perform any work or where plant or machinery is used and the fee payable to the State in respect of such registration;

(e) whereby provision is made for the continuation of any registration under this Act;

(f) as to the registration of plant and machinery and the fee payable to the State in respect of such registration;
(g) as to the establishment of one or more committees for the administration of a provision of the regulations, the constitution of such committees, the functions of such committees, the procedure to be followed at meetings of such committees, the allowances which may be paid to members of such committees from money appropriated by Parliament for such purpose and the person by whom such allowances shall be fixed;
(h) prescribing the records to be kept and the returns to be rendered by employers and users and the person or persons to whom such returns shall be rendered;
(i) as to the designation and functions of health and safety representatives and health and safety committees and the training of health and safety representatives;
(j) as to the activities of self-employed persons; and
(k) as to any other matter the regulation of which is in the opinion of the Minister necessary or desirable for the effective carrying out of the provisions of this Act.

(4) No regulation shall be made by the Minister except after consultation with the Council, and no regulation relating to State income or expenditure or to any health matter shall be made by the Minister except after consultation also with the Minister of State Expenditure and the Minister for National Health and Welfare, respectively.
(5) In making regulations the Minister may apply any method of differentiation that he may deem advisable: Provided that no differentiation on the basis of race or colour shall be made.
(6) A regulation may in respect of any contravention thereof or failure to comply therewith prescribe a penalty of a fine, or imprisonment for a period not exceeding 12 months, and, in the case of a continuous offence, not exceeding an additional fine of R200 or additional imprisonment of one day for each day on which the offence continues: Provided that the period of such additional imprisonment shall not exceed 90 days.
(7) A regulation made under section 35 of the Machinery and Occupational Safety Act, 1983 (Act No. 6 of 1983), which was in force immediately prior to the commencement of this Act and which could have been made under this section, shall be deemed to have been made under this section.

The regulations in the Act are identified by their titles (*not* by numbers). All the regulations are available online.

- Asbestos Regulations, 2001
- Certificate of Competency Regulations, 1990
- Construction Regulations, 2014
- Diving Regulations, 2009
- Driven Machinery Regulations, 1988
- Electrical Installation Regulations, 2009
- Electrical Machinery Regulations, 1988 [*most people use portable electrical tools*]
- Environmental Regulations for Workplaces, 1987 [*re lighting, ventilation, noise protection, fire safety & precautions, in the work environment*]
- Explosives Regulations, 2003
- Facilities Regulations, 1990 [*workers require seating, drinking water, and sanitary facilities*]
- General Administration Regulations, 2003 [*management's administrative procedures with regard to health and safety in the workplace*]
- General Machinery Regulations, 1988 [*ensuring health and safety with regard to driven machinery*]
- General Safety Regulations [*refers to PPE, first aid facilities, use of ladders, limited access to high risk areas, welding equipment, stacking and storage, and eviction of intoxicated persons*]
- Hazardous Biological Agents Regulations, 2001
- Hazardous Chemical Substances Regulations, 1995
- Incorporation of Safety Standards into Electrical Installation Regulations, 2009
- Lead Regulations, 2001 [*e.g. in welding or use of lead- based paint is used*]
- Lift, Escalator and Passenger Conveyor Regulations, 1994 [*buildings with lifts*]
- Major Hazard Installation Regulations [*e.g. hazardous chemical stores*]
- National Code of Practice: Evaluation of Training Providers for Lifting Machine Operators [*e.g. maintenance workshops*]
- Noise-induced Hearing Loss Regulations, 2003 [*e.g. noise zones such as workshops and construction sites*]
- Pressure Equipment Regulations, 2009 [*e.g. fire extinguishers, diving and gas cylinders, and compressors*]
- Regulations for the Integration of the Occupational Health and Safety Act, 1995 [*refers to integration of Labour Laws*]

• GENERAL ADMINISTRATIVE REGULATIONS 2003

Deals with various administrative issues regarding health and safety in the workplace. Provides for health and safety committee, negotiations, and consultations before designation of health and safety representatives, reporting of incidents and occupational diseases, recording and investigation of incidents, and some related matters. It requires that

- Employers ensure that a copy of the Act is available in the workplace.
- Any person has the right to access the Act.
- Workers report incidents to supervisors and management

• EXAMPLES OF FACILITY REGULATIONS

Sanitary facilities must have signs for males and females; be clean and hygienic; have water, toilet paper, soap or alcohol-based hand sanitizer and disposable paper towel or blower; and be in a working condition.

Safekeeping facilities must be kept sanitized; available to workers to keep personal belongings in them; and workers must use them.

Change rooms must have signs for males, females, and persons with disabilities; not be connected to areas which can contaminate them and must be kept clean and hygienic and regularly sanitized.

Dining rooms where provided they must be situated outside the work area.

Drinking water must be provided where existing water is unfit for human consumption and proper informative signage is to be provided.

Proper seating that does not cause back pains or poor posture must be provided to workers who can do their work in a sitting position while ensuring social distancing of at least 1.5m between workstations.

• EXAMPLES OF ENVIRONMENTAL REGULATIONS FOR WORKPLACES

Temperature

In cold conditions the correct protective clothing is required for construction workers working outside in winter. When working with vibrating tools in winter workers must wear padded gloves and hearing protection. In temperatures below 0° Celsius workers must be medically certified fit and wear the right protective clothing.

On the other hand, in hot conditions particularly where hard manual labour is involved workers must be medically certified fit, must be provided with enough water and first aid must be available for treatment of workers suffering from heat stroke.

Lighting

A lighting survey must be done, dirty, broken or inoperable lighting must be reported, emergency evacuation lighting must be in working condition at all times, and an evacuation procedure must be developed and periodic drills must be done to ensure that workers are able to execute it.

Housekeeping

A storage facility should be provided for whatever is not in use. Ventilation equipment must be kept in a working condition and clogged-up and dirty covers and filters must be removed and replaced.

Fire precautions

Firefighting equipment must be strategically placed, sign-posted, unobstructed, regularly serviced (check service dates) and in a good condition. All workers must know how to use them.

• NOISE INDUCED HEARING LOSS REGULATIONS

The condition and positioning of proper hearing protection signs are important. All workers and visitors entering and remaining in noisy areas must wear the required hearing protection regardless of how long they remain in a noise zone and the hearing protection equipment must be stored in the facility provided when not in use.

• GENERAL SAFETY REGULATIONS

Personal protective equipment (PPE)

Signs must be posted where PPE must be worn. All workers must know how to use or wear it and properly maintain it. PPE which is a measure of last resort must be in a good and clean condition.

First Aid

Workers must know who the first aiders are. Their names must be on the first aid box and they must be readily available to attend to anyone needing their services. The contents of the first aid box should comply with Regulation 3(9). The content must be checked regularly. It is advisable for the first aid box to be sealed to prevent abuse and pilfering of the contents.

Flammable liquid store

Any flammable liquid store must have a sign clearly demarcating it as a "FLAMMABLE LIQUID STORE". It needs to be properly ventilated to the outside atmosphere with a fan that must be switched on. The store may not contain more than 110% of the liquid to be stored in it. The correct type of firefighting equipment must be provided and strategically placed.

Stacking

All stacking must be carried out under supervision of an experienced person. Any pallets used for stacking must not be broken and whatever is stacked on them must be stable. Access to and from any stacks must be safe.

Welding and flame cutting

All welding and flame cutting activities must be screened off. Electric leads must be insulated. The correct protective equipment must always be worn such as protection of eyes, hands, legs, feet and respirators when welding is done particularly in confined spaces. No welding must be done on closed containers which are explosive and could ignite. When welding inside metal vessels always ensure there is a stand-by person in case of an emergency.

Intoxication

No worker is allowed to be intoxicated at work or partake of or offer intoxicating substances to other persons.

Ladders

Ladders must have non-skid devices and be in a good condition.

They must only be used if the following are in place:

- Hooks at upper end to ensure stability, or
- Held by a person depending on the height at which the work is to be performed, or
- Lashed, or secured by any other means to the structure being worked on.
- Ladders are NOT allowed to be lashed together.

Any ladders leaning against an object or structure:

- Are not allowed to be longer than 9 m or they will become unstable.

Wooden ladders must not be painted and inspected regularly

• OTHER RELATED LEGISLATION

Basic Conditions of Employment

These cover issues such as working time or hours, maximum permitted hours of work

- o Taking working time schedules into consideration, when providing and maintaining a safe and healthy working environment
- o Arrangements should be considered for the special needs of workers such as, for example:
  - Perform critical tasks safely without risk to health
  - Health assessment and counselling
  - Services should be available at night if work activities occur at night

        - Daily and weekly rest periods
        - Annual leave
    - Protection of children
    - Protection against discrimination
    - Protection of women
        - Protection before and after birth

Labour Relations Act

This Act deals with, for example:

- Protection against victimisation
- Protected disclosures
- Dismissal for non-compliance with health and safety rules
- Dismissal for endangering the worker and others
- Work-related injuries
- Training
- HIV and AIDS and incapacity

Employment Equity Act

This Act covers for example the following:

- Elimination of unfair discrimination in recruitment processes and in the workplace
- Implementation of affirmative action to redress inequalities of the past
- People with disabilities
- Prohibition of medical testing as condition for employment
- Prohibition of HIV testing without permission of the worker

Skills Development Act

This act deals with, inter alia, health and safety training and claiming levies

- REVIEW QUESTIONS

1. What are the responsibilities of employers in terms of the Act?
2. Why are health and safety committees important in an organization?
3. Describe how safe work or operating procedures and codes of practice contribute to health and safety at work.
4. What is the role of the CEO with respect to health and safety?
5. When is an inspector likely to issue a prohibition notice?

- REFERENCES

Anonymous. (Nd). Loose notes on safety, health and environment management
Bester, Andre., and Haupt, Theodore. Conrad. (2007): "The Role of the CEO in Construction Information Systems (IS) Performance: A Pilot Study" Acta Structilia, vol. 14, no. 1, pp. 56-80, ISSN 10 23 0564
Deacon, C.H. and Smallwood, J., 2017. *The effect of the integration of design, procurement, and construction relative to health and safety* (Doctoral dissertation, Nelson Mandela Metropolitan University).
Department of Labour (1997). Basic Conditions of Employment Act No 75 of 1997. Department of Labour.
Department of Labour (1998). Employment equity Act No 55 of 1998.
Department of Labour (1995). Labour Relations Act No 66 of 1995. Department of Labour.
Department of Labour (1993). Compensation for Occupational Injuries and Diseases Act No 103 of 1993.
Department of Labour (1993). Occupational Health and Safety Act No 85 of 1993.
DOL (n.d.). [online] Sizananitraining.co.za. Available at: http://www.sizananitraining.co.za/What%20every%20worker%20should%20know%20about%20Health%20and%20Safety%20in%20the%20Workplace.pdf.
DOL (n.d.). What Every Worker and Employer Should Know About. [online] Labourguide.co.za. Available at: https://www.labourguide.co.za/health-and-safety/739-what-every-worker-and-employer-should-.
Hermanus, M.A., 2007. Occupational health and safety in mining-status, new developments, and concerns. *Journal of the Southern African Institute of Mining and Metallurgy*, *107*(8), pp.531-538.
Lewis, John. and Thornbory, Greta. (2010). Employment law and occupational health. Chichester, West Sussex: Wiley-Blackwell.

Mpambane, S. and Haupt, Theodore. Conrad. (2007): "An Investigation into the Effectiveness of the Inspectorate in South African Home Building Industry" In Haupt, T. (Ed.) The Built Environment, Proceedings of the 2nd Built Environment Conference, Port Elizabeth, June 17-19, pp. 187 - 196, ISBN 978-0-620-38777-4

Pillay, Kersey. & Haupt T.C. (2008): "The Cost of Construction Accidents: A Pragmatic Study," In Haupt, T.C. (Ed) The Built Environment 3, Proceedings Of The 3rd Built Environment Conference, Cape Town, July 6-8, pp. 268-283, ISBN 978-0-620-41382-4

Scott, Gorden. (2011). *An assessment of health and safety management in selected rural hospitals*. [online] Dspace.nwu.ac.za. Available at: https://dspace.nwu.ac.za/bitstream/handle/10394/8437/Scott_GLS.pdf?.

Windapo, Abimbola. and Oladapo, Adebayo. A., 2012. Determinants of construction firms' compliance with health and safety regulations in South Africa. In *Procs of 28th Annual ARCOM Conference, 3-5 September 2012, Edinburgh, UK* (Vol. 2, pp. 433-444). Association of Researchers in Construction Management (ARCOM).

CHAPTER 9

RECORDKEEPING, REPORTING AND DOCUMENT CONTROL

RECORD KEEPING

The purpose of record keeping is to ensure the proper and systematic recording of information and observations collected during, for example, a SHE compliance audit or an inspection. Good record keeping and filing procedures will ensure that all supporting documentation and observations are kept for future reference. In the case of an audit, all audit information should be stored in a file and a new file opened for each compliance audit. Each file should contain several subsections to store the audit information in an orderly manner.

• KEEPING MINUTES AND RECORDS

Minutes are an official record of, for example, actions and decisions that a committee took at a meeting. They are not a record of everything that was said. They serve a historical purpose. Equally important, they serve a legal purpose, documenting the adherence of the committee to the proper procedures and the charter of the committee. In the case of any dispute, minutes may be admitted into evidence in a court of law provided that all the formalities have been complied with and that the signatures of all attendees are included. Therefore, minutes should at least be signed by the chairperson and the secretary and correctly dated. It is a general presumption, that minutes once signed are a conclusive evidence of the meetings held. It is useful that when minutes are approved, the secretary annotates the original file copy with any corrections in the margin or retypes the minutes entirely to include the corrections. The secretary then writes "Approved" on the minutes and adds both his or her initials and the date to the record.

Records and minutes are kept as written proof that any hazard or any hazardous situation, that might have developed in the workplace, has been identified and recognised and that remedial measures will be taken. The employer or a designated person must note and acknowledge the records and minutes by countersigning each of these documents when they are received.

The Act explicitly holds the CEO responsible in his PERSONAL CAPACITY and these records can be considered as legal evidence in a court case, especially where a worker or someone else has been killed or might have been killed or seriously injured because of an incident or accident in the workplace. These documents may in fact prove innocence or guilt. These records must be signed in acknowledgement of the reported problem and acceptance of the risk should a time lapse occur before rectification or remediation measures could be implemented.

• COMPANY POLICY

A company policy is a guideline to help employers to deal with worker accountability, health, safety, and their interactions with the customers of the organization. It helps foster worker wellness, fair treatment, and ensures that an organisation is following all applicable and relevant laws and regulations. A company policy will accommodate the implementation of legal requisites such as, for example, reporting and recording of incidents in the workplace. Additionally, the management of the organization may deem it necessary to implement additional procedures.

Information available in incident reports may assist management to, for example:

- ✓ formulate future policy to incorporate and implement agreed upon proposals to improve the management of safety, health and environmental protection matters within the organization and as a result of the activities and services of the organization,
- ✓ identify trends developing such as, for example, the repeated breakdown of a particular machine or piece of equipment or the recurrence of incidents in the workplace,
- ✓ motivate proposals made to management by presenting facts with supporting documents as evidence,
- ✓ assist researchers by granting them access to information from incident and accident reports,
- ✓ communicate effectively with elected health and representatives, health, and safety committees and union representatives.

• PROPERTY DAMAGE INCIDENTS REGISTERS AND REPORTS

The property damage incidents register, and reports are very important records. For example, registers and reports that must be available and up to date always include:

- ✓ Incident register
- ✓ Registers for service lifts, hooks, chains, slings, and ropes (steel and sisal) as well as all lifting tackle.
- ✓ Registers for the monthly health and safety inspections that include health and safety representatives.
- ✓ Registers for all types of machine guards in the workplace.
- ✓ Registers for all approved electrical installations, earth leakage equipment and distribution boards.
- ✓ Registers for inventory, maintenance, service and disposal of pressure vessels.
- ✓ Fire protection equipment registers for inventory, maintenance and service of hydrants, hoses and extinguishers.
- ✓ Registers for inventory, maintenance, and service of portable, extension and fixed ladders.
- ✓ Registers for inventory, maintenance, and service of portable electrical and power tools.

• REPORTING OF INCIDENTS

In terms of the Act and Section 24 incidents that should be reported and investigated include the following types of incidents:

- When a person dies
- When a person becomes unconscious
- When a person suffers the loss of a limb or part of a limb
- When a person Is injured or becomes ill, or is likely to die or suffer permanent physical defect
- When a person is unable to work for 14 days or longer because of a work-related incident
- When a "major incident" occurs

These must be reported by telephone, fax, or similar means of communication to an inspector within the prescribed period. The following occurrences must also be reported to the Provincial Director. When lives were endangered by:

- Dangerous spilled substances
- Uncontrolled release of a substance under pressure
- Flying, falling, uncontrolled moving object
- Machinery that ran out of control

Further, Regulation 8(1) of the General Administrative Regulations states that an employer must, within seven days after an incident referred to in section 24(1)(a) of the Act, give notice thereof to the provincial director using the prescribed forms of WCL1 or WCL 2. The employer or user should keep record of all section 24 incidents and any other incident where medical treatment or first aid is involved. This must be done in the form of the prescribed "Annexure 1" form. Records need to be kept for a period of at least three years.

The incident or accident investigation should officially start within a period of 7 days and finalised as soon as is reasonably practicable, or within the contracted period in the case of contracted workers. An employer must ensure that the record of the incident is examined by the health and safety committee.

• REPORTING OF INJURIES

Injuries to workers are classified into

- ✓ Fatal
- ✓ Disabling
- ✓ Non-disabling
- ✓ First aid only incidents.

These include cases where the health of a worker is affected by conditions in the workplace such as, for example, poor lighting, ventilation, control of harmful dusts and vapours, vibrations, and radiation. All occupational injuries or alleged occupational injuries that entail medical expenses and/or absence from work for more than *three* days must be reported within *seven* days in the prescribed manner. The delay to report an accident or alleged accident is a criminal offence. The Commissioner may also impose a penalty on the employer which could be the full amount of the claim.

- PROCEDURE WHEN REPORTING AN INJURY TO COMMISSIONER

STEP 1: Complete "PART A" of form W.CL.2 “Employer's report of an Occupational Injury”, sign it and provide date where indicated.

STEP 2: Detach "PART B" where perforated and forward it without delay to the doctor or hospital concerned. In minor cases, "PART B" must accompany the worker.

STEP 3: Complete "PART A", page 2 in full.

STEP 4: Forward the completed form W.CL.2 "PART A", pages 1 and 2 without delay to:

Compensation Commissioner
P.O. Box 955
Pretoria
0001

The employer is liable for the payment of compensation to the affected worker for the first three months from the date of the occupational injury. The compensation paid by the employer shall be reimbursed by the Commissioner.

When the First Medical Report W.Cl.4 was not posted together with the report of the employer of an occupational injury, it must be obtained without delay and submitted to the Commissioner. In cases of prolonged absence, a Progress Medical Report W.Cl.5 must be obtained monthly from the medical practitioner and submitted. When the worker resumes work, a Resumption Report W.Cl.6 must be completed and submitted to the Commissioner together with the Final Medical Report W.Cl.5.

The Employer's Report of an Occupational Injury must not be kept in abeyance awaiting the medical report. The medical reports must not be kept in abeyance until the condition of the worker has stabilized, improved or when he has returned to work to resume his or her duties.

- DOCUMENTS TO BE SUBMITTED IN CASE OF A FATALITY

If the death of a worker was the consequence of an accident in the workplace documentary proof indicating the cause of death must be submitted by the employer without any delay:

If the worker leaves a widow/widower and/or children under the age of 18 years as dependents the following documents must be submitted, namely:

- A marriage certificate.
- Birth certificates/baptismal certificates/sworn statements regarding the age of the widow/widower.
- Birth certificates/baptismal certificates/sworn statements regarding the ages of all the children.
- A Claim for Compensation W.Cl.3 (page 1) and (page 2).
- A statement by the dependent widow/widower of the deceased worker (W.Cl.32).
- A specified burial account and if already paid for the receipt or acceptable proof of payment.

- REPORTING OF INJURIES TO THE RELEVANT PERSONS

✓ If a worker is injured during work hours, either he, if he can do so, or a fellow worker should immediately notify his supervisor or site manager.
✓ If the injury is of a serious nature and the worker must be taken to the hospital/doctor, the supervisor/site manager and the next of kin of the affected worker must be notified as soon as possible.
✓ If the injury is of a serious nature the Department of Labour must be notified.
✓ If the injury was caused as a result of actions that could possibly be criminal in nature, the South African Police Services must also be notified.

- INVESTIGATING AND RECORDING OF ACCIDENTS OR INCIDENTS

It is important to arrive at some definite conclusion as to the cause of an accident and how to prevent a future recurrence of the event. The supervisor or employer is responsible for immediately initiating an investigation into any accident or incident that may have occurred and to ensure that remedial action is taken. Note that the intent of any investigation should never be to find the worker to apportion blame or seek discipline or retribution. The intent should always be to introduce appropriate measures to prevent a recurrence of the incident. Too often the intent is to ‘find the culprit’ and discipline him or her. Should accidents have occurred where specific injuries as expressly required in the Act have been sustained, it is legally necessary to report such to the Department of Labour and to conduct a full investigation, with the cooperation of the Department. A full report of the incident must be compiled and kept for the required minimum period of 3 years.

Regulation 9(1) of the General Administrative Regulations states that an employer must:

- ✓ keep a record in the form of Annexure 1 (below) for a period of at least three years,
- ✓ allow access to the record for inspection by an inspector, of all incidents reportable in terms of section 24 of the Act and of any other incident which resulted in the person concerned having had to receive medical treatment other than first aid,
- ✓ investigate incidents or cause them to be investigated by a person appointed by him or her, by a health and safety representative or a member of a health and safety committee within 7 days from the date of the incident and finalised as soon as is reasonably practicable, or within the contracted period in the case of contracted workers,
- ✓ cause the findings of the investigation to be entered in Annexure 1 immediately after completion of such investigation,
- ✓ cause every record to be examined by the health and safety committee for that workplace or section of the workplace at its next meeting and shall ensure that necessary actions, as may be reasonable practicable, are implemented and followed up to prevent the recurrence of such incident.

• INCIDENT REPORTS

The following information must be included in the Incident Report:

- ✓ Full details of the injured:
    - o Name
    - o Surname
    - o Date of birth
    - o ID Number
    - o Worker number
    - o Contact details address and telephone and mobile numbers
- ✓ Full details of any witnesses.
- ✓ Employer/Organization business and contact details.
- ✓ A detailed description of what happened as seen by as many reliable witnesses as possible.
- ✓ A detailed description of the actions taken by everyone involved.
- ✓ A detailed report written by the health and safety inspector of the organization/employer detailing what actions have been taken to prevent such an event from recurring.
- ✓ Dated signatures of everybody involved on all written incident report pages as these are legal documents which might be used in legal proceedings.

The following is an example of an incident or injury report:

• EXAMPLE OF INCIDENT / INJURY REPORT

Status:

Worker		Contractor	
Visitor		student	

Details of Injured Person:

Surname		First Name	
Phone (H)		Phone (W)	
Cellular No		Sex	
Date of Birth		1st Language	
Address:			

Experience in Job: Work Status

0-3 months		4-12 months	
3-5 years		5 years +	
1-2 Years		Temporary	
Permanent		Part time	

Details of Witnesses

Surname		First Name	
Phone (W)		Phone (H)	
Address:			
Surname		First Name	
Phone (W)		Phone (H)	
Address:			

Details of Incident or Accident

Date		Time of Injury	am/pm
Activity Engaged In			
Location of incident / accident:			
Describe how and what happened			

Details of Injury (the assistance of a supervisor may be required to complete this section)

Nature of injury / illness (e.g. burn, sprain, cut, etc	
Mechanism (e.g., fall, grabbed by person, muscular stress)	
Location on body (e.g., back, right thumb, left arm, etc)	
Agency (e.g., furniture, another person, hot water, etc)	

Treatment Administered

Was First Aid administered?	
Treatment:	
Referred to:	
First Aid Officer (Name):	
Signature of First Aid Officer	

The following sections must be completed by the supervisor / manager

Did the injured person stop work?	*If yes, state the time and date*
Outcome: Treated by Doctor: First Aid Officer:	
Workmen's Compensation Claim Completed:	
Returned to normal duties	
Hospitalised:	
Returned to alternative duties:	
Incident or accident investigation:	
Name of Supervisor	
Date	
Signature	

Remedial Actions

☐	Conduct task analysis	☐	Reinstruct persons involved
☐	Improve design / construction / guarding	☐	Conduct hazard / systems audit
☐	Improve staff skills mix	☐	Add to inspection programme
☐	Develop / review tasks / procedures	☐	Provide debriefing and/or counselling
☐	Improve procedures / communication / reporting procedures	☐	Improve work environment
☐	Request maintenance	☐	Improve security
☐	Review OHS policy / programmes	☐	Improve personal protection
☐	Temporarily relocate workers involved	☐	Replace equipment / tools
☐	Improve work congestion / housekeeping	☐	Other (Specify)
☐	Improve work organisation	☐	Investigate safer alternatives
☐	Develop and/or provide training	☐	Request material safety / data sheets

What, in your own words, has been implemented or planned to prevent recurrence:

Remedial Actions Completed

Signed (Supervisor)	
Title	
Date	
Review comments (Health and Safety Committee Meeting:	
Reviewed by Manager (Name)	
Signed	
Date	

(Source: Adapted from Anonymous (nd.)

• PROCEDURE WHEN REPORTING AN OCCUPATIONAL DISEASE

An employer must within 14 days after he gained knowledge of an alleged occupational disease complete an "Employer's Report of an Occupational Disease", W.Cl.1 and forward it to the

Compensation Commissioner
P.O. Box 955
Pretoria
0001

The commencement of the disease shall be the date on which the doctor first diagnosed the illness.

In the case of an alleged occupational disease, the employer must irrespective of the W.Cl.1 also submit the following documents:

- A First Medical Report for an Occupational Disease W.Cl.22.
- A Claim for Compensation for an Occupational Disease W.Cl.14.
- A Progress medical report W.Cl.26 must be submitted monthly until the condition of the worker condition has become stabilized, where after a Final medical report (W.Cl.26) must be submitted.

In cases of deafness as a result of excessive noise at the workplace of the worker, audiograms must also be submitted which must include baseline audiograms taken when the worker first started working for the organization.

• DOCUMENT CONTROL

The organization should maintain up-to-date documentation that is sufficient to ensure that its SHE management system can be adequately operated.

Typical inputs include the following items:

- details of the documentation and information systems the organization develops to support its SHE management system and SHE activities, and to fulfil the requirements of the preferred standard SHE management system,
- details of responsibilities and authorities, and
- information on the local environments in which documentation or information is used, and constraints that this can put on the physical nature of documentation, or the use of electronic or other media.

The organization should review its documentation and information needs for the OHS management system, before developing the documentation necessary to support its OHS processes. In deciding what documentation is required the organisation should determine where there is any risk that a task, through lack of written procedures or instructions, will not be performed in the required manner. All documents and data containing information required for the operation of the OHS management system and the performance of the OHS activities organization should be identified and controlled.

The organisation should consider items such as the following:

- The details of the document and data systems that support its OHS management systems and OHS activities, and which enable it to fulfil the requirements of, for example, OHSAS 18001,
- The OHS details of its assigned responsibilities and authorities. Written procedures should define the controls for the identification, approval, issue and removal of OHS documentation, together with the control of OHS data.

These procedures should clearly define the categories of documents and data to which they apply. Documents and data should be available and accessible when required, under routine and non-routine conditions, including emergencies. This could include ensuring that up-to-date plant engineering drawings, hazardous material data sheets, procedures, and instructions, are available to those persons who require them in an emergency.

The organisation should establish procedures for identifying any documents of external origin required for planning and implementing its OHS management system. The distribution of these documents needs to be controlled to ensure that the most current information is used in making decisions impacting OHS. For example, the organisation should establish procedures for managing the safety data sheets developed for hazardous substances used by the organization.

Responsibility for this task should be assigned. The person charged with this task should ensure that all persons in the organization are kept informal of any relevant changes to such information that affects their duties or working conditions. The development of the document control process of an organization will typically result in items such as the following:

- A document control procedure, including assigned responsibilities and authorities,
- Document registers, master lists or indexes,
- A list of controlled documentation and its location,
- Archive records (some of which should be held in accordance with legal or other time requirements).

Documents should be reviewed from time to time to ensure that they are still valid and accurate. This can be performed as a dedicated exercise, and could also be necessary:

- As part of a review of risk assessment of processes,
- As part of a response to an incident
- As part of the management of change procedure, and
- Following changes in legal and other requirements, processes, installation, workplace layout.

Obsolete documents retained for reference can present a particular concern, and great care should be taken to ensure that they do not return into circulation. However, it is sometimes necessary to retain obsolete documents as part of the records relating to the development or performance of the OHS management system.

- CONTROL OF RECORDS

The organisation shall establish and maintain records as necessary to demonstrate conformity to the requirements of its OHS management system, and the results achieved. The organisation should establish, implement, and maintain procedures for the:

- Identification
- Storage
- Protection
- Retrieval
- Retention
- Disposal of records

Records shall be and remain:

- Legible
- Identifiable
- Traceable

Records should be maintained to demonstrate that the organisation is operating its OHS management system effectively and is managing its OHS risks. Records that can demonstrate conformance to the requirements include among others:

✓ Records of the evaluation of compliance with legal and other requirements
✓ Hazards identification, risk assessment and risk control records
✓ Records of the monitoring of OHS performance
✓ Calibration and maintenance records for equipment used to monitor OHS performance
✓ Records of corrective action and preventive action
✓ Reports of OHS inspections
✓ Training and associated records that support evaluation of compliance
✓ OHS management system audit reports
✓ Participation and consultation reports
✓ Incident reports
✓ Incident follow –up reports
✓ OHS meeting minutes
✓ Health surveillance reports
✓ Personal protective equipment (PPE) maintenance records
✓ Reports of emergency response drills
✓ Management review records

The integrity of records and data should be maintained to facilitate their subsequent use, for example, for monitoring and review activities, for the identification of trends for preventive action.

The Basic Conditions of Employment Act, No. 75 of 1997, states that various documents relating to workers should be kept for future reference.

Document	Retention period
Written particulars of worker must be kept after termination of Employment	3 years
Worker's name and occupation	3 years
Time worked by each worker	3 years
Remuneration paid to each worker	3 years
Date of birth of any worker under 18 years of age	3 years

There is also a reference that an employer who keeps records in terms of this section is not required to keep any other record of time worked and remuneration paid as required by any other employment law.

The Compensation for Occupational Injuries and Diseases Act, No. 130 of 1993, provides for compensation for disablement caused by occupational injuries or diseases sustained or contracted by workers in the course of their employment or for death by these injuries at their place of work.

The Act states that certain records relating to the earnings should be retained.

Document	Retention period
Register, record or reproduction of the earnings, time worked, payment for piece work and overtime and other prescribed particulars of all the workers	4 years

The Occupational Health and Safety Act, No. 85 of 1993 was enacted to provide for the health and safety of workers at work and for persons using plant and machinery as well as other hazardous employment conditions. Certain documents have to be kept based on the Administrative Regulations.

Document	Retention period
A health and safety committee shall keep record of each recommendation made to an employer in terms of issues affecting the health of workers and of any report made to an inspector in terms of the recommendation	3 years
Records of incidents reported at work (Annexure 1 of the General Administration Regulations, 2003)	3 years
Records of assessments and air monitoring, and the asbestos inventory (Asbestos Regulations, 2001)	Minimum 40 years
Medical surveillance records (Asbestos Regulations, 2001)	Minimum 40 years
Records of risk assessments and air monitoring results (Hazardous Biological Agents Regulations)	40 years
Medical surveillance records (Hazardous Biological Agents Regulations)	40 years
Records of assessments and air monitoring (Hazardous Chemical Substances Regulations, 1995)	30 years
Medical surveillance records (Hazardous Chemical Substance Regulations, 1995)	30 years
Records of assessments and air monitoring (Lead Regulations, 2001)	40 years
Medical surveillance records (Lead Regulations, 2001)	40 years
All records of assessments and noise monitoring (Noise-induced Hearing Loss Regulations, 2003)	40 years
All medical surveillance records, including the baseline audiogram of every worker (Noise-induced Hearing Loss Regulations, 2003)	40 years

Other documents and records that should be kept by the organization include:

Document	Retention period
Appointments	
16(1)	40 Years
16(2)	40 Years
GMR 2	40 Years
GMR 2.7	40 Years
SHE Rep	5 years after termination
First Aider	5 years after termination
Committee Chairman	5 years after termination
Committee Member	5 years after termination
S 8 (2)(i)	5 years after termination
Lifting Tackle and Equipment Inspector	5 years after termination
Lifting Equipment Operator	5 years after termination
Lead Controller	5 years after termination
HCS Coordinator	5 years after termination
Document	**Retention period**
Appointments	
Permit To Work Coordinator	5 years after termination
Ladders Inspector	5 years after termination
Portable Electrical tools inspector	5 years after termination
Grinder Inspector	5 years after termination
Registers	
Emergency Lighting	5 Years demonstrate history/until decommission
Air Conditioner	5 Years demonstrate history/until decommission
Extraction Systems	5 Years demonstrate history/until decommission
Hygiene facility insp. – Ablution	5 Years demonstrate history/until decommission
Hygiene facility insp. – Canteen	5 Years demonstrate history/until decommission
Stacking Inspection	5 Years demonstrate history/until decommission
Ladders (Portable)	5 Years demonstrate history/until decommission
Ladders (Fixed)	5 Years demonstrate history/until decommission
Scaffolding	5 Years demonstrate history/until decommission
Goods Hoist	5 Years demonstrate history/until decommission
Lifting Tackle – Chains Blocks	5 Years demonstrate history/until decommission
Lifting Tackle – Steel Wire Ropes	5 Years demonstrate history/until decommission
Lifting Tackle – Fire Ropes	5 Years demonstrate history/until decommission
Hydraulic Jack Inspection	5 Years demonstrate history/until decommission
Air Compressors	5 Years demonstrate history/until decommission
Refrigeration Installations	5 Years demonstrate history/until decommission
Gas Welding and Cutting	5 Years demonstrate history/until decommission
Motor vehicles/Vehicles/Forklifts	5 Years demonstrate history/until decommission
Portable Electric Equipment	5 Years demonstrate history/until decommission
Earth leakage	5 Years demonstrate history/until decommission
General Electric Installation	5 Years demonstrate history/until decommission
Electrical Equipment – Hazardous location	5 Years demonstrate history/until decommission
Hand Tools	5 Years demonstrate history/until decommission
Explosive Powered Tools	5 Years demonstrate history/until decommission
Self-contained Breathing Apparatus	5 Years demonstrate history/until decommission
Gas Masks	5 Years demonstrate history/until decommission
Safety Harness	5 Years demonstrate history/until decommission
Fire Equipment Inspection	5 Years demonstrate history/until decommission
First Aid Box and Equipment	5 Years demonstrate history/until decommission

External reports and documents	
Boiler permit	Indefinite
Electrical Certificate of Compliance (COC)	Indefinite/ until changes are made and new COC issued
Earth Leakage Test by electrician	1 year until new test is done
Automatic Sprinkler Inspection Bureau certificates	Indefinite
Pressure Vessel Tests by Authorised Inspection Authority	12 years to illustrate service history
Load test certificates	12 years to illustrate service history
Noise Survey done by Authorised Inspection Authority	40 years – as long as medical records
Hazardous Chemical Substance survey done by Authorised Inspection Authority	40 years – as long as medical records
Air quality survey done by Authorised Inspection Authority	40 years – as long as medical records
Lighting survey done by Authorised Inspection Authority	40 years – as long as medical records
Document	**Retention period**
External reports and documents	
Material Safety Data Sheets	As long as the substance is utilised on site
Audit Reports done by external parties	5 years

The Electronic Communication and Transaction Act, No. 25 of 2005, regulates electronic communication and to prohibit the abuse of information. There are certain principles stated for the electronic collection of personal information and also the timeframe that this information must be kept.

Document	Retention period
Personal information and the purpose for which the data was collected must be kept by the person who electronically requests, collects, collates, processes or stores the information	As long as information is used, and at least 1 year thereafter
A record of any third party to whom the information was disclosed must be kept for as long as the information is used	As long as information is used, and at least 1 year thereafter
All personal data which has become obsolete	Destroy

✓ In determining the appropriate controls for records the organisation should take into account any applicable legal requirements, confidentially issues (particularly those relating to personnel), storage/access/ disposal/back-up requirements, and the use of electronic records.

For electronic records the use of antivirus systems and off-site backup storage should be considered.

• REVIEW QUESTIONS

1. Why is a document management and control system an important part of a health and safety management system?
2. What is the legal status of minutes of meetings?
3. What are some of the key documents that should be safely and securely stored?
4. Why would records about exposure to asbestos need to be kept for at least 40 years?
5. How would you protect documents that are stored in an electronic format?

• REFERENCES

Anonymous (nd). Loose notes on safety, health and environment management

Bsigroup.com. (2013). *BS OHSAS 18001 - Occupational Health & Safety | BSI.* [online] Available at: https://www.bsigroup.com/en-GB/ohsas-18001-occupational-health-and-safety/.

Department of Environment and Conservation NSW (2006). [online] Environment.nsw.gov.au. Available at: https://www.environment.nsw.gov.au/resources/licensing/cahandbook0613.pdf.

Gilbertson, Jan. 2005. Using near miss data as a safety improvement tool. *Health and Safety at Work.* www.healthandsafetyatwork.com/content/using-near-miss-data-safety-improvement-tool. Accessed August 15, 2020

Roughton, J. and Mercurio, J. (2002). Journey to a Safety Culture. *Developing an Effective Safety Culture*, pp.64-95.

South African Institute of Chartered Accounts. 2011. Guide on the Retention of records. ISBN 0-86983-386-3

Southern Africa Labour and Development and South African Medical Scholarships (1978). The Economics of Health Care in Southern Africa: A Review Article. *Social Dynamics*, 4(2), pp.109-160.

Southern Africa Labour and Development Research Unit (Saldru) & South African Medical Scholarships Trust (SAMST) (1978) The Economics of Health Care in Southern Africa: A Review Article, Social Dynamics, 4:2, 109-160, DOI: 10.1080/02533957808458224

PART 2

MANAGEMENT OF OCCUPATIONAL SAFETY

This section comprises of seven chapters that introduce important aspects of the management of occupational safety in general industry in South Africa. Chapter 10 deals with the need for inspections and their benefits as part of improving the management of safety in the workplace. Housekeeping, stacking, storage and lockout procedures are the focus of Chapter 11. Safety signage in terms of their design, meaning and use and the critical importance of proper machine guarding are discussed in Chapter 12. The focus of Chapter 13 is on lifting equipment, ladders, scaffolds and conveyors. Chapter 14 deals with working safely with equipment under pressure and includes the safe use of gas and other substances stored under pressure, their storage and safe transportation. In Chapter 15 the design, proper use and maintenance of key forms of personal protective clothing and equipment are discussed bearing in mind that they are measures of last resort after the hierarchy of reducing exposure to hazards has been followed. The challenges of working in confined spaces are the focus of Chapter 16.

CHAPTER 10

WORKPLACE INSPECTIONS AND AUDITS

WORKPLACE INSPECTIONS

The employer is morally, ethically, and legally obliged to provide workers with safe and healthy working conditions. To know how unsafe and health threatening working conditions are or could become, the employer needs to constantly monitor performance against health and safety standards and procedures as prescribed by law and the policies of the organization. This knowledge can be achieved by conducting regular health and safety inspections in the workplace and recording the results. In fact, workplace health and safety inspections can be a vital part of injury and accident prevention efforts if done well. Inspections can help reassure workers that the workplace is safe with no threat to their health and help the organisation demonstrate that it cares about the wellbeing of its workers and stakeholders. In order for this to happen, a more robust process is necessary. Workplace inspections help prevent incidents, injuries and illnesses. Through a critical examination of the workplace, inspections help to identify and record hazards for corrective action. Health and safety committees can help plan, conduct, report and monitor inspections. Regular workplace inspections are an important part of the overall occupational health and safety program and management system.

• CONSIDERATIONS:

As tools and equipment get older, they get worn out or damaged. These can be very dangerous if not repaired or replaced. Materials, equipment, and procedures constantly change and ignoring such changes can jeopardize worker health and safety. Lack of the necessary skills to perform a certain job can cause injury. Regular safety and health inspections will help to expose possible training needs.

Health and safety inspections may vary in purpose from routine maintenance inspections to inspection after a major incident like a fire in an apartment complex or multiple vehicle collisions.

• PURPOSE:

Inspections are important as they allow employers to:

- listen to the concerns of workers and supervisors
- gain further understanding of work activities and tasks
- determine underlying causes of hazards
- It enables one to identify potential hazards before they become a serious problem
- It acts as an early warning system.
- To identify health and safety hazards in the workplace with the aim of removing or controlling them.
- To check if relevant standards and laws are being obeyed.
- To check if agreed improvements are carried out.
- To establish the cause of injuries, death, occupational diseases, property damage and pollution.
- To record all information and report to the appropriate authorities in the required format.
- To provide information at meetings where safety, health and environmental protection matters are discussed.
- To make information available for research purposes.
- To assist in the evaluation of the SHE policy and procedures, including the completion of all registers, records, and schedules required by law.
- From the information contained in reports of inspections the employer will be able to formulate corrective and/or remedial actions. The employer will be able to monitor steps taken to eliminate hazards or control the risk such as engineering controls, administrative controls, policies, procedures, and personal protective equipment.

• BENEFITS:

There are both tangible and intangible benefits to workers in the organization such as, for example:

- Workers will feel that their health and safety are important to the employer which will arguably establish trust in and goodwill towards the employer.
- Regular inspections bring healthier and safer working conditions and practices, with less time lost due to incidents and accidents, resulting in higher productivity, which in turn enhances profits and increases the sustainability and longevity of the organization.
- Savings to the employer by decreasing time lost due to incidents and accidents, not only the cost of the incident in terms of property damage, but the time to complete the investigation and subsequent reporting. Those cases that are litigated in court could cost the employer substantially if found guilty of negligence.

• TYPES OF WORKPLACE HAZARDS:

Types of workplace hazards include:

- Safety hazards such as those caused by inadequate machine guards, unsafe workplace conditions, unsafe work practices.
- Biological hazards caused by organisms such as viruses, bacteria, fungi and parasites.
- Chemical hazards caused by a solid, liquid, vapour, gas, dust, fume or mist.
- Ergonomic hazards caused by physiological and psychological demands on the worker, such as repetitive and forceful movements, awkward postures arising from improper work methods, and improperly designed workstations, tools, and equipment.
- Physical hazards caused by noise, vibration, energy, weather, heat, cold, electricity, radiation and pressure.
- Psychosocial hazards that can affect mental health or well-being such as overwork, stress, bullying, or violence.

• IMPROVING EFFECTIVENESS OF HEALTH AND SAFETY INSPECTIONS:

Health and safety inspections can be improved in several ways, some of which include:

- Every inspection must examine who, what, where, when and how. It is important that particular attention is given to items and issues that are or are most likely to develop into unsafe or unhealthy conditions. This development results from stress, wear, impact, vibration, heat, corrosion, chemical reaction, or misuse. Areas must be included where no work activities are done regularly, such as parking lots, rest areas, office storage areas and locker rooms.
- All workplace elements must be included and inspected, namely the people, the environment, the equipment, and the process. The environment includes such hazards as noise, vibration, lighting, temperature, and ventilation. Equipment includes materials, tools and apparatus for producing a product or a service. The process involves how the worker interacts with the other elements in a series of tasks or operations.
- Frequent health and safety inspections by themselves may have very little impact in improving the SHE culture in the workplace
- Additional factors are necessary to complement routine inspections such as:
  - Implementation of relevant and applicable labour laws
  - Awareness of everyone in the workplace of labour laws and their legal implications
  - Encouragement of and support for workers to report any potential danger at the workplace without fear of victimization by their employers
  - Insurance companies should be dissuaded to not provide insurance cover to workplaces with equipment that is not examined and certified fit for use

• WORKPLACE INSPECTIONS

Documents for workplace inspections would include:

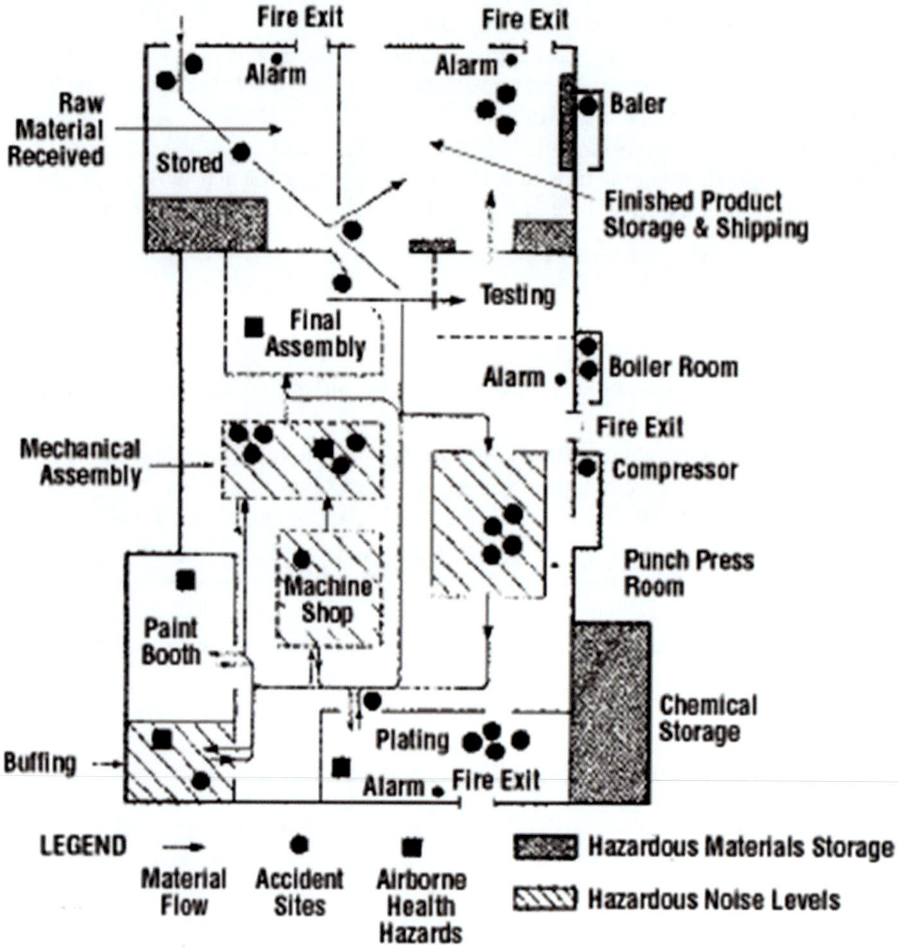

- ✓ A floor plan of the workplace as shown in the graphic in order to
  - mark areas where people work
  - write down the number of people working in each area
  - work processes and machinery in each area
- ✓ A health and safety checklist but be careful to avoid tick fever
- ✓ Materials Safety Data Sheets (MSDS) from management
- ✓ Inspection sheets to use during the actual inspection
- ✓ Process flow
- ✓ Information on chemicals
- ✓ Storage areas
- ✓ Work force size, shifts and supervision
- ✓ Workplace rules and regulations
- ✓ Job procedures and safe work practices
- ✓ Manufacturer's specifications
- ✓ Personal Protective Equipment (PPE)
- ✓ Engineering controls
- ✓ Emergency procedures - fire, first aid and rescue
- ✓ Incident and investigation reports
- ✓ Worker complaint reports regarding particular hazards in the workplace, including psychosocial hazards
- ✓ Recommendations of the health and safety committee
- ✓ Previous inspections
- ✓ Minutes of meetings where the results were discussed
- ✓ Maintenance reports, procedures and schedules
- ✓ Regulator inspection reports or other external audits (insurance, corporate specialist)
- ✓ Monitoring reports (levels of chemicals, physical or biological hazards)

✓ Reports of unusual operating conditions
✓ Names of inspection team members and any technical experts assisting in the inspections

- SAMPLE WORKPLACE INSPECTION CHECKLIST

Work Area: ……………………………………………………...

Date of Inspection: …………………………………………

Person/s Inspecting: ……………………………………….

Criteria	Area Standard C = Commendable S = Satisfactory U = Unacceptable N = Critical	Actions Required	By Whom	By When	Action Completed
Corridors / Stairs					
No blind corner					
Hand rails accessible					
Anti-slip tread on stairs					
Stairs in good condition					
Floors					
Even, visible steps, etc.					
In good condition (no trip hazards)					
No spills					
Work Areas					
Clean and tidy					
Equipment, paperwork put away					
Storage					
Items stored correctly					
Storage designated to minimise lifting problems					
Walking area clear					
Electrical					
Residual current devices fitted for moveable equipment					
Equipment checked and has current inspection tags					
No cords on floor or across walkways					
Equipment					
In good condition (in unsafe taken out of service)					
In use or stored appropriately					
Suitable for purpose used					
Maintenance checks / records up to date					
Ventilation					
Air vents, filters, extraction fans clean					
Servicing records kept up to date					

Gas Cylinders					
Cylinders secured					
Stored outside (minimum inside)					
Cylinders in use secured on trolley					
Hazardous Substances					
Material safety data sheets available for all substances					
All containers clearly labelled					
Stored appropriately					
Manual Handling					
Unnecessary manual handling eliminated					
Staff trained in manual handling					
Staff trained in use of mechanical aids					
Lighting					
Light fittings clean, working					
Work areas well lit					
Night lighting adequate					
Security lights working					
Worker Amenities					
Toilets, hand basis clean, soap available					
Lunch room / canteen clean					
Safety Signs					
OHS Policy displayed					
First aid, protective and fire equipment, signs, etc. posted					
Waste Disposal					
Bins regularly emptied					
Clean					
Food scraps in vermin proof bins					
Infectious waste disposed of appropriately (if applicable)					
Fire And Emergencies					
Extinguishers in place					
Serviced, not blocked					
Exits clearly marked, clear					
Exits, emergency lighting work					
Action cards, emergency numbers displayed					
Smoke detectors tested					
Fire blankets accessible					
Workers know procedures (ask a sample of staff)					
First aid kit available, well stocked and clean					
Records kept of first aid provided					

Grounds					
Paths even / obstruction free					
Garden sheds locked					
Easy access to grounds for staff / visitors					

(Adapted from Anonymous (nd))

It is suggested that the following rating system be considered:

Commendable: if the policies and practices meet requirements. This would be indicated by evidence that:

- all criteria are satisfied
- there is evidence of continuous improvement
- there is no area of major health or safety risk nor major concern about staff / visitors well-being, and
- there is a consistent level of achievement of the outcome.

Satisfactory: if the policies and practices generally meet the requirements and there are only minor deficiencies which can be rectified within an agreed timeframe. There would:

- be evidence of an adequate corrective plan currently being implemented to rectify the deficiency within a reasonable time, and
- no major areas of health or safety risk nor major concern about residents' well-being.

Unacceptable: if any major health or safety risks or concerns about the well-being of workers or visitors exist but the organization can provide evidence that an adequate corrective plan is being actioned and will be achieved within an acceptable timeframe; if the policies and practices of the organization do not meet the expected outcome because there are major deficiencies other than a major health or safety risk or concern about the well-being of workers which, despite corrective action, will take considerable time to rectify. **Unacceptable** *means an adequate corrective plan is being actioned and it will be achieved within an acceptable timeframe.*

Critical: if a major health or safety risk or concern about the well-being of workers or visitors exists for which there is no evidence of an adequate plan of corrective action currently being implemented. Immediate corrective action is required. **Critical** *means no corrective plan is in place and immediate corrective action is* required.

An alternative rating approach could be using a prioritization approach by assigning a priority level to the hazards observed to indicate the urgency of the corrective action required. For example:

A = Major - requires immediate action
B = Serious - requires short-term action
C = Minor - requires long-term action

INSPECTION SYSTEMS

• INTRODUCTION

The key elements in achieving and maintaining high standards of safety, health and environment are a commitment by and visible and tangible involvement of management, effective training, systems of work that are safe, regular monitoring and a committed work force. In manufacturing operations, regular checks enable the manager and supervisor to assess whether they are likely to meet objectives or if there are deviations so that they can initiate corrective action. In much the same way regular checks on health and safety matters highlight problem areas and enable action to be taken before an incident occurs. Health and safety checks can be either reactive or proactive.

Reactive checks are those that occur after the fact, where the incident has already occurred or your response to reports, which highlight that, something is not according to plan. The most common checks are incident statistics used to measure incident performance. They record the failures and give an indication of where corrective action needs to be taken. This does, however, have limited use due to the delays in corrective action taken and the fact that the accident has already occurred.

The proactive approach, on the other hand, endeavours to identify, evaluate and control hazards and risks before they develop to the stage of causing an accident.

One of the more common methods employed to identify potential hazards before they manifest themselves, is the health and safety survey or inspection, where the work area is inspected for any hazards and these are recorded, and after the inspection the hazards are assessed, and a plan of action formulated to put them right.

A well-managed inspection programme may, for example:

- Identify potential problems that were not identified during design or task analysis.
- Identify equipment deficiencies.
- Identify improper worker actions.
- Identify effects of changes in processes or materials.
- Identify inadequacies in remedial actions.
- Provide management with self-appraisal information.
- Demonstrate management commitment and involvement.

• PRINCIPLES OF INSPECTIONS

When conducting health and safety inspections, it is useful to follow these basic principles as suggested by the Canadian Center for Occupational Health and Safety (https://www.ccohs.ca/oshanswers/prevention/effectiv.html):

- Draw attention to the presence of any immediate danger - other items can await the final report.
- Shut down and "lock out" any hazardous items that cannot be brought to a healthy and safe operating standard until repaired.
- Do not operate equipment. Ask the operator for a demonstration. If the operator of any piece of equipment does not know what dangers may be present, this is cause for concern. Never ignore any item because of not having the knowledge to make an accurate judgement of health and safety.
- Look up, down, around and inside. Be methodical and thorough. Do not spoil the inspection with a "once-over-lightly" approach.
- Clearly describe each hazard and its exact location in the rough notes of the inspection. Allow "on-the-spot" recording of all findings before they are forgotten. Record what has or has not been examined in case the inspection is interrupted.
- Ask questions, but do not unnecessarily disrupt work activities. This interruption may interfere with efficient assessment of the job function and may also create a potentially hazardous situation.
- Consider the static or stop position and dynamic or in motion conditions of the item being inspected. If a machine is shut down, consider postponing the inspection until it is functioning again.
- Consider factors such as how the work is organized or the pace of work and how these factors impact safety.
- Discuss as a group, "Can any problem, hazard or accident generate from this situation when looking at the equipment, the process or the environment?" Determine what corrections or controls are appropriate.
- Do not try to detect all hazards simply by relying on senses or by looking at them during the inspection. Equipment may have to be monitored to measure the levels of exposure to chemicals, noise, radiation or biological agents.
- Take a photograph if unable to clearly describe or sketch a particular situation.

• INFORMAL INSPECTIONS

Informal inspections are simply the conscious awareness of workers as they go about their regular activities. Properly promoted and utilized, they may spot many potential problems as changes occur and work progresses. Supervisors and workers continually conduct ongoing inspections as part of their job responsibilities. Such inspections identify hazardous conditions and either correct them immediately or report them for corrective action. The frequency of these inspections varies with the amount and conditions of equipment use. Daily checks by users ensure that the equipment meets minimum acceptable health and safety requirements.

Disadvantages of informal inspections include:

✓ They are not systematic.
✓ Miss things that require extra effort to find.
✓ Workers are preoccupied and could miss things.
✓ Workers forget to follow up.

However, if done correctly, informal inspections remain an important factor, as workers are normally the first to note deficiencies. Workers should therefore be educated in the identification of hazards.

• PLANNED OR PERIODIC INSPECTIONS

As valuable as informal inspections are, they are not enough. They do not meet all the needs for inspections. Pre-operation checks involve inspections of new or modified equipment or processes. Often these are done after workplace shutdowns.

Inspections that are regular, planned inspections of the critical components of equipment or systems that have a high potential for causing serious injury or illness are referred to as periodic inspections. These inspections are included as part of standard preventive maintenance procedures or hazard control programs and could be conducted daily, weekly, monthly, or annually or at other stipulated intervals. Legislation and regulations may require that only suitably qualified or competent persons should be permitted to regularly inspect certain types of equipment, such as lifts, boilers, pressure vessels, scaffolding, and fire extinguishers.

Therefore, planned inspections should include:

- ✓ Critical parts/items inspections.
- ✓ Housekeeping evaluations.
- ✓ General inspections.
- ✓ Management audits.

It is useful to discuss with the group doing the inspection the planned inspection route before undertaking the inspection. Review where various inspection team members are going and what they are looking for. To avoid arm waving, shouting and other unsatisfactory methods of communication, it is preferable to "huddle" before going into noisy areas

Always wear the proper personal protective equipment (PPE) where necessary. Do not enter the area if PPE is not available and the unavailability as a deficiency during the inspection. When PPE is available, re-inspect the area.

• CRITICAL PARTS/ITEMS INSPECTIONS

Critical parts or items may be defined as the components of machinery, equipment, materials, structures or areas more likely than other components to result in a major problem or loss when worn, damaged, abused, and misused or improperly handled. Effective inspection programmes ensure that all these parts/items are identified, evaluated, and kept in proper condition.

No supervisor or manager should leave these inspections to chance. They must manage the system to prevent losses due to critical parts/items. This involves making an inventory, setting up inspection schedules and auditing the inspections.

• HOUSEKEEPING INSPECTIONS

Housekeeping inspections are vital in effective planned inspections and include both cleanliness and order. Housekeeping means in practice, more than clean and neat, that things are where they ought to be for maximum productivity, quality, safety, cost control.

Good housekeeping may:

- ✓ Eliminate accidents and incidents.
- ✓ Prevent wasted energy.
- ✓ Maintain best use of valuable space.
- ✓ Keep stores inventory to a minimum.
- ✓ Help control property damage and waste.
- ✓ Guarantee good workplace appearance.
- ✓ Encourage better work habits.
- ✓ Impress visitors.
- ✓ Reflect good management.

• GENERAL INSPECTIONS

A general inspection is a planned walk through an entire area of the workplace, site, or plant. It is comprehensive and inspectors look at anything and everything. Some advantages of general inspections are:

- ✓ Inspectors devote full attention to inspection.
- ✓ Inspectors prepare themselves to be observant and perceptive.
- ✓ Checklists are used as guides.
- ✓ Inspectors look beyond normal eye level.

✓ Reports of findings and recommendations are made to increase hazard awareness, corrective actions and accident prevention measures.

• FREQUENCY AND DURATION OF INSPECTIONS

The frequency and duration of inspections will depend on several factors that include:

- the frequency of planned formal inspections may be set in your legislation
- past incident records
- number and size of different work operations
- type of equipment and work processes - those that are hazardous or potentially
- number of shifts - the activity of every shift may vary
- new processes or machinery
- legislative requirements for type of operations of the organization

High hazard or high risk areas should receive extra attention.

It is often recommended to conduct inspections as often as committee meetings. Do not conduct an inspection immediately before a committee meeting but try to separate inspections and meetings by at least one week. This time allows for small items to be fixed and gives the committee an opportunity to focus on issues requiring further action.

It is not always possible to know precisely how long each inspection will take. Several factors may affect the duration of an inspection. These include, for example, what is found, how many questions are asked, and how large and complex the work area is. Ineffective inspections are those that only allow for a quick look. Since the purpose is to keep the workplace free of hazards time should not be an influencing factor. The inspection schedule should state:

- when to inspect each area or item within the workplace
- who carries out the inspection
- what degree of detail to inspect each area or item

• COMPOSITION OF THE INSPECTION TEAM

Health and safety committee members are obvious choices to be members of the inspection team to carry out formal inspections, especially if they have received the appropriate training or certification. Further, they are required to participate in terms of the Act.

Other criteria for selecting the inspection team could be:

- Extent of knowledge of regulations and procedures
- Extent of knowledge of potential hazards and possible mitigating interventions
- Extent of actual experience with and knowledge of work procedures involved

The inspection team could comprise of engineers, maintenance workers, occupational hygienists and practitioners, health and safety professionals, work supervisors or managers. They may be called upon to help with certain aspects of the inspection, or to help explain equipment or processes given their expert knowledge and experience.

Large workplaces may have more than one inspection team. The various teams can have separate areas to inspect.

• INCLUSION OF SUPERVISORS

As part of their daily duties at work, supervisors are responsible for taking action to prevent incidents, accidents, illnesses, injuries, and fatalities. Supervisors have an advantage in health and safety inspections because of familiarity with workers, equipment, and environment. This familiarity is also a disadvantage because it can interfere with their sense of objectivity. If the supervisor is not on the inspection team, before inspecting a department or area, the team should contact the supervisor in charge, but the supervisor should not act as a tour guide.

If the supervisor of the work area being inspected does not accompany the inspection team, if possible, consult the supervisor before leaving the area. It is important for each finding and recommendation to be discussed with the supervisor. Make the supervisor aware of any items that he can immediately correct. Record these on the report as corrected. This type of documentation and records serve as reminder to check the condition during the next inspection.

It is important for the supervisor to be aware that the inspection is not targeted at some other sinister agenda such as reduction in workers and disciplinary actions. The supervisor should explain clearly to the workers that this is the case, and the inspection is part of management commitment to improve working conditions and ensure a safe and healthy workplace.

• INSPECTION STEPS

While there are many different types of inspections, the steps for each type are similar. The steps are:

1. Prepare
   - ❑ pre-inspection planning
   - ❑ application of checklists
   - ❑ review of previous inspection reports.
2. Start with positive attitude.
3. Plan the inspection
   - ❑ area of responsibility
   - ❑ inspection route.
4. Determine what to look for.
5. Know what to look for.
6. Make checklists.
7. Review previous inspection reports.
8. Get tools and materials.

Steps to effective inspections:

- ❑ Refer to the map and checklist.
- ❑ Accent the positive.
- ❑ Look for off the floor and out of the way items.
- ❑ Take immediate temporary actions.
- ❑ Describe and locate each item clearly.
- ❑ Classify the hazards.
- ❑ Report items that seem unnecessary.
- ❑ Determine the basic causes of substandard actions and conditions.

Generally, identify any deviations from accepted work practices. Use statements such as: "a worker was observed operating a machine without a guard." Information derived from inspections must not be used for disciplinary measures.

Some common poor work practices include:

- using machinery or tools without authority
- operating at unsafe speeds or in other violation of safe work practice or procedure
- removing guards or other safety devices, or making the devices ineffective
- using defective tools or equipment or using tools or equipment in unsafe and unhealthy ways
- using hands or body instead of tools or push sticks
- overloading, crowding, or failing to balance materials or handling materials in unsafe and unhealthy ways, including improper lifting
- repairing or adjusting equipment that is in motion, under pressure, or electrically charged
- failing to use or maintain, or improperly using, personal protective equipment or health and safety devices
- creating unsafe, unsanitary, or unhealthy conditions by improper personal hygiene, by using compressed air for cleaning clothes, by poor housekeeping, or by smoking in unauthorized areas
- standing or working under suspended loads, scaffolds, shafts, or open hatches
- discussion with or observation of workers who may be overloaded, fatigued, working in conflict with others, or working in isolation or working alone

• DEVELOP REMEDIAL ACTIONS

Based upon certain critical factors, remedial actions should be proposed.

1. Potential severity of loss.
2. Probability of loss occurrence.
3. Cost of control.
4. Likely degree of control.
5. Control alternatives.
6. Justification for this control measure.

Take follow-up actions. The best ideas for remedial actions are of no significance if they are not implemented or if they do not work out as planned.

The inspection report is vital as it:

- ✓ Gives feedback to the management of the organization.
- ✓ Can reveal information that can help identify similar problems in other areas.
- ✓ Communicates information better and prompts people to remember.
- ✓ Documents all actions so that effort is not duplicated.
- ✓ Prompts follow-up actions and provides continuity.

• REPORTING ON HAZARDS AND UNSAFE CONDITIONS

When witnessing a worker performing unsafe acts in the workplace or notice unsafe conditions in the workplace these conditions must be reported. The purpose of a health and safety program is to prevent accidents, injuries, and losses from occurring. Inspections assist in achieving this goal. After the inspections have been done, the health and safety representative must report his/her findings, so that action can be taken if necessary, to prevent accidents from happening or recurring in the workplace.

• REPORTING PROCEDURE

The reporting procedure would depend on the type of inspection:

Continuous and frequent inspections will usually be carried out by the people doing the work, using an inspection sheet, and they will report to their supervisors by means of a completed inspection sheet. Inspection of a vehicle before leaving the premises is an example.

Planned maintenance will be scheduled in advance and the people responsible for the maintenance will also complete job cards or inspection sheets. Maintenance of machinery and vehicles are examples of this kind of inspection. When a bus or truck is taken to the workshop for a service, the mechanics will do the maintenance and affect any repairs that are necessary. The job will then be signed off by the mechanic and the foreman.

Regular inspections will usually be the duty of the health and safety representatives as it is the responsibility of the health and safety committee to make sure that there is an effective health and safety program in place. Any deviations from the health and safety program must be reported to the health and safety committee.

• REPORTING DISCREPANCIES

Once a hazard has been identified, it must be classified according to the severity, frequency and exposure, so that priority can be given to managing the risk. High priority hazards require immediate attention, while hazards with a low score will not demand the same urgency. If the priority is high, alert the supervisor and workers, as well as management and the health and safety committee about the hazard immediately. Discuss and find solutions that will not compromise the health and safety standards immediately. Note that even low priority hazards must be brought to the attention of the workers and their supervisor immediately. It could be that the workers involved are not aware that they are not adhering to the proper procedures. Discuss the matter with them, in order to ensure their cooperation.

• SUGGEST CORRECTIVE ACTION THAT CAN BE TAKEN

Possible corrective actions that can be taken with workers and their supervisors should be thoroughly discussed. Health and safety must not in any way be compromised by new procedures or corrective interventions.

• REPORT DISCREPANCIES TO THE HEALTH AND SAFETY COMMITTEE

All discrepancies as well as any corrective action that was taken must be reported to the health and safety committee. The submission of the report should always be in writing as a record.

• FOLLOW UP

It is important to follow up to ensure that the corrective action was taken and that new procedures are being implemented. Workers tend to fall back in old bad habits and sometimes need reminding to do the work in a healthy and safe manner.

• METHODS OF REPORTING

Each organisation will have their own system in place for reporting procedures. Normally, interdepartmental communication will take place in the form of a memorandum. It is important to note that any verbal communication between H&S representatives and workers and management should be followed up by some form of written communication. When reporting on health and safety matters, adhere to the *pro forma* inspection sheet that was used, as all the information will be found there.

• CONTENTS OF FINAL REPORT

All unfinished items from the previous health and safety inspection report should be noted such as any hazardous, unhealthy, and unsafe circumstances that still exist from the previous report. The information from regular inspections must be reviewed to identify where corrective action was needed. Whether these actions have been taken must be established. Older reports should be used to identify trends. Analysis of inspection reports may show the following:

- priorities for other corrective action
- need for improving safe work practices and procedures
- insight about why incidents are occurring in particular areas
- need for education and training in certain areas
- areas and equipment that require more in-depth hazard analysis

All the unsafe and hazardous conditions observed during the inspection must be written down. For each hazard the following should be done, namely:

1. At the top of the page, enter the date of the inspection
2. Enter the details of the inspection team – names and titles
3. State the department or the area that has been inspected
4. Give details of the hazard - State exactly what has been detected and accurately identify its location. Instead of stating "machine unguarded," state "guard missing on upper pulley #6 conveyor belt in North Building."
5. Classify the hazards that have been observed
6. Note the recommended method of controlling these hazards and risks and corrective action
7. Report issues in a concise, factual way. Management should be able to understand and evaluate the problems, assign priorities and quickly reach decisions.
8. State a date by when corrective action should be taken

This should be done for all the hazards, numbering them consecutively, ensuring that every member of the inspection team reads and approves the report.

Keeping health and safety inspection records is important. Previous inspection records show what has been previously been identified as needing attention. Inspection records highlight what a previous inspection team focused on and what areas the team did not include in their inspection. Do not simply repeat or copy previous inspection results. Previous inspection reports should be used to identify or pinpoint issues, and then determine whether recommendations were implemented. Importantly, the effectiveness of any changes must be evaluated.

The health and safety committee can review inspections, identify trends, and monitor the progress of the recommendations. This analysis can be used as part of the continual improvement process for the occupational health and safety program or management system.

• GUIDELINES FOR WRITING A REPORT

Good language must be used and references to details must be clear and logical. Reports are normally read by senior people who have a busy work schedule. They often simply do not have the time to read a long, detailed report and are interested primarily in the conclusions of the person who wrote the report has made. The following example is provided:

REPORT – Worker that fell two storeys
From: H.J. Lobise (Health and safety inspector)
To: G. Mogase (Head of Health and safety comity)
Date: 24 September 2010.
O.B. Reference: Number 657 of 2000.

On 24 September 2010, at 15h00, a worker (George Mbali) fell from the second storey of the construction site in 33rd avenue. Upon first assessment by paramedics, it would appear that Mr. Mbali has fractured his left leg and right ankle. A statement from the paramedic is yet to be taken. On first glance it would appear that Mr. Mbali did not hook his carabineer onto the safety line which resulted in him falling. This is to be confirmed through further investigation.

Conclusions. *There are indications that some workers are not following proper health and safety protocol, resulting in injury.*

Recommendations. *Initiate further investigations into the matter. Refresh worker training on health and safety protocol depending on investigation findings.*

Take note of the following:

- ✓ A report heading.
- ✓ From: Person and their position.
- ✓ To: Person and their position.
- ✓ Date of report.
- ✓ OB reference number.
- ✓ Body of report. Only a reference to details is made, for example the statement from the patrol security officer was not given, but it was mentioned that it was complete.
- ✓ Conclusions.
- ✓ Recommendations.

The following is an example of a workplace inspection report:

Inspection Location: __________________ Date of Inspection: __________________

Department/Areas Covered: __________ Time of Inspection: _____________

Observations					For Future Follow-up		
Item and Location	Hazard(s) Observed	Repeat Item Y / N	Priority A/B/C	Recommended Action	Responsible Person	Action Taken	Date

Copies to: __________________ Inspected by: ____________________

(Adapted from Anonymous (nd))

Example of Health and Safety Representative: Inspection form

Record that an inspection by a safety representative or representatives has taken place.

Date and time of inspection:	
Area or workplace inspected:	
Name(s) and signature(s) of health and safety representatives taking part in the inspection	
Name(s) and signature(s) of employer (or their representative) taking part in the inspection (if appropriate)	
Record of receipt of inspection form by the employer (or their representative)	
Signature Date	

(Adapted from Anonymous (nd))

This record does not imply that the conditions are safe and healthy or that the arrangements for welfare at work are satisfactory.

Example of Health and Safety Representative: Report form

Notification to the employer (or their representative) of conditions and working practices considered to be unsafe or unhealthy and of arrangements for welfare at work considered to be unsatisfactory.

Date and time of inspection:	
Particulars of matter(s) notified to employer or their representative (include location where appropriate)	
(This section below to be completed by the employer)	
Remedial action taken (with date) or explanation	
if not taken. This information to be relayed to health and safety representative(s)	
Name(s) of health and safety representative(s) notifying matter(s) to employer or their representative	
Signature of employer or their representative	
Signature(s) of health and safety representative(s): Date:	
Record of receipt of form by the employer, or their representative(s):	
Signature: Date:	

(Adapted from Anonymous (nd))

Note that as before, this report does not imply that the conditions are safe and healthy or that the arrangements for welfare at work are satisfactory in all other respects.

- CONTROL MEASURES (CORRECTIVE ACTION)

As the consequences of not managing hazards and risks have far-reaching effect on safety, health, and the environment, as well as on the production and profits of an organization, it is of vital importance to implement measures to control exposure to hazards. These measures include:

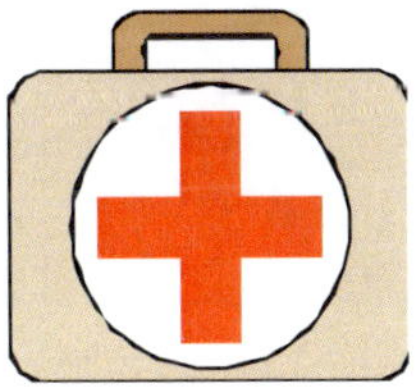

✓ **Removal** – remove the hazard completely – the best solution, if viable.
✓ **Replacement** – replace (or repair) the hazardous object or activity.
✓ **Transfer** – move the hazard to a place where it will not threaten the health or safety of workers.
✓ **Tolerate** – live with it and manage it as far as possible e.g., using public roads or transport.

An alternative health and safety checklist adapted from TUC Health and Safety Inspection Checklist:

	YES	NO
Asbestos		
Are any areas containing asbestos identified, marked and an up-to-date record kept?		
Is all the asbestos in good condition/sealed in and monitored?		
Has the risk of exposure to asbestos dust and fibres been assessed, and an up-to-date written record kept?		
Are there arrangements in place to inform any contractors about any asbestos presence, or locations where it is not known if that area is clear of asbestos?		
Is there a plan for specialist removal of asbestos?		

	YES	NO
Chemicals		
Are all containers clearly labelled with contents, hazards warnings and the precautions to be taken?		
Are there safety data sheets for all chemicals including cleaning and other materials?		
Is training provided in safe use of chemicals and on what to do in an emergency (spillage, poisoning, splashing etc.)?		
Cleanliness		
Are all work surfaces, walls and floors kept tidy and regularly deep cleaned?		
Electrical safety		
Are all electrical equipment, fittings, and tools regularly checked and maintained?		
Is access to live high voltage equipment restricted to authorized people only?		
Fire precautions		
Are there separate storage arrangements for flammable materials?		
Are bins regularly emptied and rubbish safely disposed of?		
Are cigarettes and matches disposed of separately from other rubbish?		
Are clear fire instructions displayed throughout the workplace?		
Have sources of ignition (portable heaters etc.) been replaced with safer alternatives?		
Are fire drills carried out regularly and at least once per year?		
Are fire alarms and smoke detectors checked and tested weekly?		
Are the alarms capable of warning workers throughout the building?		
Are there other forms of fire warning for the hearing-impaired?		
Are all workers given information, instruction and training on fire risks		
and precautions, as well as what to do in the event of a fire or fire alarm?		
Is emergency lighting provided and tested regularly?		
Are fire escape routes clearly signed, kept clear and wide enough to prevent a crush, and do they lead quickly and directly to a safe area?		
Are fire doors and exits clearly marked, always kept clear on both sides, never left open, and do they open easily and quickly in the direction of escape and lead quickly to a safe area?		
First aid, accidents and illnesses		
Is there a first aid box and is it fully equipped and accessible to workers?		
Is there a trained first aider or appointed person on the premises?		
Is it clear who the first aider(s)/appointed person is/are and how they may be contacted?		
Is a clean and properly equipped first aid room available?		
Are all accidents, near misses and illnesses caused by work reported and recorded in an accident book?		

Is safe social and other distancing of at least 1.5m practiced?		
Is screening for COVID-19 done?		
Are workers showing symptoms prevented from accessing the workplace?		
Are workers showing symptoms of infection referred for testing?		
Gas safety		
Are gas appliances regularly checked and serviced by qualified people?		
Do workers know what to do if there is a gas leak?		
Lighting		
Is the lighting bright enough, especially over workstations?		
	YES	NO
Are stairs and corridors etc. properly lit?		
Are light bulbs replaced promptly?		
Are windows clean on both sides and free from obstructions?		
Lifting and manual handling		
Have all workers who are at risk from lifting or moving been trained in manual handling?		
Is mechanical equipment used whenever possible, have workers been trained in its use, and is there enough space to use it?		
Where mechanical assistance is not possible, are workers trained in safe lifting techniques and is there enough space to use them?		
Are heavy items stored at a convenient or adjustable height to suit the user?		
Is the weight of loads known and clearly marked, and are they small and light enough?		
Are unbalanced, uneven, slippery, sharp or too hot or too cold loads avoided?		
Are loads securely packed to avoid them shifting or spilling?		
Are work surfaces at a comfortable or adjustable height to suit the user and at compatible heights to reduce lifting from one to another?		
Is frequent or prolonged stooping, stretching, or reaching above shoulder height, or sideways twisting of the body avoided?		
Are lifting and handling needs included in patient/client care plans?		
Do uniforms, protective equipment and other clothing that is provided allow easy movement?		
Machinery and equipment		
Are all workers trained to use, clean and adjust equipment safely?		
Is all equipment regularly inspected and maintained?		
Is there a procedure for reporting faulty equipment and for taking it out of use until repaired?		
Are all guards in place on machinery?		

Are potentially dangerous machines only operated by properly trained workers aged 18 and over?		
Noise		
Are noise levels below the recommended maximum (rough guide – you should be able to talk with someone a metre away without shouting)?		
Have the causes of noise been tackled?		
As a last resort, are suitable earmuffs or plugs provided, and are they regularly checked, cleaned and maintained, and stored in a clean and safe place?		
Overcrowding		
Is there enough space for workers to work safely?		
Is safe social and other distancing of at least 1.5m practiced?		
Are numbers of workers controlled in terms of DOH and COGTA guidelines?		
Protective clothing		
Is proper and appropriate protective clothing provided free of charge?		
Is it effective, comfortable and well fitting?		
Is it replaced as soon as they are worn out or damaged?		
Are clean overalls provided regularly?		
Slips, trips and falls		
Are floors and stairs in good condition, free from obstructions and nonslip?		
Are spills cleared up immediately?		
Is non-slip footwear provided free of charge where needed?		
	YES	NO
Do all staircases have securely fixed handrails?		
Are trailing leads and cables secured or covered?		
Is there enough storage space?		
Has a risk assessment been done on all work from heights and are the measures required being implemented?		
Stress		
Do risk assessments include stress?		
Has your employer done a stress audit?		
Are there measures in place to avoid or minimise the risk?		
Has the employer introduced the Stress Management Standard?		
Temperature (working indoors)		
Is the temperature comfortable all year?		
Does the temperature reach at least 16°C within one hour of starting work?		
Can breaks be taken away from hot areas?		

Temperature (working outdoors)		
Is warm clothing provided in cold weather?		
Are there facilities for warming up and making hot drinks when cold?		
In hot conditions, is cool drinking water provided and can breaks be taken in the shade?		
Can the work be organised so that it takes place in the shade or not during mid-day when the sun is at its strongest?		
Toilets, wash and rest facilities		
Are there enough toilets, and are they clean and in good repair?		
Are washing facilities (hot water, soap and towels) provided?		
Are sanitary disposal facilities provided in women's toilets?		
Are lockers (or something similar) provided for workers?		
Is there a rest room, and is it clean, properly lit, and ventilated?		
Are there suitable facilities for pregnant and nursing mothers to rest?		
Are there facilities for workers to eat meals?		
Ventilation		
Are fumes, steam and stale air removed?		
Is there a supply of fresh air without draughts?		
Are special precautions taken when working in confined spaces?		
VDUs (computer users)		
Are workstations and seating fully adjustable, and are workers trained and encouraged to adjust?		
Does seating give proper back support?		
Are footrests provided where needed?		
Is furniture and equipment checked and maintained regularly, with faulty items taken out of service and replaced?		
Can users easily read screens, and are they flicker and glare free?		
Is the pace of work comfortable and can breaks be taken?		
Are users offered full free eye tests?		
Are reports of aches, pains, numbness or tingling in limbs investigated?		
Is the work free from awkward postures, movements and very repetitive work?		
Is training provided about the use of equipment, methods of work, and how to avoid repetitive strain injury (RSI)?		
Are transparent screens provided between workstations and as barrier preventing contact with others?		
	YES	NO
Violence		

Has a risk assessment on violence or the threat of violence been conducted (physical, verbal abuse or intimidation)?		
Are they encouraged to report all incidents, including intimidation?		
Are there preventive measures in place to avoid or minimise the risk?		
Are there panic buttons, do they work, and are they quickly and reliably responded to?		
If name badges are worn are only first names given?		
Is there a policy and procedure for home visits, lone working and community-based working, and is this work avoided if it is unsafe?		
Is information about potential incidents or clients which may present a risk passed on?		
If workers have to make home visits or work through an isolated or threatening area, do they have alarms?		
Can public waiting areas be improved to reduce tension and stress?		
Can appointments be arranged to avoid long waiting times?		
Do interview rooms allow easy escape whilst giving privacy to the client?		
Are workers trained in what to do and how to diffuse potentially/violent situations?		
Is counselling and support for the victims and witnesses of violence provided?		

• COMPLIANCE AUDITS

An effective audit process is an essential management tool in any organization. All well-structured and conscientiously conducted audits provide an objective view of the actual status in the organization, identification of weaknesses, recognition of successes, evaluation of compliance, determination of the adequacy of policy and procedures against statutory requirements, world-class standards and social and ethical norms. Though businesses differ in size and the industries in which they operate – for example, a factory will have different regulations to a storage and distribution centre – an audit will in general look to assess the following factors:

- The strengths and weaknesses of the current system
- How the system performs within the aims of the organisation
- If the organisation is fulfilling its legal obligations within its industry
- If a proper performance reviewing system is in place.

A Health and Safety Compliance Audit is one of the most important factors in South African workplaces, to ensure compliance with the Act and Regulations, standards and by-laws. An audit is a methodical, independent, and documented assessment of a business' system and processes, in which it is measured against regulated criteria to make sure health and safety standards are being upheld. Organizations should have a management system in place to ensure safety processes continue to be maintained. In addition to ensuring compliance, health and safety audits also review the health and safety documentation of the organization and determine whether its record-keeping systems are adequate or need to be more robust. They will also look beyond the current activity in the workplace and evaluate the health and safety training program of the organization.

A health and safety audit is a pro-active management approach to measure health and safety performance of the organization. This type of audit helps the development of criteria for further improvements in organizational strengths along with the identification and control of organizational weaknesses. A health and safety audit provides a basis for taking effective planning decisions. An audit is important and essential because it will help to:

- Reduce the risk of personal traumas or injury
- Prevent material loss from a business
- Promote worker morale
- Ensure customer confidence.

Broadly speaking, health and safety audits serve two purposes:

1. They are conducted to determine whether the organization is following the applicable health and safety legislative and regulatory framework.
2. They identify and highlight any weaknesses in the organizational health and safety programs and processes and provide guidance for interventions to improve health and safety plans or corrective actions that should be undertaken within various timeframes.

Reasons for having a health and safety audit include:

- To determine if the health and safety programs and procedures in the organization are working.
- To verify that both workers and management are committed to and engaged in the health and safety programs.
- To verify that processes are following the policies and regulations of the organization.
- To determine if health and safety program activities are being documented properly.
- To discover and identify potential hazards.
- To evaluate the effectiveness of existing management controls.
- To check the health and safety of the workplace and the equipment being used routinely.
- To evaluate the adequacy of the health and safety training and performance of supervisors and workers.

A health and safety audit is a more structured and thorough process than mere technical inspections or unplanned spot-check inspections.

• TYPES OF AUDITS

Audits are conducted in two ways – internally as well as externally. While internal audits are conducted by persons within the same organization, external auditors are appointed for external audits. As far as purpose, content, sequence, and depth of investigation are concerned, both types have almost similar lines of actions. The main differences lie in approach. Practically, internal audits are subjective in that they are about examining one's own organization, own system, and own workforce although organizations claim that they authorize internal auditors with total independence. Auditors are typically selected from neighbouring departments. They might have been trained as health and safety auditors, yet they remain internal only. They could find it difficult to disclose the weakness of management and put it in writing in an auditor's report. Rather, practically, they are given the task of devising ways and means to hide the weaknesses and highlight the strengths of the organization. The consequence could be a false sense of security and well-being. Internal auditing is practically meant as preparation for third party external auditing.

Third party independent audits are usually done by expert agencies comprising of teams well-trained lead auditors, technical experts, legal specialists, and others. For the audit findings to be optimized organizations being audited should allow the team maximum freedom to judge neutrally and independently without exercising any influence from any party. Objectivity becomes the key motto, and their disclosures are to be taken as bare facts, based on which further lines of correction and improvements are to follow.

There are two basic audit type approaches: *general* and *specific*.

The *general SHE audit* is a facility-wide audit that focuses on basic hazards and their associated controls. A general audit would include, for example, a review of the interior and grounds of the facility itself that includes, for example, overall housekeeping, condition of features such as steps, pathways and walkways, parking areas, and lighting. This type of audit approach is designed to be broad-based and focuses on the entire facility being audited.

On the other hand, the *specific SHE audit* identifies SHE hazards in a specific department, section, branch or operation, or on a single piece of plant or equipment. This type of audit is particularly useful in high hazard operations or where there is a high frequency of accidents within a specific department, section, branch or operation. The specific audit is detail-oriented and of necessity time-consuming because of its nature. A few examples of specific SHE audit items include evaluating worker hand-arm vibration exposures (HAVS); inspecting compressed gas cylinders, hoses, and nozzles; and identifying appropriate fire protection measures in a cutting and welding operation. To be effective, the audit, whether of the general or specific type, must be continuous and aligned with the day-to-day operations of an organisation.

An ongoing audit process is an excellent mechanism by which management can obtain measurable and meaningful data about the SHE programs of the organization or enterprise In contrast, a one-time, single audit such as an annual SHE audit is simply ineffective in that it only provides a only snapshot of the overall status of SHE programs at a point in time. Organizations that have effective SHE audits use both general and specific types alternately.

A SHE audit can involve a walk-through of the facility, interviewing management or workers, and reviewing the documentation of the organisation.

• HEALTH AND SAFETY AUDIT PARTICIPANTS

Given the importance of management commitment and involvement in the organizational health and safety effort, senior management must support and participate in the health and safety audit process. They should endorse the process verbally and in writing to all workers in the organization. In this way workers become aware that the senior management of the organization is serious about health and safety audits and is committed to allocating appropriate resources.

There is no industry standard that indicates who should conduct health and safety audits. Variables in the size and type of business, number and expertise of workers, and special hazards relative to operations characteristic of each business will dictate which workers are assigned to the audit program. A large organization may use a health and safety director or manager to implement and oversee the entire program. In other organizations, a team approach is used, mixing facility and line managers, supervisors, engineering personnel, and workers from various departments. Finally, an outside organization can conduct the audit. Insurance carriers, government agencies, engineering firms, and health and safety consultants are commonly used.

To determine who is best for the job, consider:

- The type of health and safety audit that will be conducted, namely general versus specific
- Experience and availability of workers
- Time requirements and constraints of workers
- Special hazards and operations at the facility
- Experience and cost of outside consultants

• HEALTH AND SAFETY AUDITS AND THEIR BENEFITS

The health and safety audit must be documented in two major portions. The first part involves checklists while the second part involves the final report. Checklists are an integral component of the overall health and safety audit. These forms should suit the organization and the type of health and safety audit which can be either general and/or specific. In the planning stages, key workers should be involved to ensure that all health and safety programs, operations, and hazards are addressed. At a minimum, include checklists for housekeeping, smoking, personal protective equipment, machinery/equipment and hand tools, fire health and safety, electrical health and safety, and chemicals.

Health and safety audits are beneficial because they:

- Promote constant review of systems to ensure that they do not become weakened by habit
- Facilitate planned improvements to programs, policies, and procedures
- Help to identify weaknesses in human resources departments
- Help to demonstrate the dedication of management to worker health and safety.

A health and safety audit ought to be taken as an important link in the chain of sustenance of a worthwhile safety and health management system for the workplace. The audit should not be taken as a mere formality but should be treated as a detailed prescription by an expert independent agency for effecting improvements in health and safety culture in the workplace and job execution.

• HEALTH AND SAFETY AUDIT CHECKLIST

The audit checklist covers general safety programs and regulatory compliance; facilities and equipment; and specific hazards and operations. This checklist can help one identify safety programs needed for the operations or facilities. However, it may not list all programs or areas which need to be considered.

Risk Management Programs	**OK**	**Not OK**	**Tested**	**Regulatory Compliance Needed?**	**N/A**
Safety Management and Regulatory Compliance					
Accident Investigations					
Accident and Injury/Illness Records					
Drug Testing / Substance Abuse					
Hazard Identification And Analysis					
Job Training (Specialized)					
Occupational Safety and Health Administration (OSHA) Compliance					
Product Safety And Recall Programs					
Safety Audits					
Safety Meetings					
Safety Program					
Facilities and Equipment					
Building Utilities (Electrical)					
Compressed Air Equipment					
Elevators, Stairs, And Walking Surfaces					
Facilities and Equipment					
Emergency Response Plan					
Equipment Safety (Policies, Maintenance, Repairs)					
Fire Prevention and Protection					
Grounds Maintenance (Sidewalks, Driveway, Parking, Fences, Lighting)					
Housekeeping (Trash, Walkways, Aisles, Floor Maintenance, Tripping Hazards)					
Indoor Air Quality					
Life Safety and Evacuation					
Safety for Visitors, Contractors, etc.					
Security					
Slip and Fall Prevention (Rugs/ Mats, Floor Care, Unmarked Elevation Changes, Etc.)					
Stairs (Railings, Treads, Lighting)					
Vehicle Preventive Maintenance					

Risk Management Programs	OK	Not OK	Tested	Regulatory Compliance Needed?	N/A
Specific Hazards Or Operations					
Back Injury Prevention					
Confined Space Entry					
Construction Safety					
Cranes (Chains, Hooks, Overhead Loads, Slings)					
Drivers (Selection, Training, Supervision)					
Electrical Safety					
Ergonomics Training					
Fall Prevention and Protection					
First Aid, CPR					
Forklift Safety and Operator Training					
Hazard Communication					
Hazardous Materials Inventory					
Hazardous Materials Training (Handling, Transportation, Storage, Disposal)					
Hearing Protection					
Industrial Hygiene Assessment and Monitoring					
Laboratory Safety					
Laser Safety					
LockOut / TagOut (LO/TO)					
Machine Guarding					
Material Handling					
Material Safety Data Sheets (MSDSs)					
Office Safety (Drawers, Electrical, File Cabinets Secured, Etc.)					
Painting Areas (Ventilation, Explosion-Proof Fixtures, PPE)					
Personal Protective Equipment					
Pre-Construction Evaluation					
Scaffold And Ladder Safety; Work Platforms					
Shift work / Work Schedules					
Thermal Conditions (heat, cold)					
Tool Safety					
Trenching					
Video Display Terminal Ergonomics					
Welding, Cutting, Hot Work					
Workplace Violence Protection Program					
Other Operations specific to the Facility:					

The second portion of the documentation, the final report, identifies the health and safety audit findings, makes observations and recommendations, and offers an overall opinion. The report should provide detail on specific suggested enhancements to remedy deficiencies and should highlight serious and "repeat" observations. The final report should be communicated to management in a timely manner. Management should take ownership of the audit results and should approve improvements to safety and health programs, processes, and equipment.

• REVIEW QUESTIONS

1. What are the benefits of conducting a health and safety inspection?
2. How frequently should a health and safety inspection be done?
3. Why would one not include supervisors in the inspection team?
4. What should be in the final inspection report?
5. Why should an organization undertake a health and safety audit?

• REFERENCES

Almond, Paul. and Esbester, Mike., 2018. Regulatory inspection and the changing legitimacy of health and safety. *Regulation & Governance*, *12*(1), pp.46-63.

Anonymous (nd). Loose notes on safety, health and environment management

CCOSH (n.d.). *Effective Workplace Inspections*. [online] Ccohs.ca. Available at: https://www.ccohs.ca/oshanswers/prevention/effectiv.html.

Dowling, Carlin., Knechel, W. Robert. and Moroney, Robyn., 2018. Public Oversight of Audit Firms: The Slippery Slope of Enforcing Regulation. *Abacus*, *54*(3), pp.353-380.

HSA (n.d.). *Safety and Health Management.* [online] Health and Safety Authority. Available at: https://www.hsa.ie/eng/Topics/Managing_Health_and_Safety/Safety_and_Health_Management_Systems/.

Lim, Ki-Taek, Kim, Jim-Woo, Hanna. and Park, Y., 2018. Review of the regulatory periodic inspection system from the viewpoint of defense-in-depth in nuclear safety. *Nuclear Engineering and Technology*, *50*(7), pp.997-1005.

Morantz, Alison., 2018. Construction Site Regulation and OSHA Decentralization. *Members-only Library*.

Walters, David., Quinlan, Michael., Johnstone, Richard. and Wadsworth, Emma., 2016. Cooperation or resistance? Representing workers' health and safety in a hazardous industry. *Industrial Relations Journal*, *47*(4), pp.379-395.

CHAPTER 11

HOUSEKEEPING, STACKING, STORAGE AND LOCK OUT

DEFINITION OF HOUSEKEEPING

Housekeeping is defined by SANS 10091:2015 as use of those procedures designed to bring about and maintain cleanliness, tidiness, and orderliness of working in a factory, plant, or workplace.

A place for everything and everything in its place

For example, good housekeeping involves the following, namely:

- ✓ Good lighting
- ✓ Good ventilation
- ✓ Clean hygiene and sanitized ablution facilities
- ✓ Clearly marked passages, aisles, corridors and storage areas
- ✓ Good stacking and storage practices
- ✓ Tidy factory and plant yards
- ✓ All refuse being kept in bins

• POOR HOUSEKEEPING

On the other hand, poor housekeeping is indicated by, for example,

- ✓ Unnecessary rubbish, waste and vegetation in the factory yard or on the site
- ✓ Poor stacking and storage practices
- ✓ Unsafe handling of flammable liquids
- ✓ Compartments not being kept in good condition

The definition of housekeeping demands that the following steps should be practiced for effective housekeeping, namely

- ✓ Passages, aisles, and walkways should be marked and kept clear of obstacles to prevent workers and visitors from injuries because of trips and/or falls.
- ✓ Waste bins should be provided. Where there are no waste bins workers tend to throw waste materials and papers onto the floor which most likely create a fire hazard in the workplace.
- ✓ All spillages must be cleared and cleaned immediately to prevent workers from slipping or being contaminated if chemicals are involved.
- ✓ Any unusual parcels or packages should be reported to security
- ✓ Everything should be placed in the correct place and all products stored correctly.
- ✓ All workplaces should be cleaned up and sanitized at the end of every shift.

• WORK AREA HOUSEKEEPING

Good housekeeping is an essential part of every job or work activity. Work areas, aisles, walkways, and equipment must be kept clear of loose materials, tools, and scraps. Materials such as lumber, loose timber, boards and pipes must be stored in an orderly and secure manner. Spills of grease, water, or oil must be cleaned up as soon as possible since any delay in doing so could result in an accident to workers. A safe and unobstructed access must always be maintained to work areas. Short cuts should be avoided. Aisles, traffic lanes, or fire exits with equipment or materials should never be blocked. Knowledge of the following is also important, namely:

- ✓ warning and prohibitory signs
- ✓ location of firefighting equipment
- ✓ identity of the health and safety representative and first-aider
- ✓ location of marked emergency exits.

In particular, the following should be noted:

- ✓ avoidance of walking under suspended loads and awareness of persons working at height
- ✓ watching out for vehicles since motor vehicle accidents dominate as causes of workplace accidents and injuries
- ✓ awareness of moving plant and equipment.

• STACKING AND STORAGE

Fatal injuries in warehouses, where most of the stacking and storage take place, are higher than the national average for all industries. If stacked incorrectly, products, raw materials and other supplies can fall and cause injuries like cuts and bruises or even more serious injuries related to crushing and pinning. Employers need to ensure that workers follow a set of standards for the storage of materials to avoid these accidents.

• GENERAL SAFETY REGULATION 8

1. No employer shall require or permit the building of stacks which consist of successive tiers, one on top of another, unless -
   a) the stacking operation is executed by or under the personal supervision of a person with specific knowledge and experience of this type of work;
   b) the base is level and capable of sustaining the weight exerted on it by the stack;
   c) all the articles which make up any single tier are consistently of the same size, shape and mass;
   d) pallets and containers are in good condition; and
   e) any support structure used for the stacking of articles is structurally sound and can support the articles to be stacked on it.
2. An employer shall not permit -
   a) articles to be removed from a stack except from the top most tier or part of that tier; and
   b) anybody to climb onto or from a stack, except if the stack is stable and the climbing is done with the aid of a ladder or other safe facility or means.
3. An employer shall take steps to ensure that -
   a) persons engaged in stacking operations do not come within reach of machinery which may endanger their safety;
   b) stacks that are in danger of collapsing are dismantled immediately in a safe manner; and
   c) the stability of stacks is not endangered by vehicles or other machinery or persons moving past them.
4. Unless a stack is otherwise supported, an employer shall take steps to ensure that tiers of stacked material consisting of sacks, cases cartons, tins or similar containers
   are secured by laying up articles in a header and stretcher fashion and that corners are securely bonded; and
   are stepped back half the depth of a single container at least every fifth tier or that, alternatively, successive tiers are stepped back by a lesser amount:
   - ✓ Provided that at least the same average angle of inclination to the vertical is achieved
   - ✓ Provided further that where the containers are of a regular shape and their nature and size are such that the stack will be stable, they may be stacked with the sides of the stack vertical if the total height of the stack does not exceed three times the smaller dimension of the underlying base of the stack
5. Notwithstanding the provisions of sub-regulation (4), freestanding stacks that are built with the aid of machinery may, with the approval of an inspector, be built to a height and in a manner permitted by the nature of the containers being stacked:
   Provided that -
   - ✓ the stacks are stable and do not overhang: and
   - ✓ the operator of the stacking machinery is rendered safe as regards falling articles

• PACKING STORES

Stored materials must not create a hazard for workers. When stacking and piling materials employers should make workers aware of the height and weight of materials, the accessibility of the stored materials to the user, and the condition of the containers where the materials are being stored. To prevent creating hazards when storing materials, employers must according to OSHA do the following, namely

- ✓ Keep storage areas free from accumulated materials that cause tripping, fires or explosions, or that may contribute to the harbouring of rats and other vermin
- ✓ Piles of timber and rubble are ideal places for snakes to hide in
- ✓ Place stored materials inside buildings that are under construction at least 6m from hoist ways, or inside floor openings and at least 10m away from exterior walls
- ✓ Separate non-compatible material
- ✓ Equip workers who work on stored grain in silos, hoppers, or tanks with lifelines and harnesses

When stacking material, it can be dangerous if workers do not follow the health and safety guidelines and safe working or operational procedures. One of the most common accidents related to storage is that of a stack collapsing. When materials are stacked too high or in an unstable arrangement, removing an item from the stack or bumping the stack can cause the rest of the materials to fall. If heavy objects are involved, this can pose a real threat to workers. Falling materials and collapsing loads can crush or pin workers, causing injuries or death.

Using an appropriate stacking method is one of the best ways to keep a stack from collapsing. The following are options:

• **Block Stacking** – Stack square items in a cube, making sure to secure them with strapping like wire or plastic shrink wrap.

• **Brick Stacking** – To ensure even more security, turn each level of a stack 90 degrees. This helps hold the items in place should the stack be bumped.

• **Pinwheel Stacking** – For even more protection than the brick pattern, turn each quadrant—not just each level—of items 90 degrees. Patterns like this help "lock" everything in place.

• **Irregular Stacking** – When dealing with irregularly shaped items, try adding sheets of plywood between each layer for added stability.

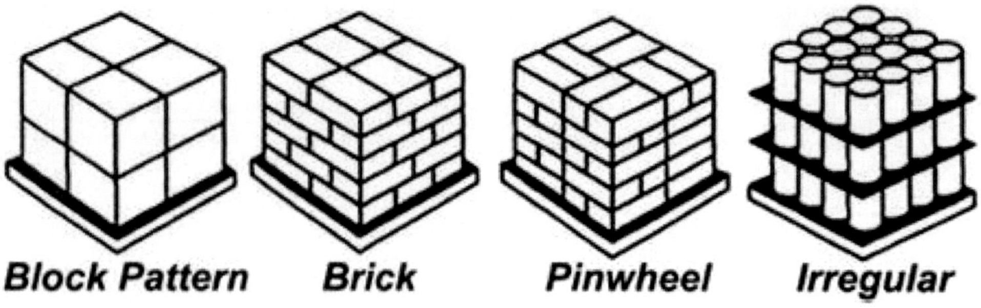

In addition to using secure stacking methods, the height and weight of the material being stacked needs to be considered to prevent a collapse. Heavy materials should generally be stacked close to the ground (if stacked on shelves) and not too high. Bricks, for example, should only be stacked 2,100mm high, and if the pile is higher than 1,200mm, the top of the pile should be tapered (50mm back for every 300mm of height above 1,200mm).

To help prevent injuries when stacking materials, workers must do the following:

- ✓ Remove all nails from timber before stacking and do not stack more than 4,800mm high (if workers will be handling it manually)
- ✓ Stack and level timber on solidly supported bracing
- ✓ Ensure that stacks are stable and self-supporting
- ✓ Avoid storing pipes and bars in racks that face main aisles to avoid creating a hazard to passer-by when removing supplies
- ✓ Use tapered stacking (a bit like a pyramid, with the layers getting narrower the higher up they are) for bagged materials and be placed in interlocking rows and cross key bags at least every ten layers
- ✓ Stack bags and bundles in interlocking rows to keep them secure
- ✓ Remove bags from the stack start from the top row first
- ✓ Store baled paper and rags inside a building no closer than 450mm to any walls, partitions or sprinkler heads
- ✓ Band boxed materials or secure them with crossties or shrink plastic fibre
- ✓ Stack barrels, drums and kegs symmetrically and if they are stored on their sides, they must be properly blocked to prevent rolling
- ✓ Block the bottom tiers of drums, barrels and kegs to keep them from rolling if stored on their sides
- ✓ Place planks, sheets of plywood dunnage or pallets between each tier of drums, barrels, and kegs to make a firm, flat, stacking surface when stacking on end
- ✓ Chock the bottom tier of drums, barrels, and kegs on each side to prevent shifting in either direction when stacking two or more tiers high
- ✓ Stack and block poles as well as structural steel, bar stock and other cylindrical materials to prevent spreading or tilting unless they are in racks
- ✓ Paint walls or posters with stripes to indicate maximum stacking heights for quick reference
- ✓ Observe height limitations when stacking materials
- ✓ Consider the need for availability of material

• ADDITIONAL CONSIDERATIONS

Fires and explosions

The risk of a fire or explosion in the workplace is real. If a material is flammable, its labelling must be checked. Have any storage instructions been provided? Combustible materials cannot be stored near open flames, devices that might spark or areas where smoking is allowed. Flammable liquids must be sufficiently separated from other materials by a firewall. Some materials may interact with others in dangerous ways. When this is the case, these materials must be stored at a safe distance from each other.

Presence of pests

Disorganized piles and stacks that are rarely touched may provide cover for pests. If materials are stored near any possible openings to the outdoors like vents, pests may get inside and take shelter among pallets, boxes, wood, or other materials. Storage areas must be kept clean and organized so that rodents and insects will not have anywhere to hide in the storage area. Piles must not be placed directly next to walls.

Trips, slips and falls

Trips, slips and falls are some of the most common accidents in the workplace. Care must be taken when stacking items in tapered piles rather than cubes where lower rows could get in the way of workers moving around the pile. The bottom layer of a stack must not protrude out too far.

Obstruction of aisles and exits

Stored goods must be kept in clearly marked areas well out of the way of pedestrian walkways. Consideration should be given to vehicles like forklifts that will need to get through an area. Extra space must be left around stacks in those situations. An example of proper stacking is shown in the figure.

Poor ergonomics

Consideration should be given to whether workers will need to frequently remove or add items to a pile. And if so will they be doing so by hand? What distance would the materials have to be carried or moved? In these cases, employers need to provide proper space for lifting and moving heavy materials. It is advisable to store heavy items above knee-height and below shoulder-height to prevent lifting injuries.

Making stacking and storage visual

Using visual cues to remind workers where to put stacks and how high stacks should be piled are simple ways to prevent mishaps. Floor tape, for example, can be placed around the corners or edges of stacks to remind workers where a stack should be placed. This is especially helpful if entire stacks are added and removed from the storage are frequently. Consider the use of floor marking tape to mark aisles, which can prevent workers from placing pallets, boxes and other materials in the way of people or vehicles. The figure is an example of demarcating areas where pallets of goods must be stacked on the floor.

Marking the maximum height of a pile on a wall or shelving unit can keep piles from growing too high. This can be done with tape, a label or paint. Making a permanent visual reminder of height limits will help workers to know whether a pile is too high or prevent them from having to get a tape or tool to measure the height.

Example of forklift warning sign

Consider using additional signage and labels throughout the workplace which are most appropriate for the kind of material being stacked or moved. Mark pedestrian traffic, vehicle traffic, loading areas and other activities related to moving supplies and products.

• SAFE USE OF SHELVES, RACKS AND PALLETS

Most businesses use shelves to store goods and materials, and racking units are often used for storage in warehouses and factories, typically in conjunction with pallets.

Shelving and racking

Shelving and racking must be suitable for their purpose and installed and maintained properly. Ensure that:

- ✓ floors are sound and level
- ✓ the manufacturer's installation instructions are followed and adhered to
- ✓ where shelving or racking needs to be fixed to a wall, the wall can support the load
- ✓ where appropriate, special safety ladders are provided to avoid people climbing on shelving
- ✓ units are spaced correctly to allow easy access for staff and, if necessary, mechanical-handling equipment such as forklift trucks
- ✓ shelves and racks are properly aligned and goods do not overhang shelves
- ✓ correct maximum loads for racking are displayed
- ✓ personal protective equipment is used where necessary
- ✓ items are stacked correctly - put the heaviest at the bottom where possible
- ✓ materials that workers frequently need are readily accessible - for example, not above head height

Check shelving and racking units regularly for any obvious signs of physical or chemical damage.

Pallets

When using pallets with racking units, avoid:

- ✓ using a pallet which cannot cope with its load
- ✓ poorly designed or constructed pallets
- ✓ using damaged pallets
- ✓ using the wrong type of pallet for the racking system used or the material or substance stored
- ✓ poor handling of pallets

Safely use any mechanical-handling equipment such as forklift trucks. Anyone operating a forklift needs to be fully trained, and people in areas where forklifts operate must be aware of the warning alarms, signs and notices and what they mean.

• STORAGE METHODS

As well as dangerous or hazardous substances, there are a range of other types of material the storage of which requires particular attention. These include:

Waste

All businesses have a duty of care to store and correctly manage the waste they create. Waste must be stored in suitable containers, making sure it does not harm the environment or human health.

When disposing of waste, or sending it to be recovered, ensure that it is handled by an authorized organization. Some waste is classified as hazardous because of its dangerous or toxic nature and is subject to additional regulation and special handling.

Timber

Sawn timber and board materials need to be stacked and stored safely. This includes ensuring that stacks of timber:

- ✓ are on firm, level ground
- ✓ do not exceed set height ratios
- ✓ are periodically inspected
- ✓ have no loose materials on top
- ✓ have no protruding nails in used timber

Glass

Storage measures should include:

- ✓ stacking glass at the correct angle
- ✓ providing appropriate personal protective equipment where necessary
- ✓ securing storage racks and ensuring these are not overloaded

Textiles

Storage also requires careful consideration in the textiles industry. For example,

- ✓ store frequently used materials at a convenient height
- ✓ plan storage areas carefully to allow for maximum possible access to materials
- ✓ dispose of obsolete stock
- ✓ consider mechanical methods of handling and moving bales and rolls

Storing food

Food and catering businesses must ensure food is correctly stored to comply with food-hygiene requirements - and protect their customers, staff and reputation. They must:

- ✓ observe temperature controls in all storage areas, including display cabinets
- ✓ store dried food off the floor
- ✓ observe use-by dates
- ✓ ensure all areas are clean
- ✓ avoid overloading refrigerated units - if they become too full, air does not circulate properly, causing food to deteriorate
- ✓ follow any storage instructions on food packaging

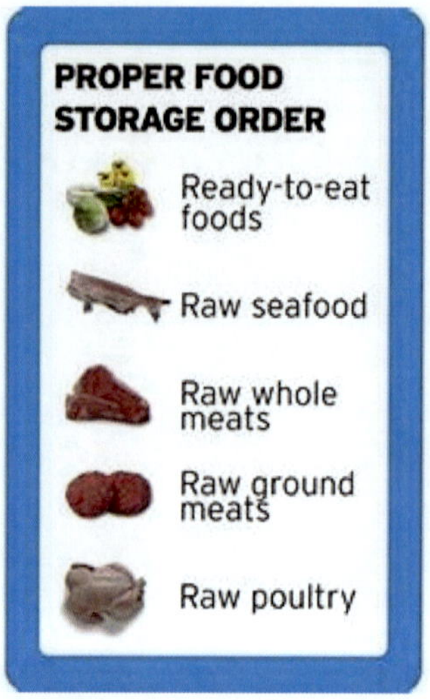

Reduce the risk of storing food by buying from reputable businesses which supply food that has been stored, processed and treated safely before it is delivered. If there are any concerns about the food when it is delivered, delivery should be refused. Assess any possible manual-handling risks caused by food storage, such as workers stacking crates of bottles or lifting heavy bags of potatoes. Ensure that all food and drink is stored in a way which minimises the risk of pollution.

• STORAGE OF DANGEROUS OR HAZARDOUS MATERIALS

Ensure that chemicals and dangerous substances are stored and handled in a way that minimises the risks posed by those substances and which limits the exposure of people to them. Assess the risks of storing and handling substances hazardous to the environment as well as human health. For instance, consider the effects of a leak to the air, water and surrounding land. This can help to avoid being prosecuted for causing pollution. Assess the risks of storing and handling dangerous substances - including the possibility of environmental damage caused by leaks and spills.

Implement any precautions needed to control risks, including:

- ✓ storing chemicals according to the manufacturer's instructions on the material safety data sheet (MSDS)
- ✓ keeping the minimum quantity of hazardous substances necessary
- ✓ storing incompatible substances separately
- ✓ taking steps to prevent release or leakage of dangerous substances
- ✓ keeping a spill kit near to storage areas, and ensuring staff are trained in what to do in the event of a spill
- ✓ cleaning up any leaks or spills that occur
- ✓ using appropriate precautions when handling substances - for example, wearing protective clothing or ensuring adequate ventilation
- ✓ ensuring workers who store and handle dangerous substances are properly trained
- ✓ checking containers used for short-term storage are properly labelled

Ensure that flammable substances are correctly stored in suitable containers and are not stored near to a source of ignition such as, for example, a heater.

Control the risks

Simple steps to control the risks of hazardous substances include:

- ✓ keeping the minimum quantity of hazardous substances necessary
- ✓ storing incompatible substances separately
- ✓ preventing release or leaks
- ✓ training workers to store and handle hazardous substances properly
- ✓ labelling storage containers properly
- ✓ storing flammable substances in suitable containers away from sources of ignition, such as boilers and heaters
- ✓ placing stores of liquid above ground where they're unlikely to be damaged, for example away from driveways
- ✓ maintaining gauges, valves and pipework
- ✓ having procedures for dealing with emergency leaks
- ✓ using a secondary containment system such as a drip tray or bund which is a storage area designed to prevent liquids escaping
- ✓ monitoring oil use - unexpectedly high use may indicate a leak
- ✓ avoid overfilling containers
- ✓ supervise deliveries

- CHEMICALS

Storage of Chemicals

Avoid the common practice of alphabetical storage of chemicals. Reactions of incompatible chemicals make spectacular demonstrations but can be a disaster in your storeroom. For example, aluminium foil and bromine react at room temperature producing enough energy to become incandescent. In addition, many incompatible concentrated acids such as acetic acid and nitric acid are often stored together. All chemicals in the store should be stored so as to avoid incompatibilities.

Chemicals could be separated in storage as follows, namely:

- ✓ Flammable or combustible and not highly toxic and compatible with water
- ✓ Flammable or combustible and not highly toxic and incompatible with water
- ✓ Oxidizers and non-flammables, compatible with water
- ✓ Oxidizers and non-flammables, incompatible with water
- ✓ Air sensitive
- ✓ Chemicals requiring refrigeration
- ✓ Compressed gas cylinders, separated as to oxidizers, reducers, corrosives, toxics
- ✓ Unstable chemicals such as explosives

Avoiding and dealing with chemical spills

Manage the workplace carefully to reduce the risk of chemical spills. If there is a chemical spill in the plant, it is important that staff know how to deal with it. If pollution is caused, prosecution and fines may follow. Pollution incidents must be reported as soon as they happen.

Plan for pollution incidents

Have a pollution incident response procedure in place for dealing with spills. Make sure that workers are familiar with the procedure and know how to implement it. If large amounts of chemicals are stored or there is a chemical warehouse, a spill alarm must be installed so that the building can be evacuated when there is a spill.

Prevent pollution from spills

Ensure that there are absorbent materials - such as sand and other containment equipment - suitable for the type and quantity of chemicals being stored and used in the workplace. Keep them close to where they might be needed, particularly in delivery areas. Ensure that workers know where they are and how to use them. Buy spill kits containing all the appropriate spill equipment for the chemicals being stored. Try to prevent liquid spills from entering drains, surface waters or groundwater. For example, use earth to block the flow of large spills, or use sand or a commercial spill kit to soak them up. Never hose down a spill, as a much worse pollution incident could result. Where chemicals are stored in a vehicle, carry a spill kit suitable for the quantity and type of chemicals involved.

EXAMPLE OF STACKING AND STORAGE INSPECTION CHECK LIST

Name of Organization:			
Date of Inspection:			
Inspection done by:			
ITEM	Yes	No	Steps to be taken
1. Is the stacking storage area demarcated?			
2. Are stacks on pallets neatly arranged?			
3. Is the base level?			
4. Is the base capable of sustaining /holding material weight?			
5. Are all material/boxes on the stack the same size?			
6. Are all pallets in use in good condition?			
7. Are containers used in good condition?			
8. Can the support structure of racks support the weight of materials packed?			

9. Are materials removed from top of stack?			
10. Do person required to climb onto stack/rack do so by means of ladder or other safe means?			
11. Are stacks that are in danger of falling immediately dismantled?			
12. Is it ensured that the stability of stacks are not endangered by workers moving past them?			
13. Are stacks packed away from light fittings and emergency exits?.			

Adapted from Qushum

• LOCKOUT SYSTEMS

• DEFINITION OF A LOCK OUT (TAG OUT)

Lockout-tagout (LOTO) or lock and tag is a safety procedure which is used in industry and research settings to ensure that dangerous machines are properly shut off and not started up again prior to the completion of maintenance or servicing work. It requires that hazardous power sources be "isolated and rendered inoperative" before any repair procedure is started. "Lock and tag" works in conjunction with a lock usually locking the device or the power source with the hasp, and placing it in such a position that no hazardous power sources can be turned on. The procedure requires that a tag be affixed to the locked device indicating that it should not be turned on.

• NEED FOR A LOCK OUT

Lockout-tagout (LOTO) or lock and tag is necessary

- ✓ To minimize/manage risk of exposure to energy sources and hazardous substances.
- ✓ To prevent accidental/unlawful machine start-up during maintenance or any other time.
- ✓ To comply with Minerals Act, Regulation 20 for the mining industry

Lockout-tagout (LOTO) or lock and tag should be done when

- ✓ Servicing and/or maintenance during normal production.
- ✓ Parking equipment in hard park area.
- ✓ When there is a break down.
- ✓ Leaving cab of a vehicle or heavy equipment
- ✓ Refuelling equipment

• LOCKOUT/TAGOUT PROCEDURE

A Lockout/Tagout procedure or method should include the following six steps, namely:

1. Preparation
2. Shutdown
3. Isolation
4. Lockout/Tagout
5. Stored energy check
6. Isolation verification

• STEP 1: PREPARATION– LOCKOUT/TAGOUT

The first step of the locking and tagging out equipment or machine for service and the maintenance is to prepare. While during the preparation phase, the duly approved or authorized person such as

- All authorized operator/drivers
- Mechanics/Service crew
- Persons who inspect equipment must investigate and obtain a complete understanding of all kinds of hazardous energy that must be controlled. The particular hazards must be identified to be able to control or monitor that energy.

• STEP 2: SHUT DOWN – LOCKOUT/TAGOUT

When the planning has been completed, the actual process of powering down and locking out machines starts. At this point the machine or equipment to be worked on must be shut down. All workers who will be affected by the shutdown must be notified.

- STEP 3: ISOLATION – LOCKOUT/TAGOUT

The next step is to isolate the machine and or the equipment from any source of energy. This may involve turning off the power at a breaker, or at the main isolator on the machine or equipment or shutting a valve.

- STEP 4: LOCKOUT/TAGOUT

With the machine or equipment separated or isolated from its energy source, the next step is to lock and tag out the machine or equipment. During this step, the approved or authorized person will attach the lockout and tagout devices to machine or equipment. This tag carries the name of the person who presented the lockout and additional information. The steps involved are:

- ✓ Attach the scissors to the lock out point
- ✓ Place the lock into the scissors and lock it
- ✓ Keep the keys on a safe place
- ✓ Team lock out
- ✓ Use one scissors and multiple lock on one machine

When two or more subcontractors are working on different parts of a larger overall system, the locked-out device is first secured with a folding scissors clamp that has many padlock holes capable of holding it closed. Each subcontractor applies their own padlock to the clamp. The locked-out device cannot be activated until all workers have signed off on their portion of the project and removed their padlock from the clamp. An example of lock out is shown in the figure.

- STEP 5: STORED ENERGY CHECK – LOCKOUT/TAGOUT

During this step it is necessary to check for any hazardous energy that might still be "stored" within the machine, or any "residual" energy. In this case, any potentially hazardous stored or residual energy must be made non-hazardous in an appropriate manner using a safe working procedure.

- STEP 6: ISOLATION VERIFICATION – LOCKOUT/TAGOUT

In the final step the approved or authorized person verifies that the machine or equipment has been suitably isolated and de-energized.

• THE EN 50110-1 SECURITY RULES

According to the European standard EN 50110-1, the security procedure before working on electric equipment comprises the following five steps, namely:

- ✓ disconnect completely;
- ✓ secure against re-connection;
- ✓ verify that the installation is dead;
- ✓ carry out earthing and short-circuiting; and
- ✓ provide protection against adjacent live parts.

• REVIEW QUESTIONS

1. Cite examples of good and poor housekeeping
2. What must be considered when stacking timber or lumber?
3. How would one safely store chemical substances?
4. How would one ensure the safe use of shelving to store materials?
5. Who should be responsible for lockout and tagout procedures?

• REFERENCES

Anonymous (nd). Loose notes on safety, health and environment management

CCOSH (n.d.). *Effective Workplace Inspections*. [online] Ccohs.ca. Available at: https://www.ccohs.ca/oshanswers/prevention/effectiv.html.

Chakraborty, Abhijeet., 2019. Construction Safety Management. *International Journal of Town Planning and Management*, *5*(1), pp.12-20.

Department of Labour (1993). Occupational Health and Safety Act No: 85 of 1993.

Dewi, Luciana. Triani., 2018. Investigation of Lockout/Tagout Procedure Failure in Machine Maintenance Process. *Jurnal Teknik Industri*, *20*(2), pp.135-140.

Emuze, Fidelis., Seboka, Linake. and Linake, Mohame., 2016. Construction work and the housekeeping challenge in Lesotho. Association of Researchers in Construction Management: 32nd Annual ARCOM Conference, 5-7 September.

Mahto, Dalgobind. G., 2016. Productivity Safety and Test of Hypothesis: An Analysis of Significance of Good Housekeeping and Safety Initiatives in Industries. *Available at SSRN 2773163*.

Materials Handling and Storage (2002). *Materials handling and storing*. Washington, DC: U.S. Dept. of Labor, Occupational Safety and Health Administration.

McKinnon, Ron. C., 2019. *The Design, Implementation, and Audit of Occupational Health and Safety Management Systems*. CRC Press.

Mining Safety (n.d.). *Mining Safety | Lock Out Machinery in the Mining Industry*. [online] Miningsafety.co.za. Available at: https://www.miningsafety.co.za/dynamiccontent/84/dynamiccontent/69/Physical-Fitness.

Nadeau, Sylvie., Kenné, Jean-Pierre., Emami-Mehrgani, Behnan. and Badri, Adel., 2016. Advances in integration of equipment lockout/tagout, determination of actual production capacity and production/maintenance planning. *Safety Science Monitor*, *19*(2).

OSHA (n.d.). *Materials Handling and Storage | Occupational Safety and Health Administration*. [online] Osha.gov. Available at: https://www.osha.gov/Publications/OSHA2236/osha2236.html.

Safe Stacking and Storage (1999). [online] Cdn.auckland.ac.nz. Available at: https://cdn.auckland.ac.nz/assets/science/for/current-students/HR/health-safety-wellness/documents/SafestackingandStorage.pdf.

Safety & Health Procedures Manual (2004). [online] Co.ontario.ny.us. Available at: http://co.ontario.ny.us/DocumentCenter/View/2292/Safety--Health-Manual?bidId=.

Sowa, Mariusz., 2018. Threats to Health and Safety at Work when Completing a Warehouse Process. *European Journal of Service Management*, *27*(1, 3/2018), pp.285-291.

Stacking and Storage of Materials. qushum.co.za/Safety/Docs/C1/V3QushumChecklist... https://www.safetyblognews.com/safe-stacking-and-storage-in-the-warehouse/

Yamin, Samuel. C., Parker, David. L., Xi, Min. and Stanley, Rodney., 2017. Self-audit of lockout/tagout in manufacturing workplaces: A pilot study. *American Journal of Industrial Medicine*, *60*(5), pp.504-509.

CHAPTER 12

HEALTH AND SAFETY SIGNAGE AND MACHINE GUARDING

Undoubtedly, health and safety signs in the workplace are one of the main means of communicating health and safety information. These signs provide information and warnings about hazards or threats which are essential to safety. It is important that workers and whoever is in the workplace at any time understand the risks in that environment and how they should be dealt with. Proper signage erected and displayed in the correct places where a potential hazard might be found in a workplace is necessary, as it is one of the main points of reference and contact for workers. These signs should clearly indicate the hazards so that it may be avoided.

The very best and effective health and safety systems use a complete set of signs that not only help to prevent dangerous situations, but effectively at the same time instruct workers in how to deal with them. For instance, a noisy area sign may be accompanied by an instruction to wear hearing protection.

Safety signs and colour are standardised in all organisations and help to protect the health and safety of workers and visitors in the workplace. SANS 1186 is the national standard for health and safety signage in South Africa. Compliance is therefore both beneficial and essential.

• ROLE OF HEALTH AND SAFETY SIGNS

Health and safety signs can play a vital role in ensuring a healthy and safe workplace. Employers who implement effective, clear visual communication may enjoy fewer accidents and injuries, increased efficiency, and safe behaviour throughout their facilities. The signs serve as a reminder, and are useful to those who are unfamiliar with a work area or process. For this reason, it is critical that health and safety signs are easy to understand. Simple graphics or words communicate more easily and quicker far better than complex instructions and allow the worker to make quick and instant decisions. Many are typically recognisable, such as the red circular prohibition sign, but others might need to be more specific depending on the situation. Good health and safety signage, inter alia,

- ✓ draw attention to health and safety hazards
- ✓ point out hazards that may not be immediately obvious
- ✓ provide general information and directions
- ✓ remind workers where personal protective equipment must be worn
- ✓ indicate where emergency equipment is located
- ✓ warn where certain actions are prohibited
- ✓ the signal word such as, for example, "Danger," "Caution," "Stop" should be readable from at 1,5m away
- ✓ signs should be placed such that workers are aware of the hazard and able to respond appropriately before being exposed to it
- ✓ the sign itself should not cause a hazard
- ✓ signs should have rounded or blunt corners, and have no sharp edges, burrs, or splinters
- ✓ a sign that warns pedestrians to look out for forklift traffic at an intersection should not block their view of incoming forklifts

Colour attracts attention and can be used extensively for health and safety purposes. For example, colour can be used as an additional health and safety measure to indicate the contents of pipes and the nature of its hazards. The choice of colour also draws attention to the severity of the hazard. For example, a dangerous hazard is communicated by the colour red. When workers are aware of the hazards around them and take the necessary precautions, the possibility of an accident or injury is minimised. However, while health and safety signs and colours are valuable in warning workers of hazards, they are not substitutes for eliminating or reducing these hazards.

- SAFETY SIGN CATEGORIES

There are three basic sign categories namely regulatory, information and warning signs.

Each category is distinguished by its **shape**	These categories are divided into sub-categories that can be recognised by their **colour**:	
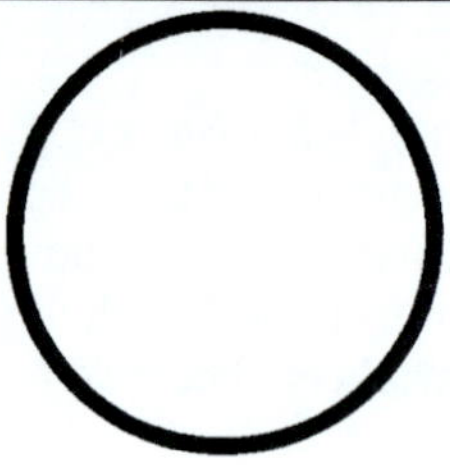 **Regulatory:** A circle indicates that an order is in force		**Permission:** allows an action
		**Prohibition:** prohibits an action
		**Mandatory:** requires an action
 **Information:** A square provides information		**Fire:** provides information on fire equipment
	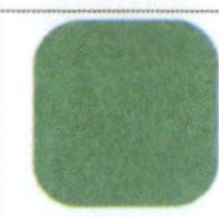	**First aid / Safety equipment:** provides information on first aid and safety equipment
		**Information:** provides general information or directions
 **Warning**: A triangle indicates caution or danger		**Caution:** indicates potential risk
		**Danger:** warns against definite danger that can cause irreversible and serious injury or death
		**Radiation:** indicates radiation risk

Basic Safety Signs

• HEALTH AND SAFETY SIGNS

Health and safety signs consist of a symbol, which is a letter or picture surrounded by a border of a definite shape and colour as shown in the following examples which also illustrate the three main categories of signs.

Regulatory symbols

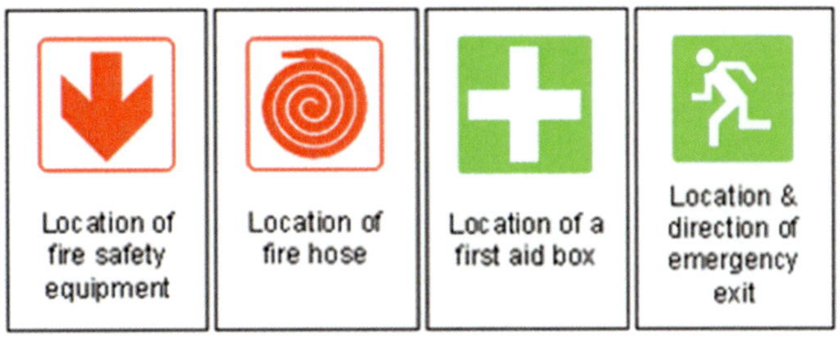

Information Symbols

Warning Symbols

Note the use of shape and colour in the examples where the shape of the sign corresponds with the category as explained and the colour code

• HEALTH AND SAFETY COLOURS

There are four main types of health and safety signage used in the workplace. Generally, the same colour schemes are used all over the world. Red signs indicate the potential of immediate danger, implying that action must be taken, or something is prohibited. Yellow or amber signs typically warn that precautionary steps should be taken to ensure safety. Blue signs are mandatory instructions that accompany danger or warning signs to ensure safety. Green signs safety signs, but does not necessarily infer any kind of danger; rather they are used to indicate escape routes or first aid.

Colour can also be used to indicate hazards or point out safety equipment. For example, colour can be applied on:

- ✓ indicator lights and buttons
- ✓ pipes
- ✓ separate work areas
- ✓ machinery
- ✓ vehicles
- ✓ walkways, floors and stairs

• INDICATOR LIGHTS, BUTTONS AND CONTROL POINTS

With the extensive daily use of colour in daily life, certain colours are associated with specific meanings. For example, red on a traffic light means stop. Many organisations take advantage of these associations to help workers easily recognise the message associated with colours and respond quickly, as in the case of indicator lights and control buttons:

Colour	Indicator Light	Button / Control
Red	Danger or alarm	Stop / off
Yellow	Caution	Caution / Intervention needed
Green	Safe condition	Start / on

Control Buttons

• PIPING SYSTEMS

There are many ways the contents of pipes are identified for health and safety purposes. A way of ensuring that workers have a definite knowledge of what a pipe actually contains is by:

- ✓ labelling the pipe with a tag that clearly identifies the material inside
- ✓ attaching the appropriate symbol to the pipe which describes the material in the pipe
- ✓ painting the pipe with a safety colour

Arrows on the pipe are often also used to show the direction of flow of the contents as a further health and safety precaution.

Colour	Classification	Examples
Red	Fire quenching materials	water, foam, carbon dioxide, etc.
Green	Liquids that are not hazardous by nature	Liquids like water
Blue	Gases that are not hazardous by nature	Compressed air
Yellow	Materials that are hazardous by nature	flammables, explosives, radioactive materials, materials with extreme temperatures or pressures

Safety Colours

• OTHER APPLICATIONS FOR SAFETY COLOURS

Safety colours are also used in work areas as well as on equipment and machinery. The colours commonly used and their meanings are listed in the table.

Colour	Meaning	Example
Yellow	Caution	Steps, guardrails, low beam storage cabinets for flammable materials; containers for corrosives or unstable materials
Orange	Warning	Hazardous parts of machines which may cut, crush or injure a worker inside movable guards or inside transmission guards for gears, pulleys, chains, etc.;
		Exposed parts (edges only) of pulleys, gears, rollers, cutting devices, power jaws, etc.
Red	Danger or stop	Containers of flammable liquid; emergency stop bars; stop buttons
	Fire protection equipment	Alarm boxes; fire extinguishers; exit signs
Green	Safety	Indicates location of first aid services; first aid kits, stretchers; emergency exit routes; safety equipment such as emergency safety showers.
Blue	Safety information	Signs requiring use of personal protective equipment; information signs; directional signs
Black	Markings for traffic routes, housekeeping	Traffic routes; passageways and floor markings

• OTHER TYPES OF SIGNAGE

Other types of signs can help workers stay healthy and safe and productive at work without referring to specific hazards in the workplace. Examples of these types of health and safety signs include:

- **Notice:** Use notice signs to deliver information about a machine, building, area, or equipment. These signs outline procedures, maintenance information, instructions, rules, and directions unrelated to personal injuries.
- **General health and safety signs:** These offer broad health and safety-related messages, typically relating to health, medical equipment, sanitation, first aid, housekeeping, and suggested general health and safety measures.
- **Admittance:** Admittance signs alert and explain the dangers and consequences associated with entering a restricted area.
- **Fire health and safety:** Fire health and safety signs point out emergency firefighting equipment and fire exits.
- **Non-hazard signs:** These communicate general health and safety facility information usually through simple text and clear symbols. These should never be used to communicate hazards, risks, or dangers because they are technically not health and safety signs but nevertheless promote a healthier and safer workplace.

• POWER TOOLS AND MACHINE GUARDING

The primary goal of all employers should be to protect all workers from injury by creating a healthy and safe working environment. Moving machine parts have the potential to cause severe workplace injuries, such as crushed fingers or hands, amputations, burns, or even blindness. Machine guards are essential for protecting workers from these preventable injuries. Any machine part, function, or process that may cause injury must be guarded. When the operation of a machine or accidental contact injure the operator or others in the vicinity, the hazards must be eliminated or controlled.

While machines and power tools can help improve production efficiency in the workplace, their moving parts, sharp edges, and hot surfaces can also cause serious workplace injuries. The list of possible machinery-related injuries is long and horrifying. Therefore, since machinery can cause injuries in many different ways when in operation, so to prevent accidents a thorough check should be undertaken before operating to ensure the machine is well maintained and all safeguarding requirements are fitted and working correctly.

Only qualified people are allowed to repair tools and equipment

When using hand or power tools it is important to ensure that:

- ✓ the right tools for the job are used
- ✓ handles are not split
- ✓ cutting tools are sharp
- ✓ power tools are not being overloaded.

Before starting machinery, it is necessary to:

- ✓ ensure that all guards are in place
- ✓ check for safety of other persons or workers in the area.

Before repairing or adjusting machinery it is important to stop and lock out the machine. A machine should not be worked on while it is in operation. Equipment or vehicles must only be operated by qualified and authorized personnel.

• MACHINE GUARDING

• SAFEGUARDING MACHINERY AND EQUIPMENT

Machine guarding is the placement of protective safety equipment, fencing and fixings around the perimeter of hazardous areas of manufacturing and engineering machinery to protect machine operators and other personnel against accidental injury or fatality.

A machine guard is a barrier which suitably encloses the moving parts of a machine. It must prevent any part of the body from reaching under, over, around or through the guard and entering the danger zone.

It is necessary to safeguard machines because the injuries caused by machines could be severe, disabling and permanent; such injuries are generally preventable; danger can often be reduced or removed by mechanical safeguards; and workers realise that unguarded machinery is dangerous. The employer should spend money on the safeguarding of machinery to show his sincerity in protecting the workers from injury.

Source: https://liftsafegroupofcompanies.wordpress.com/2016/02/02/5-potentially-fatal-machine-guarding-problems/

- PURPOSE OF MACHINE GUARDING

The purpose of machine guarding is to protect the machine operator and other workers in the work area from hazards created by ingoing nip points, rotating parts, flying chips and sparks. Some examples of this are barrier guards, light curtains, two-hand operating devices.

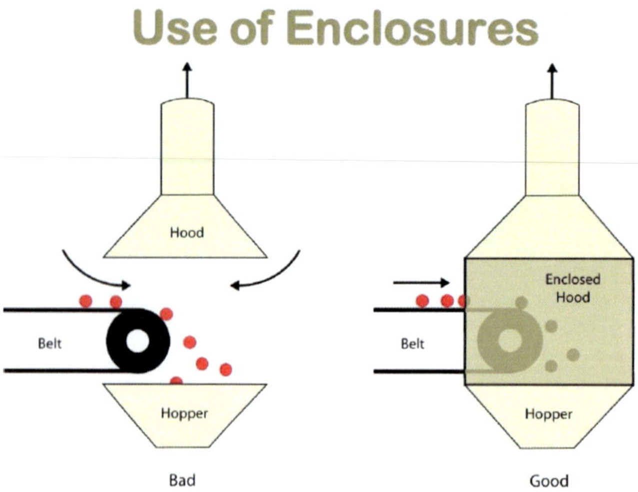

- NEED FOR MACHINE GUARDING

Machine guards are required where any type of machine activity could result in an injury ranging from cuts, burns and minor abrasions to more severe incidents such as lacerations, crushing or amputation. A large range of machine activities may create hazards to operators working with and around machinery. These can include moving parts, points of operation which is the area on a machine where work is performed, and pinch or shear points. Machine operators and other workers can also be at risk of being injured while working in the vicinity of machinery when the machine is being operated which include injuries from splinters or debris, sparks or sprays.

Guards must not create potential hazards and must be attached to the machine where possible. If guards cannot be attached to the machine, attach elsewhere.

The following is a list of machines that usually require point of operation guarding:

- Guillotine cutters
- Power presses
- Milling machines
- Power saws
- Jointers
- Portable power tools
- Forming rolls and colanders
- Conveyor belts

Machines that expose a worker to injury must be guarded. The guarding device must:

- Be in conformity with any appropriate standards.
- If specific standards are not available, the machine construction should prevent the operator from having any part of his/her body in the danger zone during the operating cycle

Special hand tools used for placing and removing material from point of operation areas must allow easy handling of the material without the operator placing a hand in the danger zone. Such tools must not replace guards required by this section.

• RESPONSIBILITY FOR MACHINE GUARDING

It is the responsibility of the machinery manufacturer and the end user to conduct a risk assessment to determine the guarding that is required to operate the machine safely and to ensure that the machine will comply with all applicable legislation. The manufacturer should consider any foreseeable use and misuse of the machinery, as well as point of operation and pinch or shear points for anyone who operates the equipment. However, it is impossible for the manufacturer to foresee any potential hazard with the equipment setup, operation and maintenance once the machine is set up within the production line or manufacturing environment of an organization. The responsibility lies with the end user to conduct a separate risk assessment on the initial installation, to recognise any potential hazards, and on a regular basis to ensure all machine guards and machine safety equipment are installed and are well maintained and working as required.

• CLASSES AND TYPES OF MACHINE GUARDS

There are **two classes** of machine guards, namely, transmission guards and point of operation guards.

There are **three types** of guards, namely:

✓ Fixed Guards - these do not move with each operation;
✓ Interlocking Guards - these guards prevent the controls of the machine from being used until the guard is moved into place; and
✓ Automatic Guards - these prevent the machine operator from coming into contact with the dangerous parts of a machine while it is moving.

• REQUIREMENTS OF MACHINE GUARDS

Machine guards should comply with the following requirements, namely:

✓ They should afford the maximum protection possible;
✓ The danger zone(s) should be blocked out during operation;
✓ The guards should be corrosion-free and fire-resistant;
✓ The guards should not produce splinters or pinch the operators;
✓ The guards should be fixtures of the machinery;
✓ The guards should not affect the efficiency of the machine; and
✓ They should be strong enough to withstand normal wear and tear.

• PREVENTATIVE MEASURES

Machine guards and safety mechanisms are designed to protect the hands and fingers of operators.

WARNING: Never remove machine guards or bypass safety mechanisms. Check lockout / tagout procedures before attempting to put your hands into machinery

Poorly maintained machinery, tools, sloppy work areas and cluttered pathways all contribute to hand injuries. Good hygiene includes hand washing or the use of a >60% alcohol-based hand sanitizer which helps to remove germs and dirt from hands. Clean hands are less vulnerable to infection and other skin problems such as contact dermatitis.

• REVIEW QUESTIONS

1. What are the basic categories of signs used in industry?
2. What are the main types of health and safety signs and what do the colours mean?
3. What other types of signs can be used in the workplace?
4. When are machine guards necessary?
5. What are some of the requirements for machine guards?

• REFERENCES

Anonymous (nd). Loose notes on safety, health and environment

Ccohs (n.d.). *(none)*. [online] Ccohs.ca. Available at: https://www.ccohs.ca/oshanswers/safety_haz/power_tools/saf_elec.html.

Graphic Products Staff, 2021. Five Ways To Meet Osha Requirements For Safety Signage, https://Www.Graphicproducts.Com/Articles/Five-Ways-To-Meet-Osha-Requirements-For-Safety-Signage/

Kumar, P., 2017. Significance of fire safety in pictorial manner. *Fire Engineer*, *42*(3), pp.13-16.

Nawi, M.N.M., Ibrahim, S.H., Affandi, R., Rosli, N.A. and Basri, F.M., 2016. Factor affecting safety performance construction industry. *International Review of Management and Marketing*, *6*(8S), pp.280-285.

OSHA (n.d.). *HAND and POWER TOOLS.* [online] Osha.gov. Available at: https://www.osha.gov/Publications/osha3080.html.

OSHA (n.d.). *Safety color codes - USDOL OSHA Public Web Site Search Results.* [online] Search.osha.gov. Available at: https://search.osha.gov/search?affiliate=usdoloshapublicwebsite&query=safety+color+codes.

OSHA (nd). Chapter 2. Methods of Machine Guarding. https://www.osha.gov/Publications/Mach_SafeGuard/chapt2.html. Accessed August 16, 2020

OSHA. 2021. Machine guarding. https://www.osha.gov/machine-guarding

OSHAcademy. 2021. Course 726- Machine guarding basics. https://www.oshatrain.org/courses/mods/726m1.html

Pathak, M.A. 2019. Study of Machinery Safety in Industries: A review.

Prince, Paul. J., 2017, January. Machine Risk Assessment and Reduction, An Ongoing Journey. In *ASSE Professional Development Conference and Exposition*. American Society of Safety Engineers.

Spellman, Frank. R. and Whiting, Nancy. E., 1999. *Machine Guarding Handbook: A Practical Guide to OSHA Compliance and Injury Prevention*. Government Institutes.

Stewart, Rebecca. Ann., 2020. Understanding how safety posters affect perception of safety culture using virtual reality.

CHAPTER 13

LIFTING EQUIPMENT, LADDERS, SCAFFOLDS AND CONVEYORS

• LIFTING EQUIPMENT

Using mechanical equipment to move and store materials increases the potential for injury to workers. Workers must be aware of both manual handling health and safety concerns and healthy and safe equipment operating techniques. Workers should avoid overloading equipment when moving materials mechanically. The weight, size and shape of the material being moved dictate the type of equipment used.

All material handling equipment has rated capacities that determine the maximum weight the equipment can safely handle and the conditions under which it can handle that weight. Employers must ensure that the equipment-rated capacity is displayed on each piece of equipment and is not exceeded except for load testing.

Although workers may be know how to use powered equipment, they must always take proper precautions especially when stacking and storing material.

• GENERAL PRINCIPLES

- ✓ Ensure that the equipment is fit for the intended purpose which implies that the equipment should be fitted with all the appropriate safety devices and they are all fully functioning and not overridden by anyone.
- ✓ All heavy lifting equipment used in the organization should be well maintained and subject to detailed examinations at least every 12 months. If lifting equipment is used to lift people, for example, using tripods and baskets, these will need to be examined at least every 6 months. The equipment should be visually inspected before every use. Furthermore, all heavy lifting equipment should be tested and certified before use.
- ✓ The load that is being lifting should not exceed the Working Load Limit (WLL). This is the maximum weight that the equipment can lift safely. The lifting equipment is only as strong as it weakest parts.
- ✓ Finally, the stability of the loads that require lifting needs to be tested before lifting. When lifting a heavy load ensure that the load is secured to the equipment correctly and securely. When lifting a load with loose parts, ensure the loose parts are secure before lifting. If this check is not carried out a potentially fatal accident or serious injuries could occur.

There are four general requirements for all lifting operations, namely:

- use strong, stable and suitable lifting equipment;
- the equipment should be positioned and installed correctly;
- the equipment should be visibly marked with the safe working load (SWL);
- lifting operations must be planned, supervised and performed in a safe manner by competent people.

• POWERED INDUSTRIAL TRUCK

When lifting items with a powered industrial truck, workers must do the following:

- ✓ Centre the load on the forks as close to the mast as possible to minimize the potential for the truck tipping or the load falling
- ✓ Avoid overloading a lift truck because it impairs control and causes tipping over
- ✓ Do not place extra weight on the rear of a counterbalanced forklift to allow an overload
- ✓ Adjust the load to the lowest position when traveling
- ✓ Follow the truck manufacturer's operational requirements
- ✓ Pile and cross-tier all stacked loads correctly when possible

• CRANES

Cranes are used for all sorts of purposes, mainly for lifting items from one place to another. For example, Tower cranes are an essential part of the construction process and are widely used to move materials around site, which can help resolve some of the health and safety problems arising from space constraints and workplace transport issues. The type of crane used will depend on a number of factors such as the size of the load, the location of the loading and unloading site, the type of load, the distance and height the load must be moved and the conditions under which the load must be moved. In order to choose the correct equipment for the task, the difference must be known between the different types of cranes. There are specific legal requirements to ensure that all cranes are installed, inspected, examined and maintained to ensure the risks relating to lifting are adequately controlled. Investigations into recent accidents have shown that enhanced standards of planning, management and execution of the crane erection and operation processes could have reduced the chance of death and injury.

Employers must permit only thoroughly trained and competent workers to operate cranes. Operators should know what they are lifting and what it weighs. For example, the rated capacity of mobile cranes varies with the length of the boom and the radius. When a crane has a telescoping boom it is extended and the radius increases. To reduce the severity of an injury, employers must take the following precautions:

- ✓ Ensure that cranes are erected and dismantled by competent people who have the necessary training and experience. Organizations should draw up written procedures for each type of crane and these procedures should be based on the instructions of manufacturers. These procedures should be available where the crane is being used and those involved in the work must be familiar with them
- ✓ Ensure that a thorough examination of the crane is undertaken after its erection by a competent person who is sufficiently independent and impartial and is not involved in the erection process
- ✓ Only allow competent people to operate the crane
- ✓ Ensure that the crane operator carries out pre-use checks at the start of each work shift to ensure that the crane has not suffered any damage or failure during the previous shift and is safe to be used
- ✓ Equip all cranes that have adjustable booms with boom angle indicators
- ✓ Provide cranes with telescoping booms with some means to determine boom lengths unless the load rating is independent of the boom length
- ✓ Post load rating charts in the cab of cab-operated cranes
- ✓ Require workers to always check the crane's load chart to ensure that the crane will not be overloaded by operating conditions
- ✓ Instruct workers to plan lifts before starting them to ensure they are safe
- ✓ Tell workers to take additional precautions and exercise extra care when operating around power lines
- ✓ Teach workers that outriggers on mobile cranes must rest on firm ground, on timbers, or be sufficiently cribbed to spread the weight of the crane and the load over a large enough area
- ✓ Direct workers to always keep hoisting chains and ropes free of kinks or twists and never wrapped around a load
- ✓ Train workers to attach loads to the load hook slings, fixtures and other devices that have the capacity to support the load on the hook
- ✓ Instruct workers to pad sharp edges of loads to prevent cutting slings
- ✓ Teach workers to maintain proper sling angles so that slings are not loaded in excess of their capacity
- ✓ Ensure that all cranes are inspected frequently by persons thoroughly familiar with the crane, the methods of inspecting the crane and what can make the crane unserviceable. Crane activity, the severity of use and environmental conditions should determine inspection schedules. In-service inspections are carried out by the crane operator, generally at weekly intervals, and records kept of these inspections
- ✓ Ensure the critical parts of the crane – such as crane operating mechanisms, hooks, air or hydraulic system components and other load carrying components – are inspected daily for any maladjustment, deterioration, leakage, deformation or other damage.
- ✓ A properly planned maintenance system is established and used. Competent people should undertake this maintenance at intervals specified by the manufacturer and records kept of the work completed including any parts that have been replaced. In general, the original manufacturers parts should be used. Where parts are sourced from suppliers other than the original manufacturer a competent engineer should assess that the parts selected meet the original manufacturers specification and are fit for purpose. Any parts replaced should be installed in accordance with the instructions of the manufacturer.

• LADDERS

Ladders are probably one of the most common tools that can be found and used in almost every industry. Falls from heights consistently rank high in terms of the frequency of incidents and accidents involving them. A third of all reported fall-from-height incidents involve ladders and stepladders. In the UK on average 14 deaths and 1,200 major injuries to workers occur each year. Many of these injuries are caused by inappropriate or incorrect use of the ladders. Where work at height is necessary the use of a ladder or stepladder must be justified in terms of whether it is the most suitable access equipment compared to other access equipment options. A risk assessment should be done and the following hierarchy of controls used, namely

1. to avoid work at height where possible;
2. then to prevent falls from height; and,

3. then to reduce the consequences of a fall.

When considering whether it could be appropriate to use a ladder or stepladder, the following factors must be considered, namely

- ✓ Use in one position for a maximum of 30 minutes;
- ✓ Use for 'light work' as they are not suitable for strenuous or heavy work. If the task involves a worker carrying more than 10 kg up the ladder or steps it will need to be justified by a detailed manual handling assessment;
- ✓ whether a handhold is available on the ladder or stepladder;
- ✓ whether it is possible to maintain three points of contact (hands and feet) at the working location. Where not possible, other measures will be needed to prevent a fall or reduce the consequences of one. On stepladders where a proper handhold is not practicable or possible a risk assessment will have to justify whether it is safe or not.
- ✓ whether the ladder or stepladder will be overloaded if the person and anything they are carrying will exceed the highest load stated on the ladder;
- ✓ whether the worker will overreach;
- ✓ whether the work to be done on the ladder or stepladder will impose a side loading.

The Occupational Health and Safety Act, 1993 General Safety Regulations Section 13A. Ladders stipulates:

1. *A worker shall ensure that every ladder is constructed of sound material and is suitable for the purpose for which it is used, and -*
   - a. *is fitted with non-skid devices at the bottom ends and hooks or similar devices at the upper ends of the stiles which shall ensure the stability of the ladder during normal use; or*
   - b. *is so lashed, held or secured whilst being used as to ensure the stability of the ladder under all conditions and at all times.*
2. *No employer shall use a ladder, or permit it to be used, if it -*
   - a. i. *has rungs fastened to the stiles only by means of nails, screws, spikes or in like manner; or*
     ii. *has rungs which have not been properly let into the stiles: Provided that in the case of welded ladder or ladders of which the rungs are bolted or riveted to the stiles, the rungs need not be let into the sides; or*
   - b. *has damaged stiles, or damaged or missing rungs.*
3. *No employer may permit that -*
   - a. *a ladder which is required to be leaned against an object for support be used which is longer than 9 m; and*
   - b. *except with the approval of an inspector, the reach of a ladder be extended by fastening together two or more ladders: Provided that the provisions of this sub-regulation shall not apply to extension of free-standing ladders.*
4. *In the case of wooden ladders the employer shall ensure that -*
   - a. *the ladders are constructed of straight grained wood, free from defects, and with the grain running in the length of the stiles and rungs; and*
   - b. *the ladders are not painted or covered in any manner, unless it has been established that there are no cracks or other inherent weaknesses: Provided that ladders may be treated with oil or covered with clear varnish or wood preservative.*
5. *When work is done from a ladder, the employer shall -*
   - a. *take special precautionary measures to prevent articles from falling off; and*
   - b. *provide suitable sheaths or receptacles in which hand tools shall be kept when not being used.*
6. *An employer shall ensure that a fixed ladder which exceeds 5 m in length and is attached to a vertical structure with an inclination to the horizontal level of 75 or more -*
   - a. *has its rungs at least 150 mm away from the structure to which the ladder is attached; and*
   - b. *is provided with a cage which -*
     - i. *extends from a point not exceeding 2.5 m from the lower level to a height of at least 900 mm above the top level served by the ladder; and*
     - ii. *shall afford firm support along its whole length for the back of the person climbing the ladder, and for which purpose no part of the cage shall be more than 700 mm away from the level of the rungs:*

Provided that the foregoing provisions of paragraph (b) shall not apply if platforms, which are spaced not more than 8 m apart and suitable for persons to rest on, are provided.

• BASIC HEALTH AND SAFETY RULES WHEN USING LADDERS

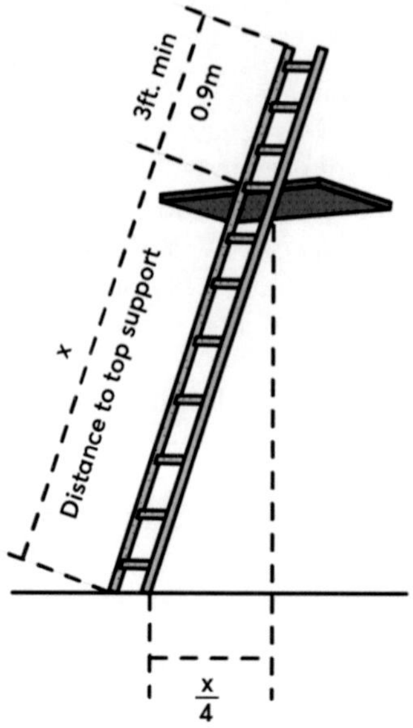

Placement of ladder against a wall
(Source: https://safesitehq.com/osha-ladder-safety/)

1. Always avoid bringing the ladder into contact with electricity or within 6 m horizontally of any overhead power lines;
2. Always use a non-conductive ladder or steps for any necessary live electrical work;
3. Avoid using ladders or stepladders where they could be struck by vehicles which can be prevented by protecting them with suitable barriers or visible plastic cones;
4. Always avoid using ladders or stepladders where they could be pushed over by other hazards such as doors or windows - secure doors (not fire exits) and windows where possible. If this is impractical, have a person standing guard at a doorway, or inform workers not to open windows until they are told to do so;
5. Do not use ladders or stepladders where pedestrians could walk under them or near them – use barriers, cones or, as a last resort, a person standing guard at the base;
6. Place the ladder so that its feet are a quarter of its length to the top support from the object it is resting against.
7. Unless the ladder is securely tied at the top it should always be held in position by another person whilst in use.
8. Wherever possible it should extend 1m above its support.
9. Ladders should be inspected at frequent intervals. Only use ladders or stepladders that have no visible defects. They should have a pre-use check each working day in accordance with the manufacturer's instructions. Ladders that are part of a scaffold system have to be inspected every seven days. Ladder and stepladder feet must be part of the pre-use check. Defective ladders should be repaired or replaced.
10. As paint conceals defects, use varnish or two coats of oil instead, to preserve wooden ladders.
11. Keep ladders clean because dirt hides defects while grease or oil cause people to slip.
12. Do not leave ladders lying on wet ground or exposed to the weather.
13. Ladders lying on floors may cause workers to trip and fall, or may be run over by vehicles and be damaged.
14. Do not use ladders horizontally as walkways, or scaffolding.
15. Tools and equipment should be hauled up by ropes.
16. Never leave a ladder where it may fall.
17. Never place a ladder in front of a doorway before taking adequate precautions.
18. Ladders should be equipped with safety feet. The feet should be in good repair and not loose, missing, splitting, excessively worn, secure and clean – the feet should be in contact with the ground.
19. Check ladder feet when moving from soft or dirty ground such as dug soil, loose sand/stone, a dirty workshop to a smooth, solid surface such as paving slabs, to ensure the foot material and not the dirt such as soil or embedded stones is making contact with the ground.
20. Store ladders in a cool place, either lying on their sides or hanging in a horizontal position from several wall brackets. Avoid warping wherever possible.
21. Use both hands when climbing up or down a ladder.

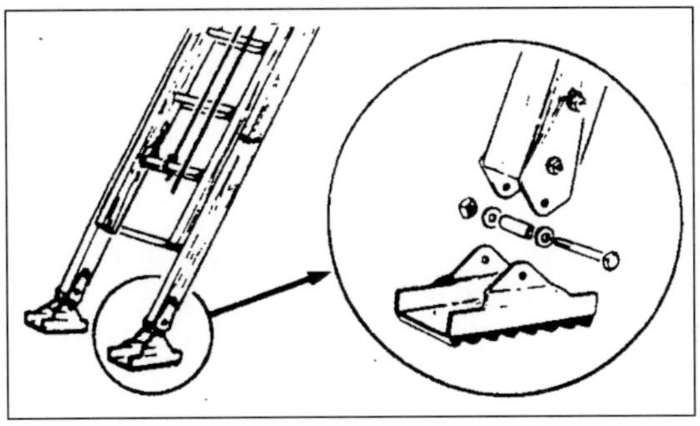

(Source: https://www.saladder.co.za/swivel-safety-feet/)

22. Only one person at a time should use a ladder.
23. Do not leave tools or other equipment on the top of a ladder. Remove these items when descending from a ladder.
24. Use the correct ladder for the job.

- LEGAL REQUIREMENTS WHEN USING LADDERS AND STEPLADDERS

Every ladder shall be –

(a) of good construction, sound material and adequate strength and suited to the purpose for which it is used.
(b) Fitted with non-skid devices at the bottom extremities of the stiles or hooks or similar devices at the top extremities; or
(c) So lashed, held or secured whilst in use as to ensure its stability under all conditions at all times.

No ladder shall be used which -

(a) has rungs fastened to the stiles only by means of nails, screws, spikes or the like, and except in the case of welded construction, rungs shall always be let into the stiles;
(b) has defective stiles or defective or missing rungs.

Except in the case of extension ladders and ladders which do not require to be leaned against an object for support, no ladder shall be used which is longer than 9m and no ladder shall have its reach extended by the tying together of two or more ladders.

Wooden ladders shall –

a. be constructed of straight grained wood
b. not be painted or covered in any way unless it has been established by the user that are no cracks or other inherent weaknesses; ladders may be oiled or covered with clear varnish or wood preservative.
c. When work is executed from a ladder, special precautions shall be taken to prevent hand tools, receptacles provided for that purpose.
d. Fixed ladders which are attached to vertical structures or fixed ladders which have an inclination to the horizontal of 75° or more and the vertical height of which exceeds 5m shall be provided with a cage extending from a point not more than 2,5m from the lower level to a height of at least 900mm above the top level served by the ladders.

The cage shall provide firm support along its whole length for the back of a person and no part of the cage shall be more than 700mm distant from the plane of the rungs.

The rungs shall be at least 150mm distant from the structure to which the ladder is secured.

- SCAFFOLDS

A scaffold is a temporary platform usually built to perform work activities on a higher elevation. To prevent a worker, material or tools and equipment from falling from the scaffold scaffolding health and safety procedures must be followed.

• HEALTH AND SAFETY REGULATIONS FOR SCAFFOLDS

Scaffold framework

Reg. 13D

1. An employer shall ensure that-

   ✓ Scaffold standards are properly propped against displacement and are secured vertically on firm foundations: Provided that putlog scaffolds shall incline slightly towards the structure;
   ✓ Steel scaffold standards with "heavy", "medium", "light" or "very light" platform loadings which shall not exceed 320, 240, 160 and 80kg/m, respectively, are spaced not more than 1,8m, 2m, 2,5m and 3m apart, respectively; and
   ✓ Ledgers are spaced vertically not more than 2,1m apart;
   ✓ Putlogs or transoms-
      ➢ which do not support a platform, are spaced at the same distances as the distances prescribed in paragraph b. in respect of scaffold standards;
      ➢ which support a platform, are spaced not more than 1,25m apart if the platform is constructed of solid timber boards;

2. No employer shall use a scaffold, or permit it to be used unless it-

   ✓ Is securely and effectively braced to ensure stability in all directions;
   ✓ Is secured at suitable vertical and horizontal distances to the structure to which work is being done, unless it is designed to be completely free-standing;
   ✓ Is inspected at least once a week and every time after bad weather by a person who has adequate experience in the erection and maintenance of scaffolds, and all findings are recorded in a register or report book.

3. No employer shall require or permit that-

   ✓ A scaffold is erected, altered or dismantled by or under the supervision of a person other than a person who has had the necessary training and experience of such work and who has been appointed by the employer in writing for this purpose.

Scaffold platforms

Reg. 13E

1. An employer shall ensure that-

   ✓ Every plank of a solid wooden scaffold platform is at least 225mm wide and 38mm thick;
   ✓ Every plank which forms part of a scaffold platform is supported at distances not exceeding 1,25m, and its ends are projected not less than 70mm and not more than 200mm beyond the last prop;
   ✓ Every plank of a scaffold platform is firmly secured to prevent its displacement; and
   ✓ Every platform is so constructed as to prevent materials and tools from falling through.

2. An employer shall ensure that every scaffold platform-

   ✓ With "heavy", "medium", "light", or "very light" platform loadings as referred to in regulation 13D (1) (b) (i) is not less than 1 125mm and not more than 1 380mm, not less than 1 125mm and not more than 1 150mm, not less than 900mm and not more than 1 150mm, and not less than 675mm and not more than 1 150mm, respectively, wide: Provided that where a platform is used only as a gangway, a platform width of 450mm shall be sufficient;
   ✓ Which is more than 2 m above the ground is on all sides, except the one facing the structure, provided with-
      ➢ substantial guard rails of at least 900mm and not exceeding 1 000mm in height; and
      ➢ toe-boards which are at least 150mm high from the level of the scaffold platform and so affixed that no open space exists between the toe-boards and the scaffold platform; Provided that the toe-boards are constructed of timber, they shall be at least 25mm thick;

Example of toe-board
(Source: https://civilengineersforum.com/wp-content/uploads/2014/12/scaffolding-safety.png)

- ✓ Is not more than 75mm from the structure: Provided that where workmen must sit to work, this distance may be increased to not more than 300mm; and
- ✓ Is kept free of waste, projecting nails or any other obstructions, and is kept in a non-slip state.
- ✓ No employer shall require or permit that a working platform which is higher than 600mm be supported on a scaffold platform, and shall provide an additional guard rail of a least 900mm and not exceeding 1 000mm in height above every such working platform.
- ✓ An employer shall ensure that convenient and safe access is provided to every scaffold platform, and where the access is a ladder, the ladder shall project at least 900mm beyond the top of the platform.

Note: To comply with Reg. 13D (3) (b) an employer must appoint in writing a supervisor, who has had the necessary training and experience, to supervise the erection, maintenance, dismantling and inspection of all scaffolding.

To comply with Reg. 13D (2) (d) a scaffold inspection register must be kept.

• TYPICAL COMPONENTS OF A SCAFFOLD

Depending on the scaffolding system being used a scaffold typically comprises of the following components, namely

1. Sole plate which is an essential component which spreads the weight of the scaffold over a greater area. While usually made of timber they may also be made from metal or concrete.
2. Adjustable base plates from which different height-adjustable base plates can be selected that come with strong and self-cleaning round threads to adjust to the ground. The can have colour and notch markings to render safeguard against over-winding.
3. Standards which are also known as uprights and are perpendicular tubes that shift the entire weight of the structure to the ground where they lean on a square base plate to distribute the weight.
4. Ledgers which are flat tubes that join between the standards.
5. Transoms which lean on the ledgers at right angles. Major transoms are positioned next to the standards; they support the standards that are in place and give support for the boards. To render additional support for the boards, intermediate transoms are placed between the main transoms.
6. Diagonal bracing which further support the basic structure comprising of vertical standards and ledgers. Additionally, their high connection standards assist special structures.
7. Decking which are also referred as planks and available, for example, in aluminium, aluminium frame with plywood board and hot-dip galvanised steel.
8. Guard rails (Hand rails and mid-rails) height approximately 1 meter.
9. Toe boards which are placed between vertical standards and are obtainable in aluminium, steel or wood. The toe board of steel lowers the fire hazard and also lasts long. Because of its design, there are no openings or gaps between the deck and the toe board. They should be minimum 150mm in height.
10. Ladders which should be secured at three locations within the scaffold.

An example of a basic putlog scaffold showing the various components:

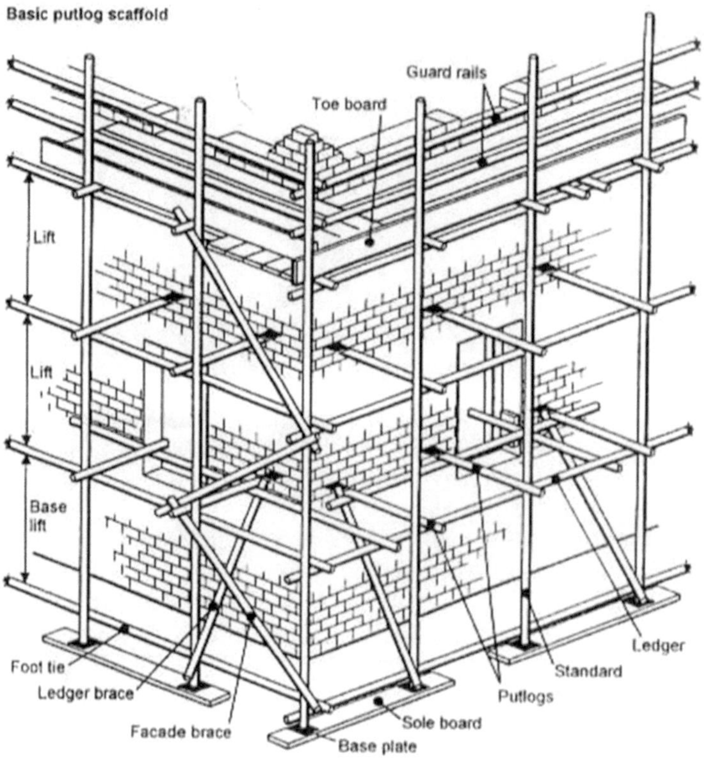

Example of basic putlog scaffold
(Source: https://scaffoldpole.com/putlog/)

• SCAFFOLDING HAZARDS

Poorly erected or maintained scaffolds can fail and collapse, and usually with catastrophic consequences. Workers could sustain severe injuries and property could sustain severe damage. Where a scaffold has inadequate foundations, tying or bracing or is overloaded, it can collapse, endangering workers and the public. Where scaffold boards or guard-rails are missing, workers can suffer severe injuries due to falls. Workers will be at risk where a healthy and safe system of work is not in place to protect them from falls.

Apart from collapse, the principal hazards for workers on scaffolds are, namely

- ✓ Falls from elevation – caused by slipping, unsafe access, and the lack of fall protection
- ✓ Struck by falling materials, tools and rubble or debris
- ✓ Electrocution from, for example overhead power lines
- ✓ Unsecured ladder within the scaffold slipping.
- ✓ Use of unsuitable, damaged and faulty materials in building the scaffold.
- ✓ Inadequately supported scaffold boards or platforms.
- ✓ Omission and/or removal of guard rails and toe boards
- ✓ Not being properly braced.
- ✓ Overloading of platform and board.

Falls may occur:

- ✓ While climbing on or off the scaffold
- ✓ When tripping over obstacles on the scaffold platform
- ✓ While working on unguarded scaffold platforms
- ✓ When slipping on the surface of the platform or boards
- ✓ When scaffold platforms or planks fail
- ✓ When worker is not properly tied off
- ✓ When incorrectly dismantling the scaffold

Fall Protection Required For Specific Scaffold Types			
Type of Scaffold	Guardrails Required	Personal Fall Arrest System Required	Both Required
Aerial Lifts		✔	
Boatswains' Chair		✔	
Catenary Scaffold		✔	
Chicken Ladder*	✔	✔	
Float Scaffold		✔	
Ladder Jack Scaffold		✔	
Needle Beam Scaffold		✔	
Self-contained Scaffold			✔
Single-point and Two-point Suspension Scaffolds			✔
Supported Scaffold	✔	✔	
All Other Scaffolds	✔	✔	
*Chicken Ladders require guardrails or personal fall arrest system or a grab-line system			

(Source: https://www.constructionsafety.co.za/special-areas/heights-scaffolding/)

- FALL PREVENTION MEASURES FOR SCAFFOLDS

The following precautions will prevent falls from scaffolds, namely:

✓ Keep workplace free of obstacles and clutter.
✓ Keep surface of platform and boards clean and dry.
✓ Avoid running cables or cords on the walkways.
✓ Use ladders in the scaffold correctly.
✓ Ensure scaffolds are erected by a competent person and follow a tagging system.
✓ Wear proper footwear.
✓ Maintain adequate illumination where scaffold is used indoors or in a poorly lit area.
✓ Use a proper safety harness if working at height (more than 1.8 meters) and always be tied off correctly.

Examples of harnesses and being tied off (Source: Author)

Examples of fall protection on scaffolds
(Source: https://safetyaspectsllc.com/onsite-safety-training/scaffolding-fall-protection/)

Stability

A scaffold is a temporary structure that is subjected to a wide range of loading during erection, use and dismantling. It should support its own dead load and live loads from construction materials, workers and tools, shock loads from material placement and wind loads. Scaffolds should be used as means of access to elevated areas of work and as working platforms. They are not support work. Where failures occur, large areas of scaffolding can collapse quite suddenly. Scaffolds can collapse because of poor construction or misuse, leading to them being loaded beyond their safe capacity to support the load. Common faults are poor foundations, inadequate tying and bracing, overloading and the removal of ties and bracing.

Scaffold stability depends on carefully following the scaffold system manufacturer's instructions and the provisions of health and safety codes and other equivalent standards.

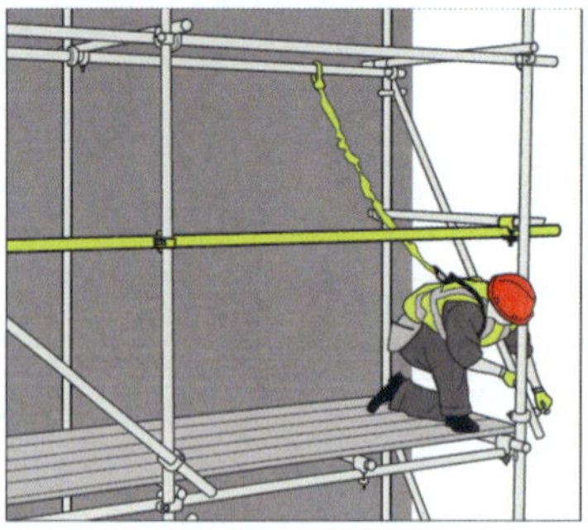

(Source: https://worksafe.govt.nz/topic-and-industry/working-at-height/scaffolding-in-new-zealand/)

In particular, the following issues should be addressed:

- ✓ The foundations should be adequate
- ✓ The scaffold should be tied to the permanent structure or to buttresses
- ✓ The scaffold should be braced
- ✓ The scaffold should not be overloaded
- ✓ Lock the wheels of mobile scaffolds

• SCAFFOLD DESIGN RATIO

The following recommended scaffolding ratio should be applied in the design of scaffolding, namely:

- For mobile scaffold: 1:3 (the height is three times the base)
- For tower scaffold: 1:4 (the height is four times the base)

• LOADING OF THE SCAFFOLD

Loading Bays

The weights of pallets of building materials such as blocks and bricks are usually in excess of the recommended load ratings of the scaffold system manufacturers. A loading bay will therefore be required where it is necessary to lift pallets of heavy materials onto a scaffold. The provision of properly constructed loading bays can avoid the excessive loading of access scaffolds and the obstruction of gangways that can otherwise occur. Loading bays are traditionally designed for a 'blanket' loading of 1 tonne per square metre (10 kN/m^2), plus an allowance of 25 per cent for light mechanical loading of materials.

Example of loading bay
(Source: https://scaffoldinghire.co.za/scaffolding-hire-south-africa/loading-bay-scaffold/)

Loading bays should be diagonally braced on all four sides or braced in compliance with the scaffold system manufacturer's recommendations. Where the internal facade bracing hinders access onto the scaffold from the loading bay the brace may be placed on the main scaffold adjacent to the loading bay or in accordance with the system manufacturer's recommendations. Issues which require consideration include:

- ✓ Standard transoms and at standard spacing and timber decking at standard spans are not usually adequate to carry the higher loadings in a loading bay. Certain systems of scaffold loading bays incorporate special load bearing transoms, often at reduced spacing.
- ✓ Where load bearing transoms are directly connected to the outside face of a scaffold, the capacity of the standards to support the combined loads imposed by the working platforms and the load bearing transoms should be assessed.
- ✓ Plan bracing should be installed from the outside corner of a loading bay to the main access scaffold and the main scaffold should be tied to the building with supplementary ties opposite these braces at intervals not exceeding 3 metres
- ✓ Where guard rails must be removed temporarily to facilitate loading, effective compensatory measures to prevent falls should be provided.

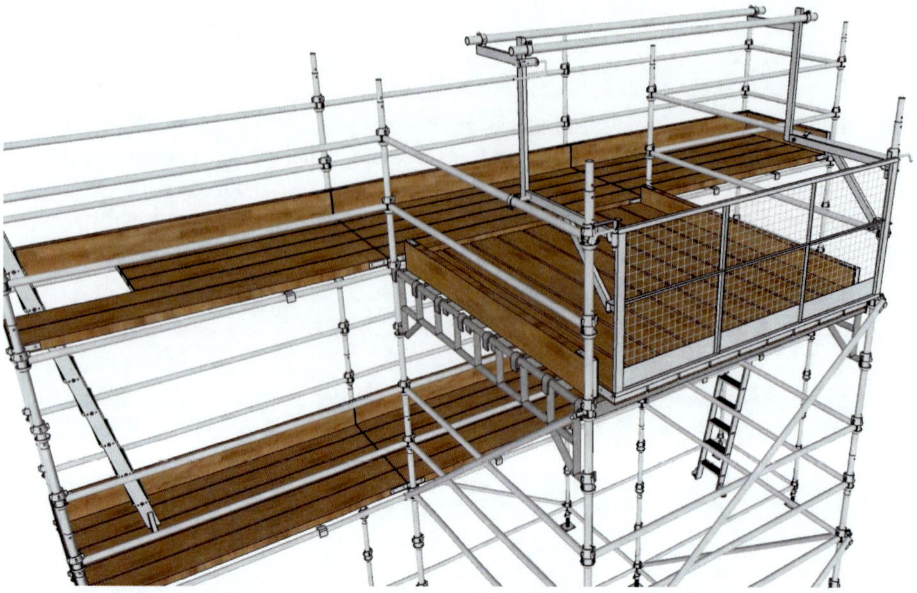

Design of loading bay
(Source: https://accesspoint.org.uk/technical-loading-bay-design/)

These measures may include movable guardrails or panels, hand holds, safety belts or harnesses affording an equivalent standard of protection as guard-rails. Temporarily unguarded openings or edges should not be left unattended and guard-rails should be replaced as soon as practicable.

• SIGNS FOR SCAFFOLDS

Easily comprehensible signs showing the safe working load, for each working lift, should be placed on scaffolds and loading bays.

Example of safe loading sign
(Source: https://www.pdsigns.ie/product/safety-construction-hazard-max-load-for-this-loading-bay-sign/)

• LOADING CHARTS

Supervisors and equipment operators such as crane and telescopic fork truck drivers, should be provided with easily comprehensible loading charts showing the weights of the typical materials used on the site. These should include, for example, the weights of the pallets of bricks and blocks, scaffold boards and standards, and mortar skips. In this way users will be able to estimate the load they are placing on the scaffold and ensure that it is less than the safe working load indicated on the signs.

BOOM LENGTH — Maximum Load Chart in pounds (lbs) with fully extended outrigger

OPERATING RADIUS (FT)	27 FT		34 FT		43 FT		52 FT		61 FT		70 FT	
	LOADED BOOM ANGLE (DEG)	LOAD RATING (LB)	LOADED BOOM ANGLE (DEG)	LOAD RATING (LB)	LOADED BOOM ANGLE (DEG)	LOAD RATING (LB)	LOADED BOOM ANGLE (DEG)	LOAD RATING (LB)	LOADED BOOM ANGLE (DEG)	LOAD RATING (LB)	LOADED BOOM ANGLE (DEG)	LOAD RATING (LB)
5	77	34,000*										
10	66	21,100*	71	17,100*	75	16,000*	78	15,700*				
15	54	15,100*	62	14,000*	68	12,100*	72	11,100*	75	10,800*	77	9,600*
20	39	11,100*	51	10,100*	61	9,100*	66	8,600*	71	8,200*	73	7,300*
25	17	7,900*	40	7,700*	53	7,100*	61	6,900*	66	6,600*	69	5,900*
30			23	6,500*	44	6,100*	54	5,600*	60	5,300*	64	4,900*
35					33	4,800*	47	4,700*	54	4,600*	60	4,150*
40					16	3,500*	38	4,100*	48	3,900*	55	3,550*
45							27	3,250*	41	3,200*	49	3,050*
50							9	2,950*	33	2,800*	44	2,650*
55									23	2,500*	37	2,350*
60											29	1,900*
65											19	1,700*

NOTE: STRUCTURAL STRENGTH RATINGS IN CHART ARE INDICATED WITH AN ASTERISK *

STOWED JIB DEDUCTIONS (POUNDS)

27 FT	34 FT	43 FT	52 FT	61 FT	70 FT
450	360	260	230	200	175

Example of loading chart for cranes
(Source: https://www.cranehunter.com/how-to-read-crane-load-chart)

• CONVEYORS

Conveyors move large amounts of materials quickly and safely. They allow workers to reduce the amount of materials handled manually. As a result, they increase work capacity and production output. Decreasing manual material handling also lessens the chance of injury to the back and hands of workers. Conveyors are safe when used correctly, but they can be dangerous, and can cause severe disablement and even fatalities, if workers fail to follow health and safety procedures when working on or around them. Materials should be placed on the conveyor so that they will be moved safely. When removing material off conveyors, workers should remain alert and in particular safeguard their hands as the moving material can create pinch points. Generally, a conveyor is simply a machine that has a pulley at each end. A belt is fitted over the pulleys, and a motor at one end of the conveyor "pulls" the belt around in a loop. The belt "runs" smoothly along, on the troughing idlers.

The material being transported may fall onto the belt from a feed chute or could be placed on it manually. It travels with the belt until it reaches the end of the belt. It then falls off the conveyor into the discharge chute or is removed manually.

In general, conveyor guards must be so designed, constructed and used that they will:

- ✓ provide positive protection preventing access to all dangerous areas during operation
- ✓ guards shall be permanently fixed or
- ✓ a physical barrier securely fixed in position by means of fasteners or other suitable devices, and which ensures that the guard cannot be altered or detached without the aid of a tool or key, or
- ✓ if a fixed barrier is not reasonably practicable, and access to dangerous areas is required during operation, an interlocked physical barrier may be used.
- ✓ If access to parts protected by guards is required for maintenance or other reasons, the guards or parts of them may be hinged or otherwise moveable, and wherever practicable they must be interlocked with the machinery.

• TYPICAL HAZARDS ASSOCIATED WITH CONVEYORS

There are many hazards associated with working at or near a conveyor, including:

- Rotating parts or pinch points can drag in, crush or entangle
- Confinement or assembly areas (the area between a fixed object and a moving one) can shear or crush
- Parts that slide or reciprocate (press down) can crush or shear
- Items can break or be ejected (thrown from) the conveyor system
- Items can fall off the conveyor
- Electrical, fire or explosion hazards

• SAFE OPERATING PROCEDURES FOR CONVEYORS

The following operating procedures apply to all conveyor installations:

- ✓ Ensure that all START/STOP and emergency controls are clearly marked.
- ✓ Keep the area clean and tidy. Good housekeeping eliminates hazards, i.e. tripping, slipping and falling.
- ✓ Lockout, isolate and danger tag or use some other control measure before working on a bogged or overloaded conveyor
- ✓ Isolate and danger tag the power source before working on a bogged or overloaded conveyor.
- ✓ Ensure that persons working near a conveyor are aware of the location of STOP/START and emergency controls.
- ✓ Wear appropriate clothing - avoid loose clothing near moving conveyors.
- ✓ DON'T walk under a moving conveyor unless the access is guarded against spillage.
- ✓ DON'T clean belts, pulleys, drum, and troughing or return idlers while a conveyor is moving.
- ✓ DON'T ride on a moving conveyor.
- ✓ DON'T repair or maintain a conveyor while in motion or in use.

Should repairs, adjustments, or belt alignment be required on a moving conveyor belt without normal guarding, it is important that the work is carried out by an authorised competent person with an attendant at the emergency stop station and that the appropriate safe work procedure is in place. When working near a conveyor:

- ✓ Wear hard hat and safety shoes.
- ✓ Tie back and tuck in long hair.
- ✓ Know the location of the emergency "shut-off" devices and how to use them.
- ✓ Do not wear loose clothing or jewellery.
- ✓ Do not climb on the conveyors.

• ADDITIONAL PRECAUTIONS FOR DIFFERENT TYPES OF CONVEYORS:

Gravity conveyors have rollers, wheels or chutes where objects move by gravity or momentum only.

- Guard pinch points on rollers and wheels and between the conveyor and receiving table.
- Provide adequate guardrails along sides to prevent any objects from falling off.
- Provide retarders such as friction areas if heavy objects are being conveyed.
- Ensure there are warning devices near the receiving areas if packages moving on the conveyors are not visible.
- Ensure draft checks such as fire doors are installed where conveyors pass through fire walls or floors.

"Powered" or "power" conveyors have belts, live rollers, slats, or buckets.

- Be positioned to prevent being hit by objects moving down the conveyor.
- Ensure that the conveyor system can when at the operating controls.
- Ensure that guards are in place for all moving parts of the drive system and in all zones where hazards such as in-running nip, drawing-in, trapping and crushing, friction burns or abrasion are present which includes above, sides, and below the conveyor.
- Guard all pinch points between the conveyor system and fixed objects.
- Locate guardrails around low level conveyors and areas where conveyors pass through the floor and/or ceiling.
- Locate emergency stop cut-off switches near the operator and along the length of the conveyor at approximately 30 metres apart or closer if possible.
- Ground belts on belt conveyors to prevent static build-up.

When working with **aerial conveyors**:

- Ensure that guards and protection plates are in place to protect workers working below from any possible falling objects.

When working with **bucket conveyors**:

- Ensure that both vertical and horizontal bucket conveyors are totally enclosed.

When working with **pneumatic conveyors**:

- Be familiar with control devices and release valves to cut off the air flow in the event of blockage.
- Shield joints and access points to prevent material from being thrown in the event of a gasket failing.
- Ensure that screening is in place at the suction end to prevent large objects from being sucked in.

When working with **portable conveyors**:

- Use only weatherproof electrical components.
- Ensure that power cables are located where they will not be walked on or run over.
- Ensure that sideboards are high enough to prevent any large items from falling and any smaller items from being thrown by the wind.
- Chock the wheels on trucks and rail cars that are being loaded or emptied by portable conveyors.
- Do not exceed the rated load capacity of the conveyor.

When working with **movable conveyors**:

- Install barrier guards, guardrails and/or mark the ground to indicate the operating area of the conveyor.

Example of a safety checklist for a belt conveyor:

Inspection Prompts – Equipment	
1.	Is the belt in a good condition?
2.	Are all the idler rollers turning and not seized?
3.	Is the housekeeping maintained around the conveyor?
4.	Are the skirts at the loading point in good condition?
5.	Is the walkway along the conveyor in a good condition?
6.	Is there adequate lighting along the conveyor?
7.	Is the full length of the conveyor protected with a trip wire?

Inspection Prompts – Equipment	
8.	Is the tension of the trip wire adequate?
9.	Is there adequate firefighting equipment along the length of the conveyor?
10.	Is the drive mechanism adequately guarded?
11.	Is the conveyor fitted with a break safe device on the incline?
12.	Is the belt running true?
13.	Are there adequate warning signs posted?
Inspection Prompts – Usage	
a.	Is the conveyor locked out as per the lock-out procedure before work commences?
b.	Are there any unauthorized persons working on or near the conveyor?
c.	Do workers working on or near the conveyor wear adequate PPE?
d.	Is the conveyor loaded correctly?
e.	Are radiation sources like mass measurement devices adequately identified?
f.	Are magnets only cleaned when the conveyor is off and locked out?
g.	Are walkway and stairs provided where it is necessary to cross a conveyor?
h.	Is there adequate guarding in place when workers have to walk underneath a conveyor?

- REVIEW QUESTIONS

1. What are basic health and safety rules when using ladders?
2. What are some of the precautions to be taken when working with cranes?
3. How can falls from scaffolds be prevented?
4. What are the design ratios for a mobile and tower scaffold respectively?
5. What are some of the hazards involved with using or working near conveyors?

• REFERENCES

Anonymous (nd). Loose notes on safety, health and environment management

Materials Handling and Storage (2002). *Materials handling and storing*. Washington, DC: U.S. Dept. of Labor, Occupational Safety and Health Administration.

Canadian Center for Occupational Health and Safety (n.d.). Available at: https://www.ccohs.ca/oshanswers/safety_haz/ladders/step.html.

Canadian Center for Occupational Health and Safety (n.d.). Available at: https://www.ccohs.ca/oshanswers/safety_haz/conveyor_safety.html

Department of Labour (1993). Occupational Health and Safety Act No: 85 of 1993.

HSA (2008). [online] Australianscaffolds.com.au. Available at: http://australianscaffolds.com.au/documents/Scaffolding_COP.pdf.

HSA (n.d.). [online] Besmart.ie. Available at: https://www.besmart.ie/fs/doc/Construction/Construction_Documents/scaffolds_PoP_v11.pdf.

HSE. Nd. HSE safety alert on the use of tower cranes. https://www.mantiscranes.ie/wp-content/uploads/2017/01/CPA-TCIG-HSE-Safety-Alert-Use-of-Tower-Cranes.pdf

HSE Books (2002). Ergonomics evaluation into the safety of stepladders. Literature and standards review Phase 1 CRR418, ISBN 07176 2302 5

HSE Books (2002). Ergonomics evaluation into the safety of stepladders. User profile and dynamic testing - Phase 2 CRR423, ISBN 0 7176 2315 1

HSE Books (2004). Evaluating the performance and effectiveness of ladder stability devices RR205, ISBN 07176 2872 1

HSE Books (2005). A toolbox talk on leaning ladder and stepladder safety Leaflet INDG403, ISBN 07176 6106 7

HSE Books (2005). Top tips for ladder safety Pocket card INDG405, ISBN 0 7176 6127 X

Kumar, Praveen., 2017. Significance of fire safety in pictorial manner. *Fire Engineer*, *42*(3), pp.13-16.

Materials Handling and Storage (2002). Materials handling and storing. Washington, DC: U.S. Dept. of Labor, Occupational Safety and Health Administration.

Nawi, Mohd. Nasrum. Mohd., Ibrahim, Sitti. Halipah., Affandi, Rohaida., Rosli, Nov. Azalina. and Basri, Fazlin. Mohamad., 2016. Factor affecting safety performance construction industry. *International Review of Management and Marketing*, *6*(8S), pp.280-285.

OSHA. 2002. *Materials Handling and Storage | Occupational Safety and Health Administration*. [online] Osha.gov. Available at: https://www.osha.gov/Publications/OSHA2236/osha2236.html.

OSHA (n.d.). *Portable Ladder Safety*. [online] Osha.gov. Available at: https://www.osha.gov/Publications/portable_ladder_qc.html.

Also https://www.cpa.uk.net/tower-crane-interest-group-tcig-publications/

Prince, Paul. J., 2017, January. Machine Risk Assessment and Reduction, An Ongoing Journey. In *ASSE Professional Development Conference and Exposition*. American Society of Safety Engineers.

Safe at Work California. Nd. Conveyors - a Safety Message. https://www.safeatworkca.com/safety-articles/conveying-a-safety-message/

Spellman, Frank. R. and Whiting, Nancy. E., 1999. *Machine Guarding Handbook: A Practical Guide to OSHA Compliance and Injury Prevention*. Government Institutes.

Stewart, R.A., 2020. Understanding how safety posters affect perception of safety culture using virtual reality.

CHAPTER 14

WORKING WITH EQUIPMENT UNDER PRESSURE

Gases under pressure in cylinders can release large volumes of gas very quickly that can result in health and/or fire hazards in the workplace depending on the properties of the particular type of gas being used. Gas can be released deliberately into the air by opening the cylinder valve, or accidentally from a broken or leaking valve. Both cases must be dealt with correctly to prevent negative consequences to both workers and property.

If gas cylinders are damaged, they can explode with great force and cause significant injury, damage and in the worst cases be fatal. Exploding gas cylinders have gone right through concrete and brick walls. These incidents are caused when an uncapped or unsecured cylinder has been incorrectly handled, knocked over or the valve breaks. Consequences of exposure to these types of incidents can be cryogenic burns or freezing or frostbite injuries from the release of refrigerated liquefied gases because these gases can be extremely cold.

Compressed gases are typically stored in heavy-walled metal cylinders designed, produced and tested in accordance with very stringent specifications. These cylinders are made in many different sizes, shapes and colours. They are permanently marked, usually on the shoulder or the top surface of the neck of the cylinder. It is advisable that cylinders be retested or inspected every five or ten years. The date of each test must be stamped on the cylinder.

Where possible, process changes or modifications must be considered that reduce exposure to the hazards presented by a particular gas. Where many cylinders of the same gas may be used in different areas of a workplace, installing fixed pipelines can often reduce the hazard and risk of exposure. These pipelines are connected to a central gas supply in a safe area which is often located externally from the work area . The need is then reduced for many sets of portable equipment supplied through flexible hoses. In the same manner, ordering cylinders equipped with flow limiting restrictors can also minimize the hazards of the sudden and unexpected failure of a process gas line. Selecting the least hazardous materials and process must always be considered so that the desired work activity can be executed effectively and safely.

• COMPRESSED GAS CYLINDERS

Due to the nature of gas cylinders and the severity of the consequences of exposure to the associated hazards, special storage and handling precautions are absolutely necessary. These hazards include oxygen displacement, explosion hazards, toxic effect of some gases, as well as the physical hazards of a ruptured cylinder. It is estimated that there are almost 200 different types of materials in gas cylinders that include, for example, atmospheric gases, fuel gases, refrigerant gases, poison gases and miscellaneous gases. Compressed gases are usually divided into six basic categories, with some gases possibly falling into more than one classification. The categories are as follows:

- ✓ Flammable Gases
- ✓ Oxygen and Oxidizing Gases
- ✓ Acid and Alkaline Gases
- ✓ Highly Toxic Gases
- ✓ Cryogenic Liquefied Gases
- ✓ Inert Gases

• GAS CYLINDER VALVES AND CONNECTIONS

Compressed gas cylinders must be connected only to regulators and equipment designed for the particular gas that the cylinder contains. A number of different standard cylinder valve outlets are available for different classes of gas. These standard connections prevent the valve connection for a flammable gas from fitting the connections for an incompatible gas such as an oxidizing gas.

Most compressed gas cylinders have valve caps or some other method of protecting the valve from damage during handling and transportation. A dust cap may be placed over the valve outlet itself to help keep it clean.

• GAS CYLINDER SAFETY-RELIEF DEVICES

Most cylinders have one or more safety-relief devices specifically designed to prevent rupture of the cylinder should the internal pressure build up to exceed the design limits of the cylinder. This pressure can become hazardly high if a cylinder is exposed to fire or heat including high storage temperatures.

There are three types of safety-relief devices that are used to relieve excessive gas pressures in a different way, namely

- **Safety- or Pressure-Relief Valves** which are normally a part of the cylinder. They are normally held closed by a spring. The force holding the valve closed is set according to the particular type of gas in the cylinder. The valve opens when the cylinder pressure exceeds the set health and safety limit. Gas is released until the cylinder pressure drops back to the set limit. The valve then closes and retains the remaining gas in the cylinder.
- **Rupture Discs,** also known as frangible or bursting discs, are usually made from metal. They burst or rupture at a certain pressure releasing the gas in the cylinder. The bursting pressure is designed so that the disc ruptures before the cylinder test pressure is reached. Because these disks cannot be reclosed the entire contents of the cylinder are released.
- **Fusible Plugs,** also called fuse or melt plugs, are used where heat could initiate an explosive chemical reaction. The fusible plug releases the gas before the hazardous reaction can begin. Fusible plugs are made of metals that melt at low temperatures. For example, acetylene cylinders have a fusible plug which melts at about 100°C which is safely below the temperature at which hazardous polymerization may occur.

Not all compressed gas cylinders have health and safety devices. Some gases are so toxic that their release through such a device would be hazardous. Cylinders for these gases are built to withstand higher pressures than normal cylinders. When these "toxic gas" cylinders are involved in a fire the entire area must be evacuated.

• IDENTIFICATION OF CONTENTS OF COMPRESSED GAS CYLINDERS

The contents of any compressed gas cylinder should be identified clearly so that they can be easily, quickly, and completely established by any worker. The content can be established in several ways such as:

- ✓ A durable label should be provided that cannot be removed from the compressed gas cylinder.
- ✓ No compressed gas cylinder should be accepted for use that does not identify its contents legibly by name.
- ✓ Colour-coding is not a reliable means of identification as the colours of cylinder vary from supplier to supplier, and labels on caps have no value because many caps are interchangeable.
- ✓ Tags should be attached to the gas cylinders on which the names of the users and dates of use can be entered.
- ✓ If the labelling on the gas cylinder becomes unclear or defaced so that the contents cannot be identified, the cylinder should be marked "contents unknown" and the manufacturer contacted regarding appropriate procedures.

• RECEIVING GAS CYLINDERS

All cylinders being delivered or brought into the plant or workspace should be inspected before storing to ensure that they are undamaged and properly labelled. Delivery of any defective cylinders should be refused. All cylinders must be free of corrosion, leakage, pitting, dents or gouges. All regulators, pressure relief valves and cylinder connections must be thoroughly checked. Cylinders should not be giving off odours, visible fumes or hissing sounds. When the cylinder was last tested should be within the required time of usually 5 or 10 years. However, some containers may need to tested within 3 years or within as long as 12 years. All cylinder labels must be intact and match other identifying markings on the cylinder. Gases that cannot be clearly identified should never be used.

By referring to compressed gases by the name on the supplier label reduces confusion, promotes recognition of the hazards involved and precautions to take, and can prevent accidental use of the wrong gas. Valve caps must always be left securely in place until the cylinder is to be used. The cylinder valve must be inspected by looking through the ports in the valve cap. Dirty, rusted or otherwise damaged valves and fixtures must not be accepted.

• TRANSPORTATION OF GAS CYLINDERS

Given the serious consequences of exposure to the hazards associated with gas cylinders, special precautions must be taken when transporting them and include the following, namely:

- ✓ Cylinders transported by truck must be fastened securely in an upright position so that they will not fall or strike each other.
- ✓ Cylinders should not be transported without their safety caps or other valve protection in place. The cap should be screwed all the way down on the neck ring of the cylinder and should fit securely. Do not lift cylinders by the cap because the cap is for valve protection only.

- ✓ Cylinders should not be transported with the regulator attached to the cylinder.
- ✓ Always use a cylinder cart to move compressed gas cylinders. Refrain from sliding, dragging or rolling cylinders on edge.
- ✓ Only one cylinder should be handled (moved) at a time.
- ✓ Avoid rolling cylinders on their sides or dragging or sliding them as they can be damaged.
- ✓ Never lift cylinders with magnets or chain or wire rope slings.
- ✓ Transport cylinders on specially built hand carts or trolleys or other devices designed for this.
- ✓ Ensure all transport devices have some way of securing cylinders to prevent them from falling.

• GENERAL MANAGEMENT AND STORAGE OF COMPRESSED GAS CYLINDERS

The following considerations are necessary when handling and storing gas cylinders, namely:

- ✓ Cylinders should not be allowed to drop nor be struck violently.
- ✓ Cylinders should be properly secured at all times whether attached to a wall, cylinder truck, cylinder rack, or post.
- ✓ Liquefied flammable gas cylinders should be stored in an upright position or such that the pressure relief valve is in direct communication with the vapour space of the cylinder.
- ✓ Caps used for valve protection should be kept on the cylinders at all times except when the cylinder is actually being used or charged.
- ✓ Cylinders should not be used for rolling, supports, or any purpose other than the transportation and supply of gas.
- ✓ Cylinders should be stored in a well-ventilated area away from flames, sparks, or any source of heat or ignition. Keep cylinders away from electrical circuits.
- ✓ Cylinders should not be exposed to an open flame or to any temperature above 50 degrees C.
- ✓ Oxygen cylinders (empty or full) in storage should be separated from fuel-gas cylinders and combustible materials by a minimum distance of 7m or by a barrier or firewall at least 1.5m high having a fire-resistance rating of at least one-half hour.
- ✓ Flammable gas cylinders should not be stored with oxygen or nitrous oxide cylinders or adjacent to oxygen charging facilities.
- ✓ Full and empty cylinders of all gases should be stored separately and identified by signs to prevent confusion and away from processing and handling areas, and from incompatible materials.
- ✓ Store compressed gas cylinders separately to minimize personal injury and damage in case of fires, spills or leaks.
- ✓ Cylinders may be stored outdoors but should be protected from the ground to prevent bottom corrosion. Where extreme temperatures prevail, cylinders should be stored so they are protected from the direct rays of the sun.
- ✓ Indoor storage areas must have walls, floors and fittings made of suitable materials. For example, use non-combustible building materials in storage areas for oxidizing gas and corrosion-resistant materials in storage areas for corrosive gas. Make sure floors are level and protect cylinders from dampness. Avoid overcrowding in storage areas or storing cylinders in out-of-the-way locations.
- ✓ If compressed gas cylinders are stored outside, use a well-drained, securely fenced area. Keep them on a raised concrete pad or non-combustible rack.
- ✓ Cylinders should not be exposed to continuous dampness, stored near salt or other corrosive chemicals or fumes. Corrosion may damage cylinders and cause their valve protection caps to stick.
- ✓ Do not ship or use any cylinder which is not provided with a legible decal that identifies its contents.
- ✓ Store compressed gas cylinders in areas which are:
  - ✓ Well-ventilated and dry.
  - ✓ Fire-resistant and supplied with suitable firefighting equipment including sprinklers, where appropriate.
  - ✓ Away from electrical circuits and ignition sources such as sparks, flames or hot surfaces.
  - ✓ Accessible at all times, but away from elevators, staircases or main traffic routes where cylinders may be dangerous obstacles.
  - ✓ Labelled with suitable warning signs.

• USE OF COMPRESSED GAS CYLINDERS

When using gas cylinders, the following noteworthy points must be considered carefully, namely:

- ✓ Know and understand the properties, uses, and safety precautions of the gas before using the cylinder.
- ✓ Always use the proper regulator for the gas in the cylinder. Always check the regulator before attaching it to a cylinder. If the connections do not fit together readily, the wrong regulator is being used.
- ✓ Before attaching cylinders to a connection, be sure that the threads on the cylinder and the connection mate are of a type intended for the gas service.
- ✓ Do not permit oil or grease to come in contact with cylinders or their valves.

- ✓ Wipe the outlet with a clean, dry, lint-free cloth before attaching connections or regulators. The threads and mating surfaces of the regulator and hose connections should be cleaned before the regulator is attached.
- ✓ Attach the regulator securely before opening the valve wide. Always use a cylinder wrench or another tightly fitting wrench to tighten the regulator nut and hose connections.
- ✓ Open cylinder valves SLOWLY. Do not use a wrench to open or close a hand wheel type cylinder valve. If it cannot be operated by hand, the valve should be repaired.
- ✓ Stand to the side of the regulator when opening the cylinder valve.
- ✓ Do not attempt to repair cylinder valves or their relief devices while a cylinder contains gas pressure. Tag leaking cylinders or cylinders with stuck valves and move to a safe, secure outdoor location.
- ✓ Close valves on empty cylinders and mark the cylinder "empty"

Provision must be made to ensure that the following actions are strictly adhered to at all times by all workers involved, namely:

- ✓ Never allow harmful gases to come into contact with the skin
- ✓ Never forget to wear eye protection in the form of chemical safety goggles when working with compressed gases
- ✓ Never breathe in harmful gases
- ✓ Never roll a cylinder to move it.
- ✓ Never carry a cylinder by the valve.
- ✓ Never leave an open cylinder unattended.
- ✓ Never leave a cylinder unsecured.
- ✓ Never force improper attachments on to the wrong cylinder.
- ✓ Never grease or oil the regulator, valve, or fittings of an oxygen cylinder.
- ✓ Never refill a cylinder.
- ✓ Never use a flame to locate gas leaks.
- ✓ Never attempt to mix gases in a cylinder.
- ✓ Never force connections, using homemade adaptors or tampering with cylinders in any way.
- ✓ Never discard pressurized cylinders in the normal trash.

• GUIDELINES REGARDING STORAGE OF CYLINDERS IN STOREROOMS

Any storerooms in which gas cylinders are being or will be stored should be constructed to be fire-proof and designed in such a way that in the event of a fire, the cylinders can be removed easily. They should be well ventilated at the top and the bottom, and must never be below ground level.

Light fittings, as well as all electric switches in stores containing flammable gases, should either be of the flame-proof type, or should be placed outside the building lighting the interior through fixed windows. 'NO SMOKING - NO NAKED LIGHTS' symbolic signs, should be posted in the area of the store.

Ventilation systems must be designed and built so that they do not result in any unintended or preventable hazard. All hoods, ducts, air cleaners and fan must be made from materials compatible with the gas used. Separate ventilation systems may be needed for some compressed gases to keep them away from systems exhausting incompatible substances. Handigas cylinders stored inside a building on an industrial user's premises should be limited to a total mass of 19kg per 600m3 of building space with a total maximum of 100kg.

• EQUIPMENT USED WITH COMPRESSED GAS

All equipment used with compressed gases must be clean, properly designed and maintained, and made from materials compatible with the particular gas used. Acetylene forms explosive compounds in contact with copper, silver and mercury or their alloys, including bronze or brass containing more than 65 percent copper. Ammonia attacks brass and can react with mercury to form an explosive compound. Do not use mercury pressure gauges in ammonia systems.

Avoid pressurizing ordinary glass equipment. Use specially designed glass equipment and protective devices. Where cylinders are connected to a manifold or header, make sure specialists properly design and install the system. Use effective flashback arrestors on acetylene and other flammable gas systems.

Always follow the correct procedures for assembling and disassembling compressed gas equipment. Check that all the connections are clean and do not leak. Check for leaks, using the gas suppliers recommended method, after assembling and before starting to use equipment. Never use old clips or twisted wire for hose connections. If a hose works loose and flails around, serious injury could result. Poor hose connections are a common cause of accidents.

Acetylene under pressure can explode. Never use acetylene outside of the cylinder at pressures over 103 kPa (15 psig) unless special precautions are taken. If an acetylene cylinder has been accidentally left on its side, set it upright for at least an hour before use. Otherwise, it will emit a burst of solvent instead of gas when the valve is opened.

Corrosive gases can "freeze" the valve stem, making it hard to open the valve. This results from the gas corroding the valve metal. Minimize "freezing" by rotating the valve stem at least once a day while the cylinder is in use. Flush the regulator or manual control valve with dry nitrogen or dry air as soon as possible after use.

• EMERGENCY PROCEDURES ASSOCIATED WITH GAS CYLINDERS

Swift response to emergencies involving chemical fires or gas cylinder leaks is necessary, critical and could be lifesaving. The following steps and actions should be taken without delay, namely:

- Evacuate the affected area immediately if no one is available who is trained to handle the problem or if it is clearly beyond their control.
- Alert other people in the area to the emergency.
- Call the fire department immediately.
- Report the problem to those responsible at work for handling emergencies.
- Ensure that appropriate fire extinguishers are available. Be aware of at least two different exit paths in the event of fire.
- Obtain first aid and remove all contaminated clothes which have been exposed to harmful chemicals.
- Know the appropriate first-aid measures before an incident occurs:
    - For contact with refrigerated liquefied gases, thaw frosted parts with lukewarm water. Do not rub affected area. Get immediate medical advice or attention.
- Provide emergency eyewash stations in the workplace and showers wherever accidental exposure to gases that can damage skin or eyes is possible.
- Only specially trained and properly equipped people should handle emergencies. Nobody else should go near the area until it is safe to do so.
- Planning, training and practicing for emergencies help workers to know what they must do.
- Use MSDSs for the gases which have specific sections on spill and leak procedures, first aid instructions, and fire and explosion hazards to develop an emergency plan.
- Understand and practice emergency procedures in order to know what to do if it becomes necessary.

• MANAGING EMPTY CYLINDERS

It is extremely important that empty cylinders are managed according to the types of gas that they contained when full. Managing them correctly is as equally important as managing full cylinders as the consequences of poor management of either could be disastrous and catastrophic.

Non-Liquefied and Dissolved Gases

The amount of material remaining in a cylinder containing non-liquefied or dissolved gas such as acetylene is directly proportional to the cylinder pressure gauge reading. As the gas is used, the reading on the cylinder pressure gauge drops. When the cylinder pressure gauge reads zero, the cylinder is not really empty. The cylinder still contains gas at atmospheric pressure. Keep a slight positive pressure in the cylinder. It is useful to regard the cylinder as empty when the cylinder pressure gauge reads about 172 kPa or when the cylinder will not deliver at least 172 kPa to the outlet pressure gauge.

Liquefied Gases

The pressure in liquefied gas cylinders remains constant at a given temperature as long as any liquid remains in the cylinder. The only way to know how much material remains in a liquefied gas cylinder is to weigh the cylinder. The empty or tare weight of the cylinder is usually stamped on its neck or valve stem. To monitor whether the cylinder is empty or not, record the net weight of the cylinder contents on a card attached to it. Note that the cylinder should never be emptied completely. Retain some material in the cylinder to maintain a slight positive pressure.

Additional general precautions:

Keeping a positive pressure in a compressed gas cylinder that is regarded as empty as just explained, helps to prevent back flow or suck back. This back flow is the drawing-back into the cylinder of contaminants or moist air from a higher-pressure system or the atmosphere. To achieve this result always keep the valves of empty cylinders closed so that a positive pressure can be maintained.

Because temperature increases or drops in atmospheric pressure can force gas out of the open valve of an empty cylinder, the cylinder must be handled as though it is full since it does contain gas. This release could result in hazardous conditions depending on the gas and how much is forced out. Temperature drops or increases in atmospheric pressure can cause air to be drawn in through the open valve. Air could cause a serious contamination and a corrosion problem inside the cylinder.

Therefore, handle the empty cylinder as follows, namely:

- Tightly close the cylinder valve before removing the gas discharge equipment.
- Clearly mark or label the cylinder "empty" or "MT."
- Place the cylinder in a storage area separate from that used for full cylinders.
- Keep incompatible materials away from the cylinder.
- Notify the gas supplier if the cylinder or any part of it is damaged or defective, contaminated, or may have been exposed to a possibly hazardous condition such as a fire or electric arc.

Extra care must be taken when scrapping unserviceable cylinders which should first be destroyed as a pressure vessel. Contact the gas supplier for advice on disposing of unserviceable cylinders or preferably allow them to remove and destroy the cylinder correctly.

• PRESSURE EQUIPMENT LEGISLATIVE REQUIREMENTS

Notice was given of the Pressure Equipment Regulations (hereinafter referred to as PER) promulgated under section 43 of the Act in Government Gazette 32395 Notice No. R. 734 dated 15 July 2009 to govern the use of pressure equipment such as gas cylinders. The PER contains the following sections, namely:

Regulation 1: Definitions
Regulation 2: Scope of application

1. These Regulations shall apply to the design, manufacture, operation, repair, modification, maintenance, inspection and testing of pressure equipment with a design pressure equal to or greater than 50 kPa, in terms of the relevant health and safety standard incorporated into these Regulations under section 44 of the Act.
2. Regulations 3, 4, 5, 9(1), 9(2) and 9(3) shall not apply to pressure equipment in use or on order prior to the publication of these Regulations, which equipment shall be designed and constructed according to the requirements applicable at the time of order.
3. The following pressure equipment shall be excluded from these Regulations:
   a) Piping for the supply, distribution and discharge of water below its boiling point at atmospheric pressure and associated pressure equipment and headraces such as penstocks, pressure tunnels, pressure shafts for hydro-electric installations and their related specific pressure accessories;
   b) Aerosol dispensers;
   c) Pressure equipment intended for the functioning of road and rail vehicles, excluding a fuel gas system;
   d) Pressure equipment comprising casings or machinery where the dimensioning, choice of material and manufacturing rules are based primarily on requirements for sufficient strength, rigidity and stability to meet the static and dynamic operational effects or other operational characteristics and for which pressure is not a significant design factor, and such pressure equipment may include –
      i. Engines, including turbines and internal combustion engines;
      ii. Reciprocating steam engines, gas turbines, steam turbines, turbo-generators, compressor engines, pumps and actuating devices;
   e) Open metal-making pots and blast furnaces
   f) Housing for electrical machinery such as switchgear, control gear, transformers and rotating machines;
   g) Tyres and flexible pressurised casings used for recreational purposes;
   h) Fixed electrical hot-water storage container of water capacity from 15 litres to 450 litres operating at a maximum pressure of 600 kPa manufactured to the requirements of SANS 151, which shall be installed in accordance with the requirements of SANS 10254.

Regulation 3: General requirements
Regulation 4: Duties of manufacturers
Regulation 5: Duties of importers and suppliers
Regulation 6: Duties of users
Regulation 7: Approval and duties of approved inspection authority
Regulation 8: Registration of a steam generator
Regulation 9: Pressure equipment marking
Regulation 10: Pressure and safety accessories
Regulation 11: Inspection and test
Regulation 12: Risk-based inspection

Regulation 13: Repairs and modifications
Regulation 14: Records
Regulation 15: Access
Regulation 16: Door interlocks
Regulation 17: Gas reticulation equipment and systems
Regulation 18: Transportable gas containers
Regulation 19: Fire extinguishers
Regulation 20: Offences and penalties
Regulation 21: Repeal of regulations and Annexures

The Vessels under Pressure Regulation, 1996, published under Government Notice No. R. 1591, dated 4 October 1996, will be repealed by the PER.

Regulation 22: Short title. These Regulations shall be called the Pressure Equipment Regulations, 2009, and shall come into effect on 1 October 2009: Provided that approved inspection authority for in-service inspections shall come into effect on 1 April 2011 on condition that the inspection shall be carried out by an authorised person.

Some extracts of the requirements of these regulations follows:

- DUTIES OF USERS 0F PRESSURE EQUIPMENT– REGULATION 6

(1) The user shall ensure that the pressure equipment is operated and maintained within its design and operating parameters.
(2) The user shall, subject to the relevant health and safety standard incorporated into these Regulations under section 44 of the Act -
(a) provide the manufacturer, repairer or modifier with comprehensive information of the operating or intended operating conditions of the pressure equipment, including the characteristics of the fluid and operating parameters of other connected pressure equipment, where reasonably practicable;
(b) ensure pressure equipment has a certificate, issued by the manufacturer, including a verification signature by an approved inspection authority when required, which certifies that the pressure equipment has been designed and manufactured in accordance with the relevant health and safety standard incorporated into these Regulations under section 44 of the Act;
(c) ensure pressure equipment has a certificate issued by the repairer or modifier, including a verification signature by an approved inspection authority when required, which certifies that the pressure equipment has been modified or repaired in accordance with the relevant health and safety standard incorporated into these Regulations under section 44 of the Act;
(d) ensure that pressure equipment has a certificate issued by an approved inspection authority before commissioning, where applicable; and
(e) ensure that a gas system has a valid certificate issued by an authorised person.

- REGISTRATION OF A STEAM GENERATOR REGULATION 5

(1) No user may use a steam generator unless such user is in possession of a certificate of registration issued in terms of sub-regulation (3) for that steam generator.

- PRESSURE EQUIPMENT MARKING – REGULATION 9

(1) Every manufacturer of pressure equipment shall cause the pressure equipment to be marked in accordance with the relevant health and safety standard incorporated into these Regulations under section 44 of the Act.
(2) Every manufacturer shall cause a data plate to be permanently fixed in a conspicuous place to any steam generator or pressure vessel with the following minimum particulars:
(a) Name of manufacturer;
(b) country of origin;
(c) year of manufacture;
(d) manufacturer's serial number;
(e) reference number, date and edition of the health and safety standard;
(f) design pressure in units of Pascal;
(g) design temperature for both minimum and maximum in degrees Celsius;
(h) capacity in cubic metres;
(i) unique mark of an approved inspection authority as applicable; and
j) the hazard category in accordance with the requirements of SANS 347.
(3) In the case of composite pressure equipment the following information shall be included in addition to that referred to in sub-regulation (2):
(a) The resin system of the corrosion barrier lining;

(b) the resin system of the structural wall; and
(c) the name and specific gravity of the medium for which the vessel was designed.

• PRESSURE AND SAFETY ACCESSORIES – REGULATION 10

(1) No user may require or permit pressure equipment to be used unless it is provided with all the pressure and safety accessories required by the relevant health and safety standard which is incorporated into these Regulations under section 44 of the Act and used in the design, construction and manufacture of such pressure equipment: Provided that alternative safety accessories other than those required by the standard may be fitted with the written approval of an approved inspection authority.
(3) Every user of a steam generator or pressure vessel shall ensure that the steam generator or pressure vessel in use is fitted with at least one pressure measuring device.
(4) Every user of a steam generator or pressure vessel shall ensure that the steam generator or pressure vessel in use is fitted with at least one safety valve and that safety valve is kept locked, sealed or otherwise rendered inaccessible to any unauthorised person.
(5) The number and capacity of the safety valve referred to in sub-regulation (4) shall comply with the requirements of the design standard for the steam generator or pressure vessel or as required in terms of sub-regulation (2).

• PRESSURE EQUIPMENT RECORDS – REGULATION 14

(1) Every user of pressure equipment shall keep a record, which shall be open for inspection by an inspector, in which the certificate of manufacture, and the results, after manufacturing, of a" inspections, tests, modifications and repairs shall be recorded.
(2) When pressure equipment is sold, the manufacturer shall ensure that it is accompanied, where relevant, with instructions for the user, containing a" the necessary safety information relating to -
(a) mounting, including the assembling of different pieces of pressure equipment;
(b) putting into service; and
(c) maintenance, including checks by the user

• TRANSPORTABLE GAS CONTAINERS – REGULATION 18

(1) No user shall use. require or permit a transportable gas container to be used, and no user shall fill. place in service, handle, modify, repair, inspect or test any transportable gas container, other than in compliance with the relevant standards incorporated into these Regulations under section 44 of the Act.
(2) The inspection and test referred to in sub-regulation (1) shall be carried out by an approved testing station.

• FIRE EXTINGUISHERS – REGULATION 19

(1) No user shall use, require or permit the use of a fire extinguisher unless designed, constructed, filled, recharged, reconditioned, modified, repaired, inspected or tested in accordance with the relevant safety standard incorporated into these Regulations under section 44 of the Act.
(2) No person shall fill, recharge, recondition, modify, repair, inspect or test any fire extinguisher unless such person is an authorised person employed by a permit holder: Provided that a permit is issued by an organisation approved by the chief inspector.

• OXY/FUEL GAS AND WELDING

Oxy/fuel gas equipment has many uses - welding, cutting, heating, straightening, and descaling. The equipment is versatile, easy to move and cheap. It is so widely used that sometimes people forget about the dangers. Many people are injured each year by the incorrect or careless use of oxy/fuel gas equipment. Some people die. The main hazards are:

- fire caused by heat, sparks, molten metal or direct contact with the flame;
- explosion when cutting up or repairing tanks or drums which contain or may have contained flammable materials;
- fire/explosion caused by gas leaks, backfires and flashbacks;
- fumes created during flame cutting;
- fire/burns resulting from misuse of oxygen;
- burns from contact with the flame or hot metal;
- crushing or impact injuries when handling and transporting cylinders.

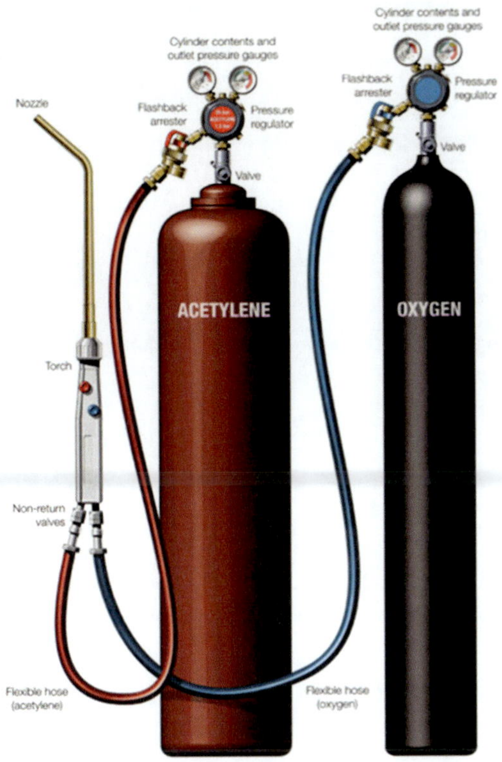

Example of oxy-acetylene connections
(Source: https://www.hse.gov.uk/pubns/indg297.pdf)

• PREVENTING FIRE FROM HEAT, SPARKS, MOLTEN METAL OR DIRECT CONTACT

The flame from an oxy/fuel gas torch will quickly ignite combustible material such as wood, paper, cardboard, textiles, rubber, and plastics. This ignition source is commonly referred to in industry as a "blowpipe". Many processes also generate sparks and hot spatter which can easily ignite these materials.

Taking the following precautions will help prevent fire:

- move the work piece to a safer location for carrying out hot work;
- remove nearby combustible materials such as flammable liquids, wood, paper, textiles, packaging or plastics;
- protect nearby combustible materials that cannot be moved. Use suitable guards or covers such as metal sheeting, mineral fibre boards or fire-retardant blankets;
- check that there are no combustible materials hidden behind walls or in partitions, particularly if the welding or cutting will go on for some time. Some wall panels contain flammable insulation materials such as, for example, polystyrene;
- use flame-resistant sheets or covers to prevent hot particles passing through openings in floors and walls such as doorways, windows, and cable runs;
- where the consequences of a fire are severe, such as work inside a confined space, a fire watch may need to be appointed during and after the work finishes. It is normal to maintain fire watch for 30 minutes after hot work finishes;
- prevent flame, heat, sparks or hot spatter from landing on the hoses;
- keep fire extinguishers nearby.

• EXPLOSION WHEN CUTTING UP OR REPAIRING TANKS AND DRUMS

If a welding blowpipe or burner is used on a tank or drum containing flammable material whether solid, liquid or vapour, it can explode. As well as flammable liquids such as petrol, diesel and fuel oil, substances such as paints, glue, anti-freeze and cleaning agents may also release flammable vapours.

Tanks and drums that are regarded as empty usually still have residual material in the bottom, and in seams and crevices. Just a teaspoon of flammable liquid in a drum can be enough to cause an explosion.

Never use an oxy/fuel gas blowpipe on a drum or tank that has contained, or may have contained, flammable material, unless it is safe to do so. It may be safer for a specialist company to carry out the work.

• BACKFIRES AND FLASHBACKS

Backfires

A backfire is when the flame burns back into the torch, often with a sharp bang. This may happen when the torch is held too close to the work piece, or if the nozzle is partly blocked. The flame may go out or it may reignite at the nozzle. Sometimes the flame burns back into the torch, and burning continues inside it. Backfires do not usually cause injury or damage, but they can indicate a fault in the equipment.

If a backfire does occur:

- shut off the torch valves, oxygen first (usually coloured blue) and then the fuel gas (usually coloured red);
- shut off the oxygen and fuel gas cylinder valves;
- cool the blowpipe with water if necessary;
- check the equipment for damage or faults, particularly the nozzle.

Flashbacks

Flashbacks are commonly caused by a reverse flow of oxygen into the fuel gas hose or fuel into the oxygen hose, producing an explosive mixture within the hose. The flame can then burn back through the torch, into the hose and may even reach the regulator and the cylinder. Flashbacks can result in damage or destruction of equipment, and could even cause the cylinder to explode.

The following precautions will help to prevent flashbacks:

- Use the correct lighting-up procedure. Purge the hoses before lighting the torch to remove any potentially explosive gas mixtures. Use a spark igniter and light the gas quickly after turning it on.
- Make sure the blowpipe is fitted with spring-loaded non-return valves.
- Use the correct gas pressures and nozzle size for the job.
- Maintain the equipment in good condition.

These measures will reduce the risk of a flashback but will not completely eliminate it. Non-return valves will not stop a flashback once it has occurred.

Protecting cylinders from flashbacks

Fit flashback arresters to both the oxygen and fuel gas hoses near to the regulators. For long lengths of hose, fit arresters on both the torch and the regulator.

The fitting of a flashback arrester is not a substitute for healthy and safe working practice. If a flashback does occur:

- if it is safe to do so, close the cylinder valves on both fuel gas and oxygen;
- if the fire cannot be put out at once, evacuate the area and call the emergency fire services.

After a flashback, carefully check for damage to the torch, hoses, regulators, flashback arresters and other components. Replace parts where necessary. If in doubt, consult the supplier.

Acetylene cylinders

Particular attention should be given to any acetylene cylinder which has been involved in a flashback or affected by fire. There is a risk that the acetylene could start to decompose, and the cylinder could explode. If an acetylene cylinder becomes hot or starts to vibrate, the area must be evacuated immediately and the emergency fire services called.

• FUMES CREATED DURING FLAME CUTTING

The fumes from welding and flame cutting metals are harmful. It may be necessary for fume extraction and/or filtering respirators which are respiratory protective equipment to reduce the risk of ill health. Dirt, grease and other contamination increase the quantity or volume of fumes generated and can introduce very toxic substances to it. Hot work on items with lead paint, chromium or chromate paint or cadmium plating is particularly hazardous.

- If flame cutting occurs outdoors and the metal is clean and unpainted respiratory protective equipment is not usually necessary. Try to work in a position where the wind blows the fume cloud away from workers and others.
- If flame cutting occurs indoors or in conditions were there is little air movement, the need for extraction and/or respiratory protective equipment depends on how much cutting is going to be done.

- Automated flame cutting machines, particularly multi-head machines, will normally need fume extraction fitted to them.
- Work on materials coated with lead or chromate paints, galvanised or cadmium plated or metal contaminated with oil, grease etc. are likely to require fume control such as extraction and/or respiratory protective equipment.

• FIRE/BURNS FROM MISUSE OF OXYGEN

Oxygen leaks increase the risk of fire. Clothing contaminated with oxygen, even fire-retardant clothing, will catch fire easily and burn very fiercely. Oxygen can cause explosions if used with incompatible materials. In particular, oxygen reacts explosively with oil and grease.

The following precautions should be taken, namely:

- never use oxygen to blow dust off clothing;
- never attempt to improve air quality inside confined spaces by releasing oxygen in the space;
- never allow oil or grease to come into contact with oxygen valves or cylinder fittings;
- only use equipment designed for use with oxygen. In particular, check that the regulator is safe for oxygen and for the cylinder pressure.

• CONTACT BURNS WHEN WELDING

The following precautions will help to prevent burns from contact with the flame, hot metal, sparks and molten slag:

- work in a safe location away from other people;
- wear protective clothing, boots, gauntlets and eye protection;
- shut off the torch when not in use. Do not leave a lighted torch on a bench or the floor as the force of the flame may cause it to move;
- clamp the work piece, avoid holding it by hand.

• CRUSHING OR IMPACT INJURY

The following precautions will help prevent injury when handling and transporting cylinders.

- Prevent cylinders from falling or being knocked over by securing them with a stout chain or strap. It is normal to chain them to a wheeled trolley or against a wall.
- Avoid moving a cylinder by tilting it on its base and rolling it. It is better to use a trolley.
- If transported in a road vehicle, the cylinders should be secured so they do not come loose in an accident.

• WELDING TRAINING

Oxy/fuel equipment is relatively easy to use but users may still need some instruction or training in:

- the safe use of the equipment, including the correct methods to purge, light and shut down;
- general safe working practices and the precautions to take;
- when and how to use the fire extinguishers;
- information about escape routes, how to raise the fire alarm and emergency procedures.

• PERMIT-TO-WORK FOR WELDING

The potential for injury and property damage are such that many companies want to control hot work very closely. One way to achieve this is to operate a written permit system for welding and flame cutting work. Sometimes called hot work permits, they aim to ensure that welding and flame cutting does not start until after the risks for that particular task are identified and measures to eliminate or control risk are in place. A permit should specify:

- what work will be done;
- how and when it is to be done;
- what safety and health precautions are needed;
- who is responsible for checking it is safe to start;
- who will check the work is done safely;
- who is responsible for confirming that work is complete and there is no longer a risk from, or to, the people doing the work.

• HOT WORK

Working with ignition sources near flammable materials is referred to as "hot work." Welding and cutting are examples of hot work. Fires are often the result of the "quick five minute" job in areas not intended for welding or cutting. Getting a hot work permit before performing hot work is just one of steps involved in a hot work management program that helps to reduce the risk of starting a fire by welding or cutting in areas where there are flammable or combustible materials.

• DEFINITION OF A HOT WORK MANAGEMENT PROGRAM

Hot work management programs are put in place to control or eliminate hot work hazards and their risks. Programs include the development of policies, procedures, and the assignment of responsibilities and accountabilities for all aspects of hot work. A program includes:

1. Policies
    a. Where hot work is permitted
    b. When hot work is permitted
    c. Who authorizes hot work
2. Procedures
    a. What must be assessed before permitting/performing hot work in an area or on a process piece of equipment or area
    b. What to do to prepare an area for hot work
    c. What to do if hot work cannot be avoided in a particularly hazardous area
    d. What hot work tools are required
    e. How to obtain a hot work permit, when they are required, and who can administer them
3. Training
    a. Workers, supervisors, maintenance individuals, fire wardens, trained fire watch individuals, and contractors all have different roles, and must be trained accordingly
4. Communications
    a. Posting procedures
    b. Posting policies
    c. Posting signs in areas that are prohibited from having hot work performed in them

• ALTERNATIVES TO HOT WORK

Hot work may be substituted with other methods such as, for example:

Instead of:	Use:
Saw or torch cutting	Manual hydraulic shears
Welding	Mechanical bolting
Sweat soldering	Screwed or flanged pipe
Torch of radial saw cutting	Mechanical pipe cutter

• GENERAL GOOD PRACTICES BEFORE PERFORMING HOT WORK

Ensure that proper hot work procedures are strictly adhered to and the following should be considered, namely:

- ✓ Ensure that all equipment is in good operating order before work starts.
- ✓ Inspect the work area thoroughly before starting by checking for combustible materials in structures such as partitions, walls, and ceilings.
- ✓ Remove any combustible materials on floors around the work zone. Combustible floors must be kept wet with water or covered with fire resistant blankets or damp sand.
- ✓ Use water ONLY if electrical circuits have been de-energized to prevent electrical shock.
- ✓ Remove any spilled grease, oil, or other combustible liquid.
- ✓ Move all flammable and combustible materials a reasonable distance away from the work area.
- ✓ If combustibles cannot be moved, they should be covered with fire resistant blankets or shields. Gas lines and equipment should be protected from falling sparks, hot materials and objects.
- ✓ Block off cracks between floorboards, along baseboards and walls, and under door openings, with a fire resistant material. Close doors and windows.

- ✓ Cover wall or ceiling surfaces with a fire resistant and heat insulating material to prevent ignition and accumulation of heat.
- ✓ Secure, isolate, and vent pressurized vessels, piping and equipment as needed before beginning hot work.
- ✓ Inspect the area following work to ensure that wall surfaces, studs, wires or dirt have not heated up.
- ✓ Vacuum away combustible debris from inside ventilation or other service duct openings to prevent ignition. Seal any cracks in ducts. Prevent sparks from entering into the duct work. Cover duct openings with a fire resistant barrier and inspect the ducts after work has concluded.
- ✓ Post a trained fire watcher within the work area during welding, including during breaks, and for at least 30-60 minutes after work has stopped. Depending on the work done, the area may need to be monitored for up to 3 hours after the end of the hot work.
- ✓ Eliminate explosive atmospheres such as vapours or combustible dust or do not allow hot work. Shut down any process that produces combustible atmospheres, and continuously monitor the area for accumulation of combustible gases before, during, and after hot work.
- ✓ If possible, schedule hot work during shutdown periods.
- ✓ Comply with the required legislation and standards applicable to the workplace.

The following is an example of a pro forma hot work permit:

HOT WORK PERMIT - EXAMPLE

1. THIS PERMIT IS VALID FOR ONE JOB ONLY
2. THIS COPY IS TO BE RETAINED BY THE PERSON AUTHORISED TO PERFORM THE HOT WORK AND MUST BE PRODUCED ON THE REQUEST OF ANY OF OUR EMPLOYEES.
3. MANAGEMENT WILL RETAIN A COPY
4. DESCRIPTION OF WORK ....
5. EXACT LOCATION OF JOB...
6. PERMIT VALID FROM .... TO ....
7. NOTE PARTICULAR HAZARDS OR HAZARDOUS AREAS AND LIST ....
8. STANDARD PRECAUTIONS TO BE TAKEN

Precaution	Management Requirement (✓)	Contractor to Confirm (✓)
Equipment in good working order?		
All combustibles removed a minimum of 5 metres from the work area.		
Fire extinguisher available		
Hosereel available		
Fire blanket provided		
Welding curtains provided		
Ventilation provided		
Barricades required		
Signs required		
Check the area for explosive atmosphere		
Post welding/cutting fire watch		

Additional Precautions	Management Requirement (✓)	Contractor to Confirm (✓)

I VERIFY THAT THE JOB AREA HAS BEEN EXAMINED AND AUTHORISE HOT WORK TO BE CARRIED OUT PROVIDING THE ABOVE CONDITIONS ARE MAINTAINED THROUGHOUT THE TERM OF THE PERMIT.

NAME:...SIGNED:...

NAME OF PERSON(S) DOING HOT WORK...
COMPANY NAME :...
SIGNED:..

DATE.......................

ON COMPLETION OF HOT WORK SIGN & RETURN TO PLACE OF REGISTRATION
I verify that the hot work has been completed in accordance with the authorised conditions outlined in part one of this form.

PERSON DOING HOT WORK:

SIGNED:...

Location of work inspected 30 minutes after job completed

SIGNED...

A written Hot Work Permit needs to be issued by a supervisor who should examine the location, methods of protection and specify necessary precautions. The Permit is issued for a specific time and area. Work should not begin in a new area without obtaining a new Permit. A permit should not be issued unless the following requirements are met:

- ✓ Adequate fire extinguishing equipment is located at the work site
- ✓ Combustible materials have been relocated at least 11 meters from the area of operation. If they cannot be relocated, they should be shielded by non-combustible welding curtains or fire resistant blankets
- ✓ Combustible floors and walls should be wet down with water if possible. A fire watch is trained and present at the site
- ✓ The fire watch should be maintained for at least 30 minutes after operations have been completed and checked again at least one hour afterward
- ✓ Work should only begin after ensuring that all equipment and surrounding conditions are safe
- ✓ Continue to work only as long as approved work conditions remain unchanged

The job is not considered complete until the responsible supervisor signs the Hot Work Permit and checks the area. Many losses occur when the Hot Work Permit was signed off without a final review of the area.

• SPECIAL CONSIDERATIONS WHEN DOING HOT WORK

- ✓ Never cut or weld around explosive atmospheres or where combustible dusts or liquids are present. Explosive material, such as flammable liquids, gases, vapours or combustible dust, may reside in the uncleaned or improperly prepared drums, tanks or other containers. Sparks can pass through cracks, or holes in walls, floors, broken windows or open doorways.
- ✓ Before initiating Hot Work at roof level or on scaffolding, be sure that the area below is secured and any sparks or hot debris will not cause a fire.
- ✓ Hot Work Permits should not be issued if any portion of the plant's fire protection systems are impaired.

• REVIEW QUESTIONS

1. What are the basic health and safety rules to be observed when working with compressed gas cylinders?
2. What must be considered when transporting and storing compressed gas cylinders?
3. What actions should be taken when dealing with empty compressed gas cylinders?
4. How can flashbacks and backfires be prevented when using an oxy/fuel gas blowpipe?
5. When would a hot work permit system be used?

• REFERENCES:

Anonymous (nd). Loose notes on safety, health and environment management

Balamurugan, P., Muthamilselvi, P. and Balashanmugam, P., 2019. Effective work permit system to minimize the hazards in eid parry (India) limited. *International Journal of Engineering, Science and Mathematics*, *8*(8), pp.20-42.

BOC (n.d.). *Storage of Cylinders*. [online] BOC Zimbabwe Gases. Available at: http://www.boc-gas.co.zw/en/health_and_safety/gases_and_cylinders/storage_cylinders/index.html.

Canadian Center for Occupational Health and Safety. Nd. Available at: https://www.ccohs.ca/oshanswers/chemicals/compressed/compress.html

Canadian Center for Occupational Health and Safety. Nd. Available at: https://www.ccohs.ca/oshanswers/chemicals/howto/gas_cylinder.html

Canadian Center for Occupational Health and Safety. Nd. How do I Work Safely with - Compressed Gases. Available at: https://www.ccohs.ca/oshanswers/prevention/comp_gas.html

Department of Labour (1993). Occupational Health and Safety Act No: 85 of 1993.

Glass, George., 2016. *Gas Cylinder Safety, Course 9518* (No. LA-UR-16-28217). Los Alamos National Lab.(LANL), Los Alamos, NM (United States).

Health and Safety Executive. 2012. Safety in gas welding, cutting and similar processes. Available at: https://www.hse.gov.uk/pubns/indg297.pdf

Kukfisz, Bozena., Ptak, Szymon., Półka, Marzena. and Woliński, Marek., 2018. Fire and explosion hazards caused by oxygen cylinders. *WIT Transactions on The Built Environment*, *174*, pp.141-151.

Rinawati, Seviana., 2018. Level of Safe Behavior with the Implementation of Hot Work Permit Approach in PT Bbb East Java. *Journal of Vocational Health Studies*, *1*(3), pp.89-96.

Supply, O.M.C., 2017. Code of Practice 7 the safe use of oxy-fuel gas equipment (individual portable or mobile cylinder supply).

VCU (n.d.). [online] Safety.louisiana.edu. Available at: https://safety.louisiana.edu/sites/safety/files/Compressedgascylindersafety.pdf.

Yan, Chin. Koy., Siong, Piong. Han., Kidam, Kamarizan., Ali, Mohamed. Wijayanuddin., Hassim, Mimi. Hassim., Kamaruddin, Mohd. Johari. and Kamarden, Hamidah., 2017. Contribution of permit to work to process safety accident in the chemical process industry. *Chemical Engineering Transactions*, *56*, pp.883-888.

CHAPTER 15

PERSONAL PROTECTIVE CLOTHING AND EQUIPMENT

Personal Protective Equipment or PPE is the most commonly used primary control measure to personal safety. Instead it should be a measure of last resort after exposure to hazards has been mitigated as far as reasonably possible

Personal Protective Equipment (PPE) includes all items of clothing and work accessories that are specifically designed to protect workers from exposure to hazards present in the workplace that could potentially cause serious injuries in the workplace itself and illnesses subsequently arising from working in that workplace. Workplace injuries and illnesses may result from contact by the worker with chemical, radiological, physical, electrical, mechanical, or other workplace hazards. Generally PPE may include gloves, safety glasses, steel toed boots and footwear, goggles or face shields, hearing protection in the form of earplugs or muffs, hard hats, respirators, or overalls, coveralls, vests and full body suits. Therefore, it is important that all PPE should be safely designed and constructed and be maintained in a clean and reliable fashion. If the PPE does not fit properly, it can make the difference between being safely covered or dangerously exposed.

Protective equipment should not replace engineering, administrative or procedural control measures used to protect workers. This means that all alternative ways of protecting workers must be fully investigated and implemented before using PPE. When engineering, work practice, and administrative controls are not feasible or do not provide sufficient protection, employers must provide personal protective equipment to their workers and ensure its proper use. Therefore, PPE should always be a measure of last resort and not first resort as is typically the case. Where PPE is a measure of first resort it creates in the mind of the worker a false sense of security. Consequently, the worker assumes that the hazard has been mitigated and that there is no threat from exposure to it.

Therefore, it is imperative that the hierarchy of health and safety hazard controls as advocated by The National Institute for Occupational Safety and Health (NIOSH) in the USA should be used to mitigate exposure to hazards. This hierarchy which is often depicted as a pyramid contains five levels and is meant to be worked from the top of the pyramid downward. The five controls in the hierarchy are:

- **Eliminating the hazard:** The most effective way to keep workers safe from a workplace hazard is to eliminate the hazard altogether. If a tool or piece of equipment represents a danger to workers and the facility the necessity of the item should be considered and where not absolutely necessary, it should not be used at all.
- **Reducing the hazard (Substitution):** Where it is not viable to completely eliminate the hazard should be substituted out with something less dangerous. Even though the hazard still exists, workers will not be in nearly as much danger.
- **Implementing engineering controls:** In cases were trying to eliminate or reduce the hazard is not viable, the next best option is to limit the exposure to the hazard by using engineering controls. Engineering controls are designed as modifications or designs to machines, equipment, ventilation systems, and processes that reduce the amount of exposure and physically protect the worker from harm.

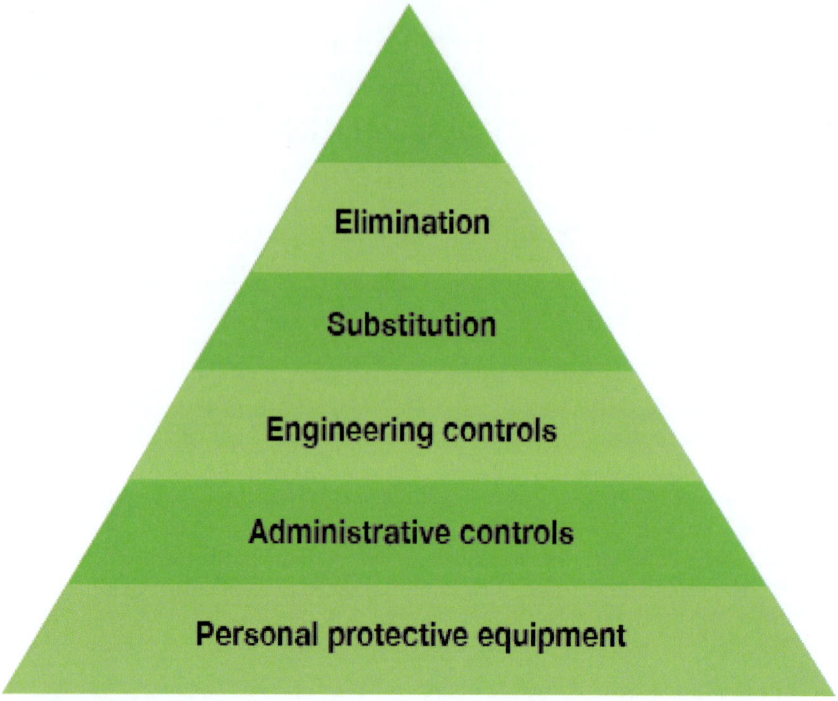

(Source: The National Institute for Occupational Safety and Health (NIOSH))

- **Utilizing administrative controls:** Where it is not feasible to implement engineering controls, the next option in the hierarchy of hazards to be considered is to use administrative controls. These controls are at risk for human error and are used as a temporary intervention rather than a sustainable, long-term solution. These controls are typically rules and regulations that are put in place to minimize exposure to hazards. Examples of administrative controls include the following:
  - **Training:** Workers should be trained to identify hazards, monitor hazard exposure, and health and safe procedures for working around the hazard. Additionally, workers should know how to protect themselves and their co-workers.
  - **Procedures:** The steps in a job process may need to be rearranged or updated to keep the worker for encountering the hazard. Developing standardized safe work practices such as a safe work or operating procedure (SWP or SOP) is an important step.
  - **Maintenance:** Having a properly designed, documented and implemented maintenance schedule for machines known to be hazardous is important in the health and safety improvement effort. Preventive maintenance will address any equipment issues before they become a problem.
  - **Housekeeping:** Sustaining a clean and clutter-free space will greatly reduce the risk of injury and can minimize the severity of an accident.
  - **Signs:** Wall and floor signs can be posted or installed to enforce administrative controls. Visual cues can remind workers which areas are prohibited from entering and, for example, when breaks need to be taken to limit heat exposure.
- **Personal protection equipment:** Finally, PPE should be considered when all other options have been explored and exhausted. Although considered "the last line of defence," or a measure of last resort, PPE can make the difference between a minor injury, a severe, life-threatening injury and even death.

Employers are also required to train each worker required to use PPE to know:

- When PPE is necessary
- What kind of PPE is necessary
- How to properly put PPE on, adjust, wear and take it off
- The limitations of the PPE itself
- Proper care, maintenance, useful life, and disposal of the PPE

A workplace risk assessment is necessary in an organization to identify the potential hazards that workers could be exposed to and select the appropriate PPE for adequate protection of the workers. Therefore, the workers should be trained with potential topics of knowledge which comprises the proper PPE for their work, when this PPE should be used, how to wear, adjust, maintain, and discard this equipment properly, and the shortcomings of the PPE.

It is critical therefore, to use or wear all the equipment, protective devices or clothing recommended on equipment, products or Material Safety Data Sheets (MSDSs) or as required by the employer, the Act or any other relevant health and safety regulation. Any broken, worn or missing PPE should be reported to the supervisor and where unsure about what PPE is most appropriate for the work activity request advice from the supervisor.

- MAINTENANCE OF PPE

All PPE should always be used and maintained in a sanitary and reliable condition. Maintenance must be carried out on, for example, respirators, safety shoes, and gloves to ensure that the equipment will always function properly and increase their lifespan and use. Damaged or broken equipment must always be replaced. Slack, worn-out, sweat-soaked, or twisted headbands of hardhats must be replaced regularly. A visual inspection of the hardhat can determine when the headband must be replaced.

- CLEANING OF PPE

PPE must always only be cleaned in accordance with the specifications of the manufacturer. Any other methods of cleaning will cause the equipment to malfunction and reduce their effectiveness.

- STORAGE OF PPE

Goggles should be kept in a case when not in use. Spectacles, in particular, should be given the same care as one's own glasses, since the frame, nose pads, and temples can be damaged by rough usage. Items such as respirators and hearing protection equipment should be placed in a clean, dust-proof container, such as a box, bag, or plastic envelope, to protect them until reissue. Only store equipment according to the specifications of the manufacturer.

Proper storage of PPE
(Source: https://nasdonline.org/7314/d002518/oregon-osha-pesticide-use-and-your-personal-protective.html

- PROTECTION PROVIDED BY PPE

PPE is designed to provide protection to specific body part. PPE only differs in the amount of protection it offers as well as the specific threat it protects against. It is therefore very important that the correct PPE is used to provide the protection required after implementing the hierarchy of control. The following table sourced and adapted from WorkSafe in New Zealand is informative of the various body parts and the range of PPE that could be used to protect them:

Body area	Potential clothing examples
Head and neck	Face shields Sunhats Hard hats Neck protection such as scarves for use during welding Hairnets
Eyes	Safety glasses and goggles Face shields Visors
Ears	Earplugs Earmuffs Semi-insert/canal caps
Hands and arms	Gloves made of latex, rubber, leather or metal Long sleeved tops Gauntlets and sleeving that covers part or all of the arm
Feet and legs	Rubber boots Thermally insulated boots Safety boots with protective toecaps and penetration-resistant mid-soles Foundry boots Chainsaw boots Anti-static, electrically conductive boots Boots with oil/chemical resistant soles Non-slip shoes
Whole body	Conventional or disposable overalls Aprons Chemical suits Cooling vests Weather-proof gear – water-proof trousers, rain coats Fire-proof clothing Hi-visibility (hi-viz) clothing

(Adapted from https://worksafe.govt.nz/topic-and-industry/personal-protective-equipment-ppe/protective-clothing/)

The following figure illustrates the areas of the body that need to be protected together with the appropriate signage to inform about what must be worn when entering a particular work zone:

Body parts protected by PPE and signage
(Source: https://innovativeengineering.com/safety/)

- HEAD PROTECTION

The head is a very delicate part of the body because in and around the head are

✓ the eyes, with which a worker sees
✓ The ears, with which a worker hears
✓ the nose, with which a worker smells
✓ the mouth, with which a worker eats and speaks
the brain, with which a worker thinks

Therefore, any Injuries to the head must be taken seriously and therefore proper head protection is critical.

Wear a hard hat for head protection
(Source: https://www.optimumsafetymanagement.com/blog/hard-hat-safety-regulations/)

- POTENTIAL HAZARDS

Accidents that cause head injuries are in the main difficult to anticipate or control. If hazards exist that could cause a head injury, these can be eliminated by wearing appropriate head protection. It has been correctly stated that the brain is the best PPE a worker has and therefore should be protected and used!

Hazards include

✓ Falling or flying objects which are a common cause of head injuries. Falling or walking into hard and solid objects can also cause head which include neck sprains, concussions and skull fractures.
✓ Accidents involving electricity result in electrical shocks and burns.

- ✓ Splashes, spills and drips of toxic liquids such as acids or caustics which can irritate and burn the eyes and skin.

• HARD HATS

Safety hard hats protect the head from impact, penetration and electrical shock. Most hard hats are water resistant, flame resistant and adjustable. It is important for protective headgear to fit correctly on the head and must be adjusted to the head size of each individual. Given that increasing numbers of women are working in areas requiring the wearing of hard hats it is possible to procure them in a variety of colours. However, in the process of making hard hats more fashionable the safety features have not been compromised or reduced.

• WEARING A HARD HAT

The most common form of head protection used in industry is the safety helmet which is generally referred to as a hard hat. The following are guidelines for its use, namely:

- ✓ Always wear a hard hat while working in areas where there are potential head hazards.
- ✓ Adjust the suspension inside the hard hat so that it sits comfortably but securely on the head.
- ✓ Inspect the shell of the hard hat for cracks, cuts and dents. Inspect the suspension system for worn or broken straps. If the hard hat needs to be repaired, have it repaired immediately or ask the employer to issue a new one.
- ✓ Never paint, scratch or drill "air holes" in a hard hat. If working at night reflective plastic tape can be applied. Never use metal tape on a hard hat because it can conduct electricity.
- ✓ Never carry personal belongings such as cigarettes, lighters or pens in the hard hat while wearing it.

• CARING FOR A HARD HAT

Because a hard hat is an important piece of PPE, the following are necessary, namely:

- ✓ Clean the hard hat at least once a month or more frequently if needed to remove oil, grease, chemicals and sweat that can collect in and around the hat.
- ✓ Clean the hard hat by soaking it in a solution of mild soap and hot water for 5-10 minutes. Rinse with clear water, wipe and let air-dry. Preferably follow the recommendations of the manufacturer for cleaning the hat.
- ✓ Always store the hard hat in a clean, dry and cool place as sunlight and heat can damage the suspension of the hat.

• HARD HAT HEALTH AND SAFETY SIGNS

The *circle* ***indicates that the sign is an*** *order* ***that must be enforced and followed by all. Because it has a*** *blue background,* ***it also indicates that one has to*** *do* ***something. The*** *symbol* ***is an indication to wear*** *head protection*

• EAR AND HEARING PROTECTION

Noise is a common problem found in many workplaces. Research has shown that high levels of noise can damage hearing. It is of increasing concern that workers are losing their hearing and going deaf because of exposure to noise levels in excess of 80dB. Losing hearing is a slow but sure process and is less noticeable than other types of workplace injuries. It is, however, a permanent handicap for those who are affected.

Noise exposure is measured in decibels (dB). The level of exposure depends upon a number of factors such as, for example:

- ✓ the loudness or volume of the noise as measured in decibels
- ✓ the expected and actual duration of exposure to the noise
- ✓ whether the worker moves between work areas which have different levels of noise
- ✓ whether the noise is generated from one or multiple sources

Hearing loss can be permanent, so wear protective equipment when noise levels are high.

The following questions can be used as a guide, namely

- ✓ Is it necessary where you are working to speak loudly, raise your voice or even shout to be heard by another worker about 1 meter away?
- ✓ Is there a ringing or buzzing noise in your ears at the end of the day or working shift?
- ✓ Is it difficult to hear people especially after leaving the workplace at the end of the working shift?

If the answer is YES to any of these questions, the noise levels in the workplace are probably too high and there is a need to wear ear protection.

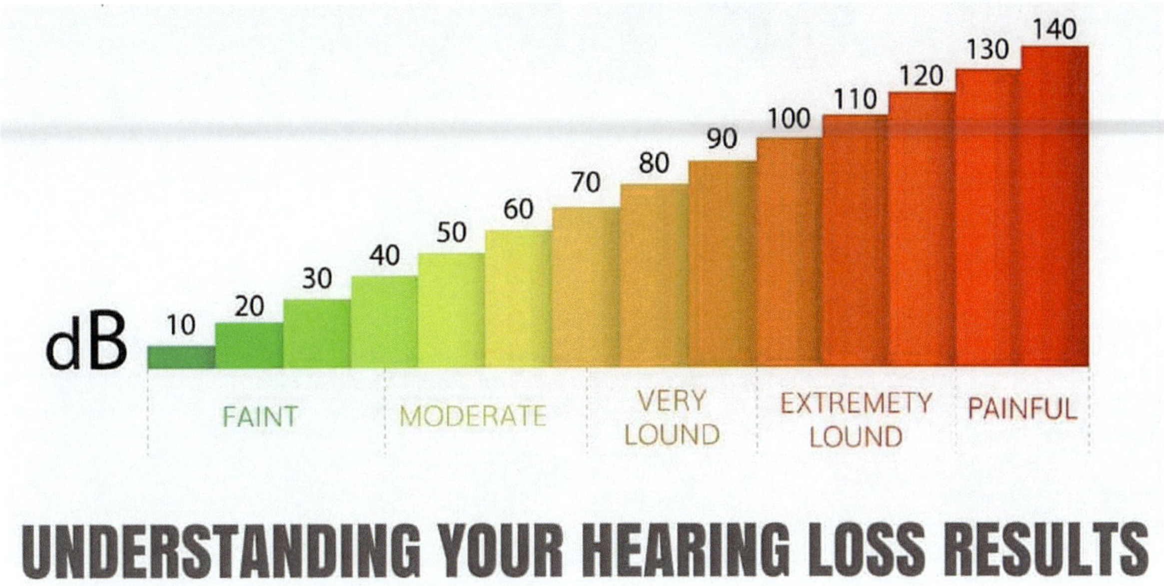

(Source: https://hearingwellnesssolutions.com/understanding-hearing-loss-results/)

One should wear a hearing protection device whenever there is an exposure to noise that is 80 decibels or greater for an 8-hour period of time.

Most hearing protection devices have an assigned rating that indicates the amount of protection provided. Depending on the level of exposure, various hearing devices can be used.

Permissible noise exposures	
Duration per day in hours	Sound level dBA slow response*
8	90
6	92
4	95
3	97
2	100
1½	102
1	105
½	110
¼	115

*A sound level meter's response is generally based on either a fast or slow exponential averaging. Fast corresponds to a 125-millisecond time constant; slow corresponds to a 1-second time constant.

(Source: https://www.professionalroofing.net/Articles/A-noise-notice--07-01-2007/1099)

• TYPES OF EAR PROTECTION

There are many types of hearing protection devices available with each type of hearing protection having its own advantages and disadvantages.

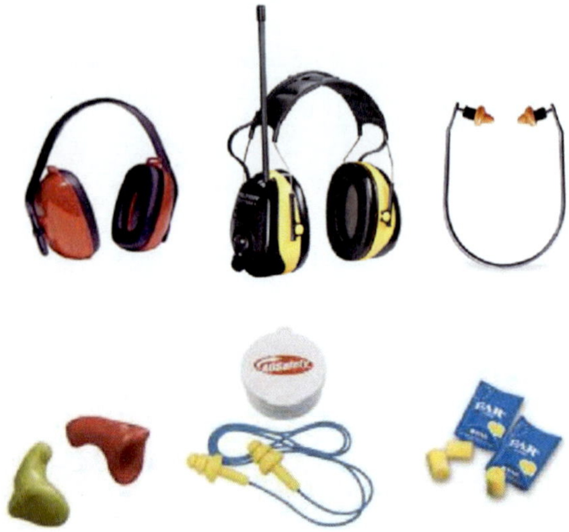

Clockwise from top left: Standard earmuffs, radio earmuffs, canal caps,

Different types of ear protection
(Source: https://www.researchgate.net/figure/Examples-of-common-personal-hearing-protection-devices_fig2_264555852)

The advantages of foam and PVC earplugs are:

- ✓ small & lightweight
- ✓ comfortable in hot environments
- ✓ easily used with other safety equipment

The disadvantages of earplugs are:

- ✓ may work loose and require occasional refitting
- ✓ require specific fitting instruction
- ✓ are frequently dirty

The advantages of earmuffs are:

- ✓ easy for the employer to supervise the wearing of this device
- ✓ one size fits all
- ✓ fits better for longer periods of time

The disadvantages of earmuffs are:

- ✓ may fit tight on the head
- ✓ uncomfortable in a warm environment
- ✓ problems occur when used with other equipment

- WEARING EAR PROTECTION

Generally, foam earplugs provide the most protection.

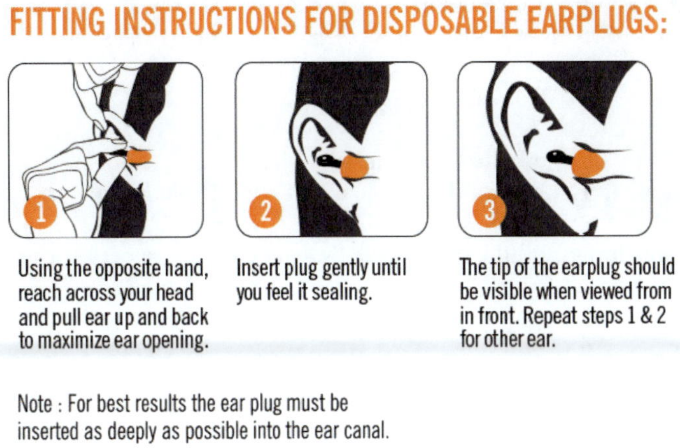

(Source: https://blog.prochoice.com.au/personal-protective-equipment/health-and-hygiene-dangers-of-disposable-ppe/)

To insert foam earplugs properly:

- ✓ Reach around the back of the head and gently pull the ear back and up.
- ✓ Roll the plug into a small roll.
- ✓ Insert the plug well into the ear canal.
- ✓ Hold the plug in place for a few seconds while it expands and forms a good seal

To insert PVC earplugs properly:

- ✓ Reach around the back of the head, and gently pull the ear back and up.
- ✓ Begin to insert the plug into the ear canal.
- ✓ Work the plug well into the ear canal using a circular motion.
- ✓ Once inserted, wait a few seconds to make sure the plug is securely fitted into the ear.

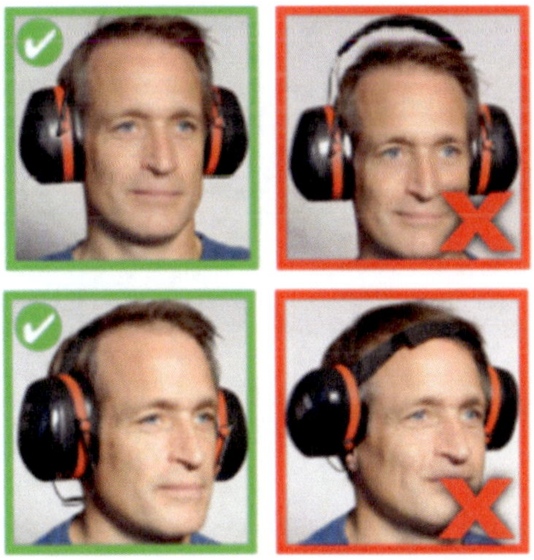

How to wear earmuffs correctly
(Source: https://www.3mnz.co.nz/3M/en_NZ/safety-nz/stories/full-story/?storyid=0cd922a8-6639-4f84-ae38-a4383f902d5c)

Earmuffs fit over the outside of the ears. Because their effectiveness is limited to the quality of the seal around the ear they are often not as effective as earplugs. For this reason, a combination of earplugs and earmuffs are used in some work environments.

- ✓ Make sure that the earmuffs fit tightly around the ears.
- ✓ Consider using earplugs if glasses or earrings are worn or there is facial hair that would prevent the earmuffs from forming a good seal around the ears.

• CARING FOR EAR PROTECTION

Foam earplugs

- ✓ When not using foam earplugs, store them in a clean, cool, dry place.
- ✓ If foam earplugs become dirty, torn or stiff, throw them away and ask the supervisor for a new pair.

PVC earplugs

- ✓ When not using PVC earplugs, store them in a clean, cool, dry place.
- ✓ If PVC earplugs become dirty, they can be cleaned with a mild solution of soap and water. Rinse, then dry them with a soft towel.
- ✓ If PVC earplugs become torn or brittle, throw them away and ask the supervisor for a new pair.

Earmuffs

- ✓ When not using earmuffs, store them in a clean, cool, dry place.
- ✓ Always inspect earmuffs for cracks especially around the foam cups. If the earmuffs are damaged, have them repaired immediately or ask the supervisor for a new pair.

• HEARING PROTECTION HEALTH AND SAFETY SIGNS

The circle indicates that the sign is an order that must be enforced and followed by all. Because it has a blue background, it also indicates that one has to do something. The symbol is an indication to wear hearing protection.

• EYE AND FACE PROTECTION

Eye and face protection comes in many types, depending upon the threat that is to be reduced. Thousands of workers have their eyesight affected and many are blinded each year from work-related eye injuries which could have been prevented if they had been provided with and used the correct eye protection.

Types of eye and face protection
(Source: http://corsafetydemo.corsafety.ca/orientation/ppe05/02ppe05.htm)

Eyes are very delicate and perform the same function to the brain as a colour television camera. But unlike a television camera that is made of plastic and metal, the eye is made of soft tissues and blood vessels. Damage to the eyes causes serious trouble that is often permanent.

• HAZARDS INVOLVING THE EYES

The major types of accidents involving the eyes that cause blindness include:

- ✓ objects striking the eye
- ✓ chemicals and other hazardous materials that get into the eye
- ✓ being struck by swinging objects such as chains and ropes

✓ looking at bright energy sources such as welding operations or lasers

Small particles of matter such as dust can also enter the eyes and damage them. This can also happen with regards to toxic chemicals in the form of gases, vapours and liquids.

Always read the appropriate MSDS before working with any hazardous material and check with the supervisor or health and safety manager to learn the type of eye or face protection that is appropriate in order to work safely.

• EYE PROTECTION MEASURES

There are a wide variety of ways to protect the eyes and circumstances when eye protection is important.

✓ Many types of machines such as compressors, conveyor belts etc. are equipped with guards, shields and screens. Always make sure that these are in place and in good working order before working on these machines. Wear eye protection.
✓ Ventilation - Operations which use or produce vapours, gases, mists, dusts, powders and other airborne particles should be ventilated. Ventilation, along with damping systems, can significantly reduce the amount of airborne particles that could be hazardous to the eyes.
✓ Lighting - Good lighting is important in work areas. Good lighting reduces eye strain and glare. It also promotes health and safety and improves productivity.
✓ Signs and warnings - Obstructions and objects that stick out should be identified and marked. Use caution when working around these objects.

(Source: https://www.dreamstime.com/caution-wear-eye-protection-transparent-background-image141031595)

✓ Eyewash stations should be located within a maximum of 30m of the work area. Eyewash basins should be provided in the ablution facilities at work. Each first aid box should include an eye wash container.
✓ Safe work practices - Apply safe work procedures and practices and do not participate in horseplay.

• TYPES OF EYE PROTECTION

Safety glasses are perhaps the most widely used type of eye protection. While they may to be similar in appearance to regular glasses, they are much stronger and more resistant to impact and heat than regular glasses. In addition, most safety glasses are equipped with side shields that give protection from exposure to hazards that may not be directly ahead.

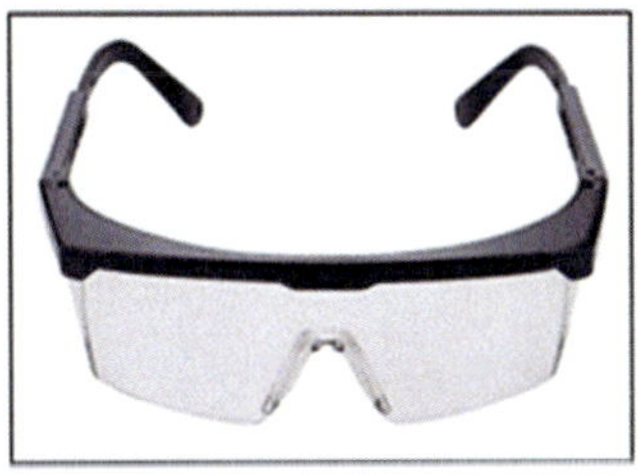

Safety glasses

Both prescription for workers with poor eyesight and non-prescription safety glasses are available. In addition, a wide variety of lens coatings are available for special work situations.

Goggles provide more protection than safety glasses because they fit closer to the face. Because goggles surround the eye area, they give more protection in situations where splashing liquids, fumes, vapours, powders, dusts and mists might be encountered. Different types of goggles are available. If it is necessary to work with chemicals, the goggles must indicate that they are chemical splash goggles.

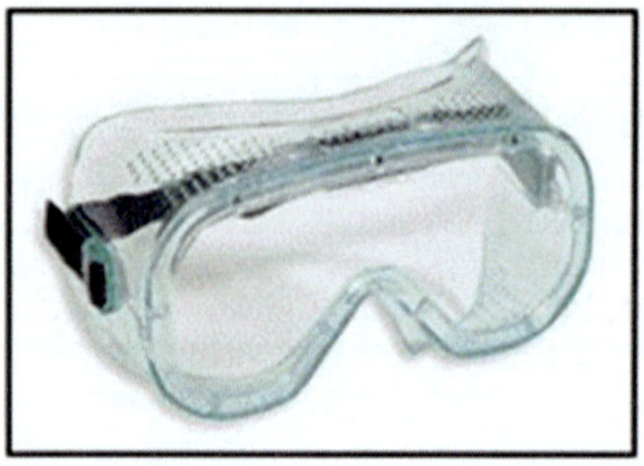

Goggles

Face shields are made of clear and transparent Perspex fastened with straps that fits around the head and offers full face protection. Many face shields can be used while wearing a hard hat. It is often used around operations with exposure to molten metal, chemical splashes or flying particles. With the recent COVID-19 pandemic face shields have been used to reduce the likelihood of infection from the virus by protecting the person wearing them from droplets emitted during speaking, coughing or sneezing.

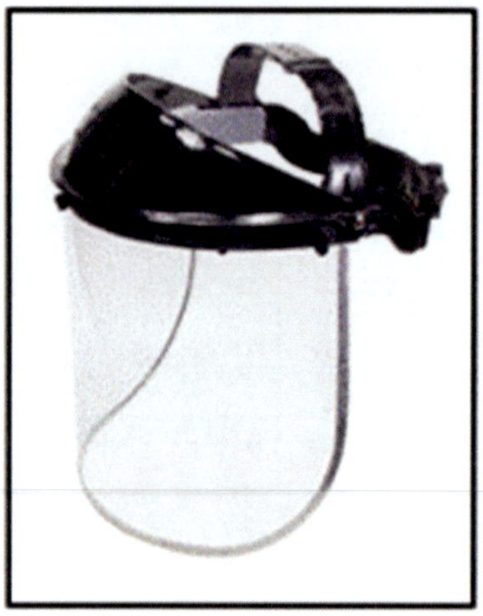

Face shield

One should always wear safety glasses or goggles when using a face shield for added protection. Face shields alone are NOT considered adequate eye protection.

- CONTACT LENSES

When wearing contact lenses, keep the following health and safety tips are useful, namely

Contact lens
(Source: https://www.sunwaymedical.com/blogpost/the-best-contact-lens-is-lvc)

- ✓ While using a full-face respirator contact lenses must not be worn.
- ✓ Wear contacts with caution when working in areas where there might be exposure to fumes, dusts, powders, vapours, chemical splashes, molten metals, or intense heat, light or glare.
- ✓ If anything gets under contact lenses, take time to remove and clean them. Follow the instructions of the optician or ophthalmologist for cleaning and caring for contact lenses.
- ✓ Some chemicals can react with contacts and cause permanent injury.

- ✓ It is wise to keep an extra pair of contacts or a pair of glasses handy in case one of the contact lenses could be lost or damaged while working.

Do not wear contact lenses in hazardous areas. Contact lenses do not provide eye protection and may reduce the effectiveness of emergency eyewash.

• CARE OF EYE PROTECTION EQUIPMENT

It is important to take proper care of eye protection equipment to prevent accidents and damage to the eye and face.

- ✓ Clean eye protection equipment usually by using mild soap and water. Special wipes that are designed for cleaning protective eye equipment may also be used. Never use abrasive soaps, rough paper, or cloth towels. These items will scratch and damage eye protection equipment.
- ✓ Always keep eye protection equipment in good working condition. If it is damaged, have it repaired or replaced.
- ✓ Store eye protection equipment in a sanitary, cool, dry area away from moisture.
- ✓ Read the manufacturer's directions and warnings before using any eye protection equipment.
- ✓ If there are any questions concerning eye protection equipment, consult the supervisor at work.

Eye safety is a choice that each and every one has.

• EYE PROTECTION HEALTH AND SAFETY SIGNS

Areas requiring eye protection will be clearly marked on hazard identification placards posted at the entrance of the workplace and / or at the specific work area.

*The **circle** indicates that the sign is an **order** that must be enforced and followed by all. Because it has a **blue background**, it also indicates that one has to **do** something. The **symbol** is an indication to wear face protection.*

• RESPIRATORY PROTECTION / FACE MASKS

Health hazards in the workplace are a major concern for both employers and workers. It is important to remember that hazardous materials only present a health hazard when they come into contact with the body. Hazardous materials can enter the body in three ways, namely via

- ✓ Ingestion by swallowing them
- ✓ Skin absorption through contact with the skin
- ✓ Inhalation by breathing them in

Of the three ways, inhalation is the most common way of contact for most materials which are health hazards. This includes inhaling dust, fumes, oil mist and vapours from solvents and various gases. Inhaling hazardous materials damage the delicate structure of the lungs and can lead to respiratory diseases. These diseases often cannot be cured and eventually could lead to death. It is important to know how to protect against the many substances which are health hazards that become airborne. The use of face masks or shield has become a mandatory form of preventing COVID-19 infection not only in public spaces but especially in the workplace.

• POTENTIAL RESPIRATORY HAZARDS

- ✓ Dust is formed whenever solid material is broken down into tiny particles.
- ✓ Vapours are substances that are created when a solid or liquid material evaporates into the air.
- ✓ Gases are materials that become airborne at room temperature. Gases may have an odour, but many do not. Some gases can be seen, but again, others cannot. Gases may be heavier than air or lighter than air, but in either case, can travel undetected for great distances.
- ✓ Inhalation of droplets containing the highly infectious COVID-19 virus

- SELECT THE CORRECT RESPIRATOR

There are many types of respirators and they have different abilities to protect against different hazards

Respirator type	Protection	No protection
Filter Respirator	Dust, Fumes, Smoke, Mist, Microorganisms, Asbestos	Chemical vapours or gases Oxygen deficiency
Chemical Cartridge / Canister Respirators	Certain gases and vapours up to a particular concentration	Oxygen deficiency Particulate matter
Air Supply Respirator	Depending on type: Particulates Chemical vapours and gases Oxygen deficiency	

The first step in selecting the correct respirator or face mask is to determine the level of hazard that is posed by the environment in which work is being done. To do this, four basic questions should be answered, namely:

- ✓ What type of contaminant is present?
- ✓ What is the form of the contaminant?
- ✓ How toxic is the contaminant?
- ✓ What is the concentration of the contaminant?

In addition to determining the level of hazard that is posed by the environment, the following must also be considered, namely:

- ✓ How long will the exposure to the contaminant be?
- ✓ What is the individual sensitivity to the contaminant?
- ✓ What are the individual requirements of affected workers?
- ✓ Are glasses being worn?
- ✓ Does the worker have a beard or other facial hair?
- ✓ Are dentures being worn?
- ✓ Will other protective equipment have to be worn as well?

Discuss with the supervisor or health and safety representative to determine the correct respirator that will be needed to work healthily and safely.

- TEST THE RESPIRATOR

The following steps should be followed, namely:

1. With one hand, hold the respirator to the face
2. While holding the respirator in place, slip the head harness over the head
3. Adjust and tighten the head harness straps until the respirator fits snugly to the face. The best way to tighten a respirator is to tighten the straps from the bottom up

The worker should demonstrate donning the respirator to the supervisor or health and safety representative.

Always check the guidelines provided by the manufacturer before donning, fitting and using the respirator

Because an airtight seal is needed between the face and the respirator it is necessary to fit test the respirator each time it is worn to ensure that no contaminant gets inside the face piece and into the lungs.

Positive-pressure check

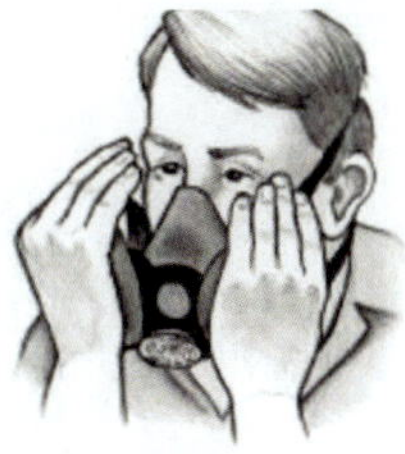
Negative-pressure check

(Source: https://ohsonline.com/articles/2014/06/26/oregon-osha-posts-respiratory-program-guide-for-small-businesses.aspx)

Positive Pressure Test

The exhalation valve of the respirator must first be closed by covering it with the hand before breathing out slowly. The face piece will typically bulge out slightly. The breath should be held for about 10 seconds. If there are no air leaks during this period from around the face piece, the respirator has a good fit. If the fit is not good, the head harness straps must be readjusted. The pressure test must be repeated until the desired result is achieved. Another way to test the seal formed by the respirator around the face is to perform a negative pressure test.

Negative Pressure Test

Close the inhalation valves of the respirator with the hands and breathe in slowly. The face piece will collapse slightly. Hold the breath for about 10 seconds. If during this time, no air leaks in from around the face piece the respirator has a good fit. If there is not a good fit, the head harness straps must be readjusted the pressure test must be repeated.

It is a good idea to perform both the positive and negative pressure tests on the respirator before using it

• MONITOR THE RESPIRATOR

When any work activity is being done wearing a respirator, not only must the seal around the face be monitored, but the efficiency of the respirator must also be monitored.

The respirator is not working when:

- ✓ The contaminant can be smelt or tasted
- ✓ Breathing becomes difficult
- ✓ When the worker becomes dizzy or feels sick
- ✓ The manufacturer's recommended service life of the filters or cartridges expires
- ✓ The respirator is damaged

WARNING: Never use or continue to use a respirator that is not working perfectly

• CARE FOR THE RESPIRATOR

After using a respirator, it should be cleaned and properly inspected. During cleaning, the following should be looked out for, namely:

- ✓ Cracks or chips in the faceplate
- ✓ Cracks or holes in the breathing tube or airlines
- ✓ Worn or frayed straps
- ✓ Worn or damaged fittings
- ✓ Bent or corroded buckles
- ✓ Improperly seated valves

If anything is found to be wrong with the respirator, it must be repaired or replaced immediately

Storing a respirator
(Source: https://pksafety.com/blog/how-to-clean-your-respirator-mask

- STORING A RESPIRATOR

If the respirator is not going to be used immediately, it will need to be stored and placed in a sealable plastic bag. The respirator must be stored in a convenient place, but away from:

✓ Dust
✓ Sunlight
✓ Heat
✓ Extreme cold
✓ Moisture
✓ Damaging chemicals

- BODY PROTECTION

Hazards that threaten the upper body tend to threaten the entire body. A variety of protective clothing such as overalls, or coveralls, coats, long pants, rubber aprons, leather aprons and disposal body suits are available for specific work conditions.

- POTENTIAL HAZARDS

When hazardous chemicals come into contact with the skin, a lot of damage is caused which can be permanent if proper care has not been taken. Chemical resistant protective clothing is designed with health and safety in mind.

- PROTECTIVE SUITS, OVERALLS OR COVERALLS

Long coats, overalls, and disposable body suits protect workers and everyday clothing from contamination.

Examples of long coat, overall and two piece suit to protect body
(Source: https://workwearsolutions.wordpress.com/2013/12/17/work-overalls-and-work-clothing/)

Coveralls and Two-Piece Suits

Rubber, neoprene, and plastic clothing protect workers from most acids and chemical splashes. The purpose of chemical protective clothing and equipment is to shield or isolate individuals from the chemical, physical, and biological hazards that may be encountered during operations involving hazardous materials. During chemical operations, it is not always apparent when exposure to the various hazards could occur. Many chemicals pose invisible hazards and offer no warning properties. Protective clothing, such as a chemical resistant apron or two-piece suit must be worn whenever the worker faces potential hazards arising from chemical exposure.

Chemical Resistant Apron and Two-Piece Suit
(Source: https://www.grainger.com/product/LAKELAND-Chemical-Resistant-Bib-Apron-6EHH6 and https://www.abatix.com/catalog/product/view/id/103775)

The apron is made out of chemical resistant material and protects the body from the front while handling chemicals. The two-piece suit protects the rest of the body, as the legs and arms fit in snugly. Note that under COVID-19 conditions there may be the need for extra protection and specialized PPE.

Wash reusable protective clothing separate from other clothing

- CARE FOR PROTECTIVE CLOTHING

It is important to take care of the protective clothing. Therefore, a suit or apron in good condition must always be used.

Before use:

✓ Inspect for damages
✓ Inspect for holes
✓ Ensure apron is long enough and it covers the legs up to the feet
✓ Ensure that the apron will not slip when handling chemicals

- HAND PROTECTION

It has been estimated that almost 20% of all disabling accidents on the job involve the hands. Without fingers or hands the ability to work will be greatly reduced. Human hands are unique. No other creature in the world has hands that can grasp, hold, move and manipulate objects like human hands. They must be protected and cared for.

- POTENTIAL HAZARDS INVOLVING HANDS

There are many ways in which hands can sustain a bad injury. These include:

✓ Exposure to tools and machines with sharp edges that can cut the hands.
✓ Working with or in proximity to staples, screwdrivers, nails, chisels and stiff wire that can puncture the hands.
✓ Getting hands caught in machinery that can sprain, crush or remove hands and fingers.
✓ Coming into contact with caustic or toxic chemicals, electrical sources or extremely cold or hot objects can irritate or burn the hands.

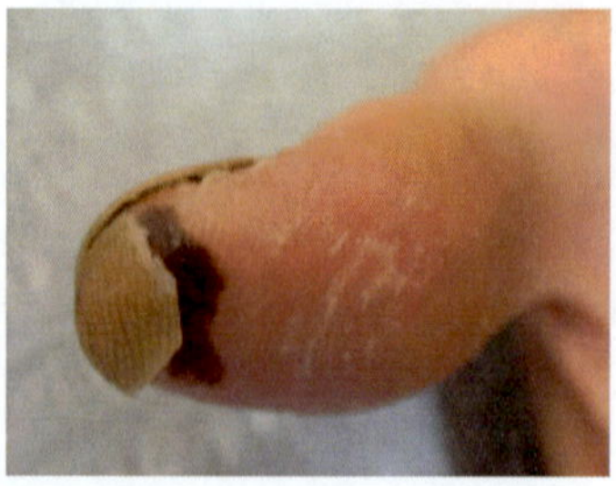

Example of toxic fluoride burn
(Source: https://dermnetnz.org/topics/chemical-burn/)

WARNING: Toxic substances are poisonous substances that can be absorbed through the skin and enter the body.

• TYPES OF HAND PROTECTION

Many types of gloves have designed to with protection of the hands in mind such as, for example:

- ✓ Metal mesh gloves that resist sharp edges and prevent cuts.
- ✓ Leather gloves that shield hands from rough surfaces.
- ✓ Vinyl and neoprene gloves that protect hands against toxic chemicals.
- ✓ Rubber gloves that protect hands when working around electricity.
- ✓ Padded cloth gloves that protect hands from sharp edges, slivers, dirt, and vibration.
- ✓ Heat resistant gloves protect hands from heat and flames.
- ✓ Latex disposable gloves that are used to protect hands from germs and bacteria. These have become popular as a result of COVID-19 pandemic.
- ✓ Lead-lined gloves that are used to protect hands from radiation sources.

Types of gloves as hand protection
(Source: https://www.123rf.com/photo_105092929_protective-gloves-multiple-types-rubber-and-cloth-materials-for-household-gardening-and-cleaning-haz.html)

• WEARING AND USING GLOVES

Because hand protection is so important the following guidelines should be followed, namely:

- ✓ Select and use the right kind of glove for the job that is going to be done.
- ✓ Select gloves that fit.
- ✓ Some gloves may be chemical specified and have a life expectancy. Throw them away after the recommended time has expired.
- ✓ Remove any rings, watches or bracelets that might cut or tear the gloves.
- ✓ Wash hands before and after wearing gloves. Practice good personal hygiene always.
- ✓ Inspect gloves before they are used. Look for holes and cracks that might leak.
- ✓ Replace gloves that are worn or torn.
- ✓ After working with chemicals, hold gloved hands under running water to rinse away any chemicals or dirt before removing the gloves.
- ✓ Wash cotton gloves as needed.
- ✓ Avoid borrowing gloves. Gloves are *personal* protective equipment.
- ✓ Store gloves right side out in a clean, cool, dry, ventilated area.

- ✓ Never wear gloves around powered rotating equipment such as drills and lathes.

• OTHER HAND PROTECTION MEASURES

Several other forms of protection including barrier creams may be used to protect hands. These include, for example:

- ✓ Water repellent creams are used to protect the hands from caustic chemicals.
- ✓ Solvent-repellent creams are used to protect hands from solvents, oils, and other organic chemicals.
- ✓ Vanishing creams protect the skin against mild acids and make cleaning up easier.

WARNING: Never substitute a barrier cream even when using gloves.

- ✓ Forearm cuffs are used to protect the forearm.
- ✓ Thumb guards and finger cots protect only the thumb or fingers.
- ✓ Mittens protect the hands while working around very cold or hot materials.
- ✓ Hand pads are often found in laboratories. Hand pads protect hands while working around very hot materials.

• HAND HEALTH AND SAFETY SIGNS

The** circle **indicates that the sign is an** order **that must be enforced and followed by all. Because it has a** blue background**, it also indicates that one has to** do **something. The** symbol **is an indication to wear hand protection.

Although the sign instructs to wear safety gloves, it does not provide information on the specific hazard exposure or the type of hand protection to use. Additional signs are often placed together with the first sign to provide further clarification of the specific hazard exposure expected.

For example, this sign is a warning sign. A triangle will always warn against a specific danger. This sign warns one of the potential risk of corrosive materials. The hand protection to use must protect one against corrosive material.

• SOME GENERAL GUIDELINES FOR USING GLOVES

The following are useful guidelines when using gloves at work, namely:

- ✓ Inspect and test new gloves for defects.
- ✓ Always wash hands before and after using gloves.
- ✓ Wash chemical-protective gloves with soap and water before removing.
- ✓ Do not wear gloves near moving machinery as they may get caught in the moving parts of the machine.
- ✓ Do not wear gloves with metal parts near electricity.

It may be necessary to wear gloves to prevent infection with the COVID-19 virus when coming into contact work surfaces that could have been infected.

Gloves are easily contaminated. Avoid touching surfaces such as telephones, doorknobs, etc. when wearing gloves.

• FOOT PROTECTION

There are two major categories of work-related foot injuries. Category one includes foot injuries from punctures, crushing, sprains, and lacerations. Category two includes injuries that result from slips, trips and falls. These do not always result in a foot injury. However, ignoring foot health and safety contributes to increasing the likelihood of them occurring. Other foot conditions unrelated to the workplace such as calluses, ingrown toenails or simply tired feet are common among workers. These can have serious consequences for health and safety at the workplace because they cause discomfort, pain and fatigue. Fatigue causes injuries affecting the muscles and joints. A tired worker who is suffering pain is also less alert and more likely to act unsafely. An accident or incident may consequently result.

Since feet are designed for mobility, maintaining an upright stance can be extremely tiring. Therefore, standing for hours every working day not only tires the feet of workers but is likely to also cause permanent damage. Continuous standing can cause the joints of bones of the feet to become misaligned such as, for example, flat feet and can cause inflammation that can lead later to rheumatism and arthritis. All working safety footwear should provide comfort without compromising protective value.

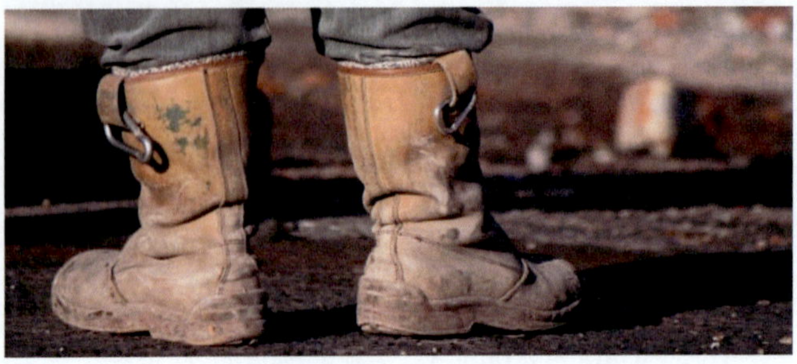

Example of a worker wearing foot protection
(Source: https://www.highspeedtraining.co.uk/hub/ppe-in-construction/)

• POTENTIAL FEET-RELATED HAZARDS

Workers may be exposed to injuries including:

- Crushing from falling objects – impact injuries,
- Crushing from rolling cylinders – compression injuries,
- Punctures from sharp objects,
- Burns or shocks from electrical hazards,
- Burns from molten metal or hot surfaces,
- Skin contact or burns from chemicals,
- Injuries from spills and splashes,
- Frostbite, blisters or burns from extremes in cold, heat or moisture,
- Poor housekeeping involving poorly maintained machinery, tools, sloppy work areas and cluttered walkways, or
- Slips and falls from wet or slippery surfaces.

• PREVENTATIVE MEASURES FOR FOOT HEALTH AND SAFETY

Ergonomically improving the design of the workplace can increase foot health and safety and reduce damage to the feet of workers. Examples of the design interventions include

- Separating mobile equipment from pedestrian traffic and installing safety mirrors and warning signs can decrease the number of incidents that might result in cut or crushed feet or toes.
- Proper guarding of machines such as chain saws or rotary mowers can avoid cuts or severed feet or toes.
- Effective housekeeping reduces the number of incidents at workplaces. For example, loose nails, other sharp objects, and littered walkways are hazards for foot injury.
- Using colour contrast and angular lighting to improve depth vision in complicated areas such as stairs, ramps and passageways reduces the hazard of tripping and falling.
- Posting safety signs in conspicuous places where safety footwear is required when there is a potential hazard from falling objects and sharp objects.

• FEATURES OF FOOTWEAR

The right type and size of footwear should always be worn. Where a steel toe cap is part of the footwear it should cover the whole length of the toes from tips to beyond the natural bend of the foot. Preferably a soft pad covering the edge of the toecap will increase the level of comfort. Should the toecap cut into the foot, then either the size or style of the footwear is incorrect.

A further consideration is that soles of safety footwear come in a variety of thicknesses and materials. They must be chosen according to the hazards and type(s) of flooring encountered in the workplace.

Additionally, the uppers of protective footwear come in a variety of materials. Therefore, the selection of footwear should take into account the hazards, and individual characteristics of the worker's foot.

A steel midsole which protects the foot against penetration by sharp objects should be flexible enough to allow the foot to bend.

It is important to be aware that no one type of non-slip footwear can prevent the wearer from slipping on every surface type.

• TYPES OF PROTECTIVE FOOTWEAR

There are a wide ranges of protective footwear options. The choice of which to wear depends on the hazards to be protected from and the work activity to be done by the worker.

- ✓ Steel toe footwear protects the toes from falling objects and from being crushed.
- ✓ Metatarsal footwear has special guards that run from the ankle to the toes and protects the entire foot.
- ✓ Reinforced sole footwear has metal reinforcement that protects the foot from punctures.
- ✓ Latex / rubber footwear resists chemicals and provides extra traction on slippery surfaces.
- ✓ PVC footwear protects the feet against moisture and improves traction.
- ✓ Butyl footwear protects against most ketones, aldehydes, alcohols, acids, salts, and alkalis.
- ✓ Vinyl footwear resists solvents, acids, alkalis, salts, water, grease, and blood.
- ✓ Nitrile footwear resists animal fats, oils, and chemicals.
- ✓ Electrostatic dissipating footwear conducts static electricity to floors that are grounded.
- ✓ Electrical hazard footwear is insulated with tough rubber to prevent shocks and burns from electricity.
- ✓ Disposable footwear includes shower slippers, clear polyethylene and non-woven booties used in dust free work areas.

Examples of the range of protective footwear
(Source: https://www.tendersontime.com/blogdetails/supply-safety-shoe-6988/)

• WEARING AND USING SAFETY FOOTWEAR

It is important that workers know when and how to use the appropriate safety footwear. The following need to be considered:

- ✓ Select and use the right kind of footwear for the job going to be performed. Footwear should meet or exceed the minimum standards set by the SABS.
- ✓ Avoid footwear made of leather or cloth working around acids or caustics. These chemicals quickly eat through the leather or cloth and can injure the feet.
- ✓ Select footwear that fit.
- ✓ Inspect footwear before they are used. Look for holes and cracks that might leak.
- ✓ Replace footwear that is worn or torn.
- ✓ After working with chemicals, hose footwear with water to rinse away any chemicals or dirt before removing them.
- ✓ Avoid borrowing footwear. Footwear is *personal* protective equipment.
- ✓ Store footwear in a clean, cool, dry, ventilated area.

Additional aspects for the proper use and maintenance of safety footwear include the following, namely:

- ✓ Leggings protect the lower legs and feet from heat hazards such as molten metal and welding sparks.
- ✓ Metatarsal guards protect the instep area from impact and compression. They are made of aluminium, steel, fibre or plastic and may be strapped to the outside of the shoes.
- ✓ Toe guards fit over the toes of regular shoes to protect the toes from impact and compression.
- ✓ Combination foot and shin guards protect the lower legs and feet.
- ✓ Safety shoes have impact resistant toes and heat resistant soles to protect the feet.

- ✓ Electrically conductive shoes provide protection against the build-up of static electricity. These shoes are worn by workers working in explosive and hazardous locations such as explosives manufacturing facilities or grain elevators. These shoes are non-conductive and will prevent the owner's feet from completing an electrical circuit to the ground. Non-conductive footwear must not be used in explosive or hazardous locations
- ✓ As with all protective equipment, safety footwear should be inspected before each use.
- ✓ Shoes and leggings should be checked for wear and tear at reasonable intervals. Look for cracks or holes, separation of materials and broken laces.
- ✓ Soles should be checked for metal or other embedded items that could present electrical or tripping hazards.

Do not wear sandals or open-toed shoes in laboratories, workshops, construction sites or other potentially hazardous areas.

- WORKPLACE SIGNAGE PERTAINING TO PPE

Safety Sign	Meaning
	**Safety Shoes** It is recommended that approved safety shoes be worn to protect your feet. Different safety shoes for different conditions such as, for example, shoes that offer protection against slipping, acids, crushing, and liquids.

Safety Sign	Meaning
	**Eye Protection** Proper eye protection must be worn when the nature of the operation presents a potential eye or face injury. Examples of these hazards include flying objects, dust, hot or splashing metals, harmful rays, caustics or acids.

  	**Respiratory Equipment** Approved respiratory equipment shall be worn when the worker is exposed to toxic chemicals or dusts, spray-painting, or other inhalation hazards.
	**Head protection** Hardhats shall be worn in all designated areas and construction areas. Hardhats offer protection against bumps and falling objects.
	**Gloves** Appropriate gloves and aprons shall be worn when handling hazardous chemicals and abrasive materials.
	**Hearing protection** Hearing protection will be worn. There are many different types of hearing protection available, each with its own noise reduction capabilities, use the type as prescribed by the employer.

Safety Sign	Meaning
	**Safety harness and fall protection.** Use a safety harness and Lanyard. This safety equipment must be used when working on scaffolding or any elevated surface that is higher than 2 meters.
	**Aprons shall be worn** Wearing of protective aprons. Workers working with or cutting and welding of metals will normally do this.

Our responsibility for maintaining workplace health and safety comes with our right to a healthy and safe workplace. Everyone is accountable (as an individual) for carrying out their responsibilities. The greater the authority, the greater the accountability.

• REVIEW QUESTIONS

1. Why should PPE not be a measure of first resort?
2. How would the wearing of contact lenses be affected by PPE?
3. When would you have to wear ear protection?
4. How would you know which type of gloves to wear?
5. How would you select the right safety footwear?

• REFERENCES

Anonymous, (nd). Loose notes on safety, health and environment management

Ayu, Baiq. Fathin., Tualeka, Abdul. Rohim. and Wahyudiono, Yustinus. Denny. Ardyanto., 2018. The analysis of factors which are related to the compliance of welder workers in using workplace personal protective equipment in Pt. pal Indonesia. *Indian Journal of Public Health Research & Development*, *9*(5), pp.47-52.

Balkhyour, Mansour. A., Ahmad, Ijaz. and Rehan, Mohammad., 2019. Assessment of personal protective equipment use and occupational exposures in small industries in Jeddah: Health implications for workers. *Saudi journal of biological sciences*, *26*(4), pp.653-659.

Canadian Center for Occupational Health and Safety. (n.d.). Foot Comfort and Safety at Work. Available at: https://www.ccohs.ca/oshanswers/prevention/ppe/foot_com.html

Canadian Center for Occupational Health and Safety. (n.d.). Hazard control. Available at: https://www.ccohs.ca/oshanswers/hsprograms/hazard_control.html

Creative Safety Supply (n.d.). What is the hierarchy of hazards? Available at: https://www.creativesafetysupply.com/qa/workplace-safety/what-is-the-hierarchy-of-hazards.html

EHS (n.d.). *Personal Protective Equipment (PPE) | Office of Environmental Health and Safety*. [online] Ehs.princeton.edu. Available at: https://ehs.princeton.edu/workplace-construction/workplace-safety/personal-protective-equipment-ppe.

Macedo, Luciana., Gorman, Terry., & Berger, Elliot. (2016) 'Assessment of the effects of various personal protective equipment (PPE) and apparel in the performance of earmuffs', Paper presented at the Australian Institute of Occupational Hygienists Inc. (AIOH) 34Th Annual Conference & Exhibition, 3 - 7 December 2016. Gold Coast, Queensland

Martinelli, Katie. 2017. PPE in construction: Safety Guidance and Checklist. https://www.highspeedtraining.co.uk/hub/ppe-in-construction/ Accessed August 20, 2020

Nath, Nipun. D., Behzadan, Amir. H. and Paal, Stephanie. G., 2020. Deep learning for site safety: Real-time detection of personal protective equipment. *Automation in Construction*, *112*, p.103085.

Nikulin, Andrei., Ikonnikov, Dmitry. and Dolzhikov, Iliya., 2019. Smart Personal Protective Equipment in the Coal Mining Industry. *International Journal of Civil Engineering and Technology*, *10*(4), pp.852-863.

Nikulin, Andrei. and Romanov, Anatoly., 2017. Control over the use of personal protective equipment by employees, head protection. *Ecology, Environment and Conservation*, *23*(1), pp.384-389.

OSHA (2010). *Eye and Face Protection - PDF Free Download*. [online] Docplayer.net. Available at: http://docplayer.net/48691681-Eye-and-face-protection.html.

OSHA (n.d.). Personal Protective Equipment. Available at: https://www.osha.gov/personal-protective-equipment.html

Princetown University Environmental Health and Safety (n.d.) Foot Protection. Available at: https://ehs.princeton.edu/workplace-construction/workplace-safety/physical-safety/personal-protective-equipment-ppe/foot-protection.html

Safety Grade (n.d.). *Personal Protective Equipment | SafetyGrade | SafetyGrade*. [online] Safetygrade.com. Available at: http://www.safetygrade.com/guides/personal-protective-equipment.

Smith, Todd. D., DeJoy, David. M. and Dyal, Mari-Amanda., 2020. Safety specific transformational leadership, safety motivation and personal protective equipment use among firefighters. *Safety Science*, *131*, p.104930.

Yarpuz-Bozdogan, Nigar., 2018. The importance of personal protective equipment in pesticide applications in agriculture. *Current Opinion in Environmental Science & Health*, *4*, pp.1-4.

CHAPTER 16

WORKING IN CONFINED SPACES

A confined space is an area in the workplace which while they are not necessarily designed for people, are large enough for workers to enter and perform certain work activities. A confined space is an area which is substantially enclosed although not always entirely, and where serious injury or death can occur from hazardous substances or dangerous conditions within the space or nearby such as, for example, lack of oxygen. It also has limited or restricted means for entry or exit and is not designed for continuous occupancy. Confined spaces include, for example, tanks including storage tanks, vessels, silos, storage bins, hoppers, vaults, pits, sewers, manholes, inspection chambers, tunnels, equipment housings, unventilated or poorly ventilated rooms, ductwork, depressions, basements and pipelines.

An example of confined space signage

OSHA in the USA uses the term "permit-required confined space" to describe a confined space that has one or more of the following characteristics, namely:

- ✓ contains or has the potential to contain a hazardous atmosphere;
- ✓ contains material that has the potential to engulf an entrant;
- ✓ has walls that converge inward or floors that slope downward and taper into a smaller area which could trap or asphyxiate an entrant; or
- ✓ contains any other recognized safety or health hazard, such as unguarded machinery, exposed live wires, or heat stress.

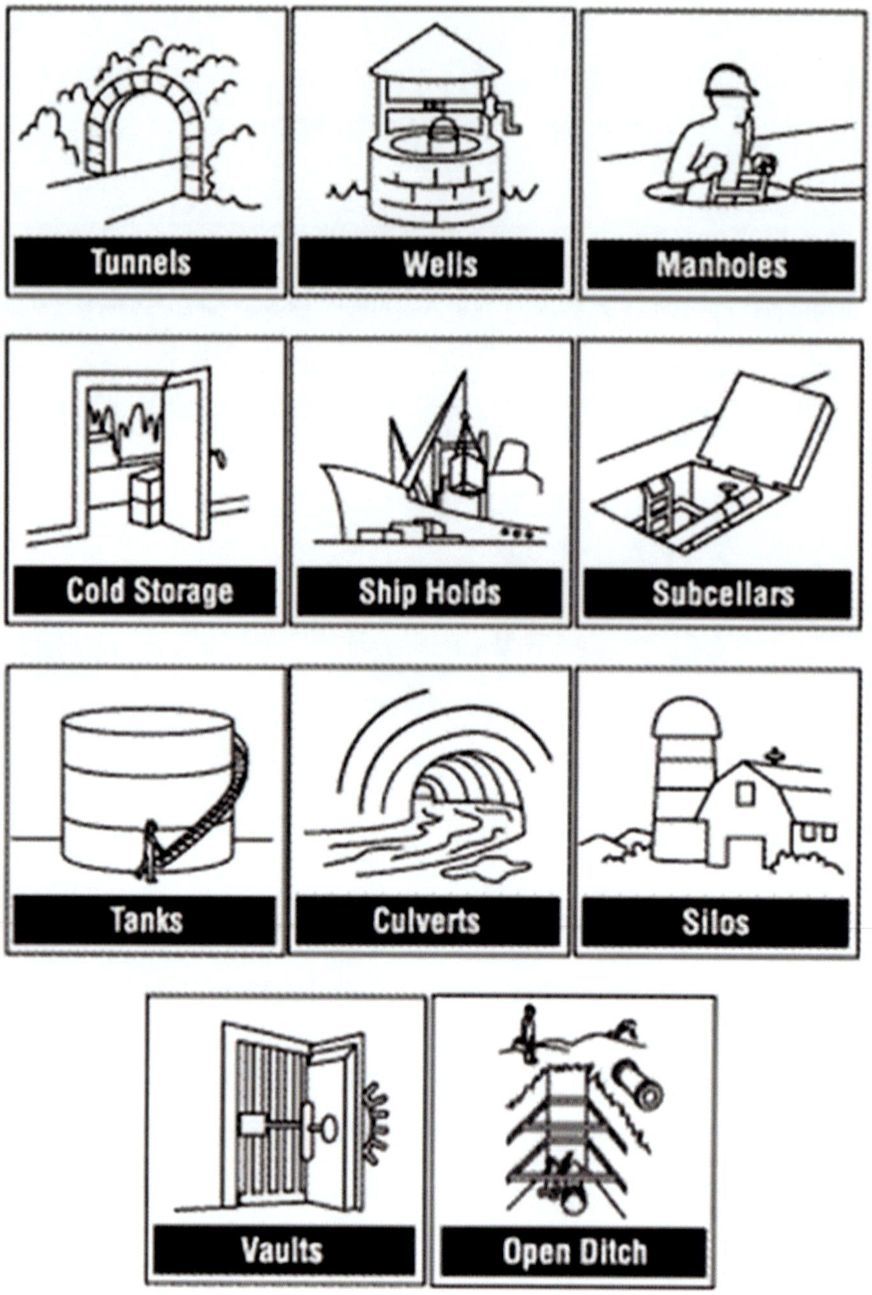

Examples of confined spaces
(Source: http://www.hrdp-idrm.in/e5783/e17327/e17791/e17825/)

• EXAMPLES OF CONFINED SPACES FOUND IN THE CONSTRUCTION SECTOR

The following are examples of confined spaces generally found on construction projects and sites:

• TRENCH

A trench is a type of excavation or depression in the ground. Trenches are generally defined by being deeper than they are wide, and by being narrow compared to their length as opposed to a simple hole. For more information, visit

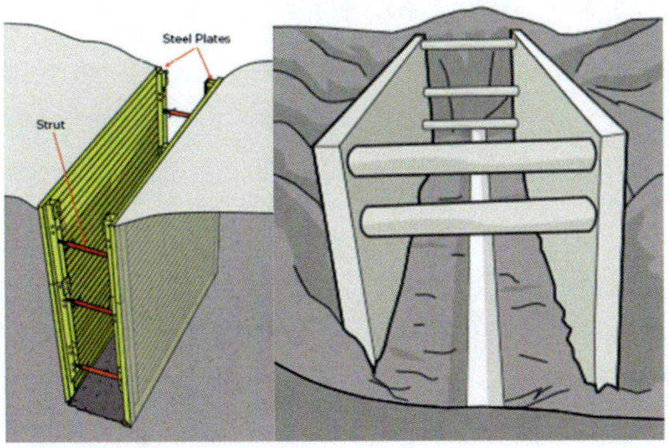

Example of a trench excavation and a trench shield or box

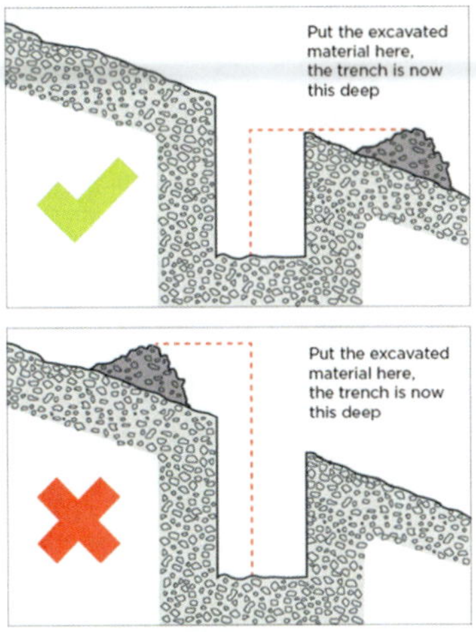

Section through a typical trench excavation

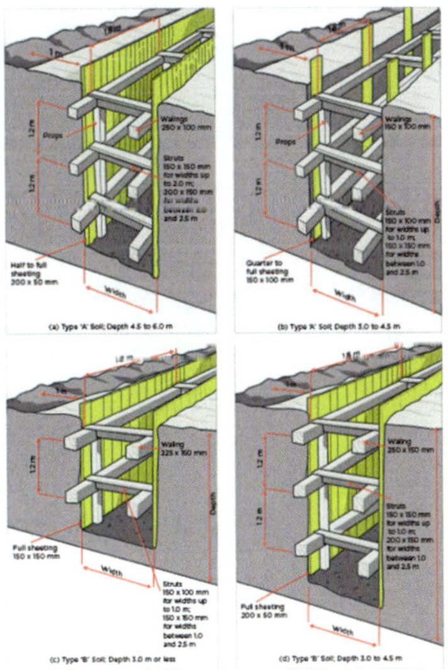

(Source: https://worksafe.govt.nz/topic-and-industry/excavation/excavation-safety-gpg/)

• MANHOLE

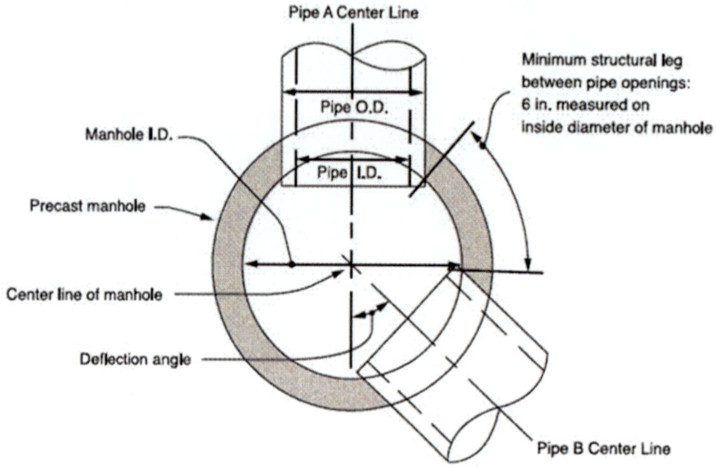

Section through a manhole
(Source: https://precast.org/2010/11/manhole-sizing-recommendations/)

A manhole is an opening to a confined space such as, for example, a shaft, or utility vault containing electrical cables or waterlines. Manholes are often used as an access point for an underground public utility allowing inspection, maintenance, and system upgrades. The majority of underground services have manholes, including water, sewers, telephone, electricity, and storm drains.

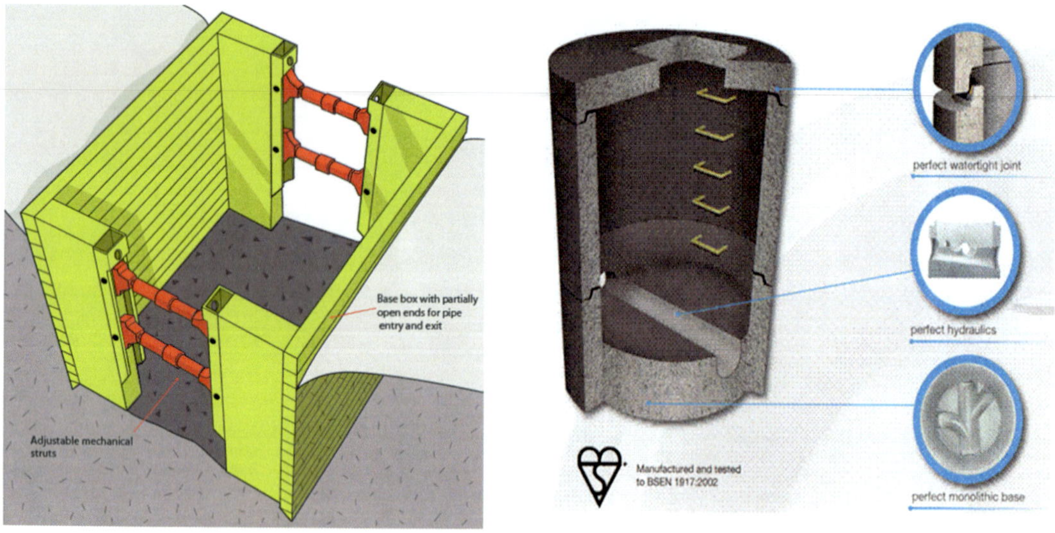

Sections through a typical manhole
(Source: https://worksafe.govt.nz/topic-and-industry/excavation/excavation-safety-gpg/)

• TUNNEL

A tunnel is an underground passageway, completely enclosed except for openings for ingress and egress, commonly at each end. A tunnel may be for foot or vehicular road traffic, for rail traffic, or for a canal. Some tunnels are aqueducts to supply water for consumption or for hydroelectric stations or are sewers. Other uses include routing power or telecommunication cables.

Example of hot work being done in a tunnel
(Source: https://www.masstransitmag.com/rail/infrastructure/article/21135720/new-york-mta-completes-one-l-of-a-tunnel-project

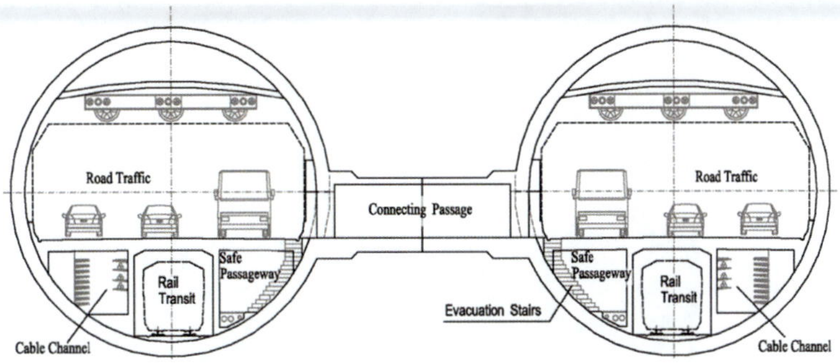

Example of cross section through a tunnel
(Source: https://www.researchgate.net/figure/Cross-section-of-the-twin-bored-tunnels-for-the-Shanghai-under-river-tunnel-after-Huang_fig13_258311283/download)

Example of a tunnel in New York
(Source: https://www.masstransitmag.com/rail/infrastructure/article/21135720/new-york-mta-completes-one-l-of-a-tunnel-project

- DUCTS

Ducts are used in heating, ventilation and air conditioning (HVAC) to deliver and remove air. These needed airflows include, for example, supply air, return air, and exhaust air. Ducts also deliver, most commonly as part of the supply air, ventilation air. As such, air ducts are one method of ensuring acceptable indoor air quality as well as thermal comfort. Ducts are also used for conveying substances such as a liquid or gas. A duct system is often called ductwork.

Example of ductwork
(Source: https://www.indiamart.com/proddetail/rectangula-gss-ducting-4216356612.html

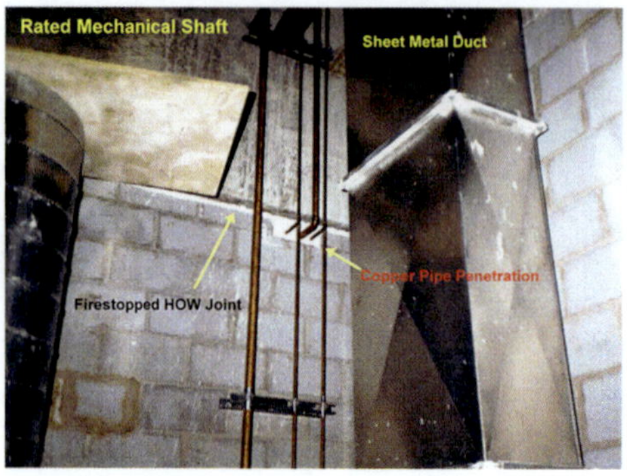

Example of a service shaft
(Source: https://en.wikipedia.org/wiki/Duct_(flow)

- WATER TABLE

The water table is the level below which the ground is completely saturated with water.

A water table refers to the surface of a body of underground water below which the soil or rocks are permanently saturated with water. The water table separates the groundwater zone or zone of saturation that lies below it from the zone of aeration that lies above it. The water table fluctuates both with the seasons and from year to year because it is affected by climatic variations and by the amount of precipitation used by vegetation.

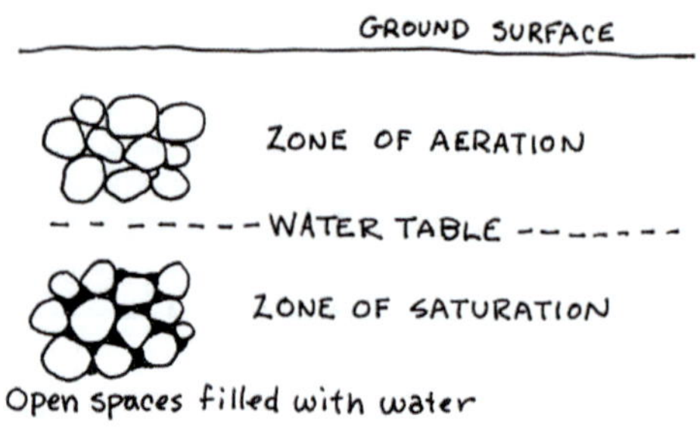

Section showing a water table
(Source: http://sncyear7geography.weebly.com/groundwater.html)

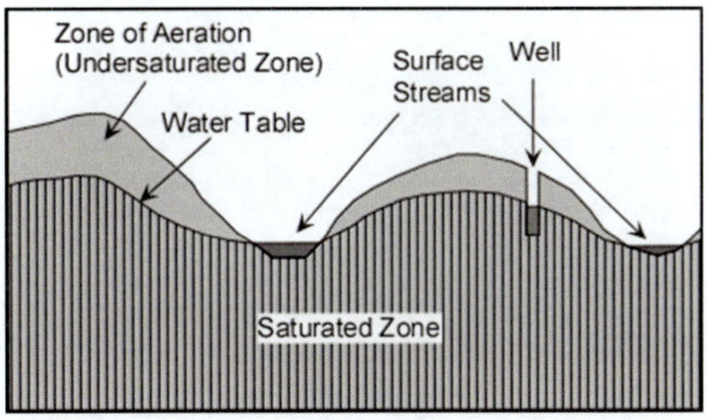

Section showing a water table
(Source: https://www.tulane.edu/~sanelson/eens1110/groundwater.htm)

- DEPRESSION

A depression in geology is a landform sunken or depressed below the surrounding area

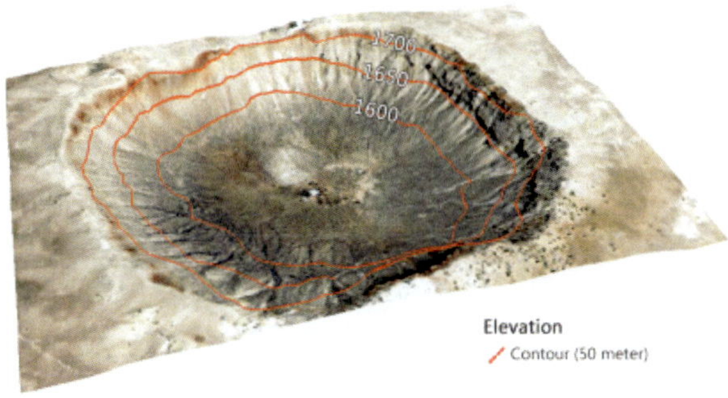

An example of a depression
(Source: https://gisgeography.com/contour-lines-topographic-map/)

Examples of depressions
(Source: https://www.gsi.ie/en-ie/programmes-and-projects/groundwater/activities/understanding-irish-karst/karst-landforms/Pages/Enclosed-depressions.aspx

- BASEMENT

A basement is one or more floors of a building that are either completely or partially below the ground floor. Basements are typically used as a utility space for a building where such items as the boiler, water hot water cylinders or tanks, breaker panel or fuse box, car park and air-conditioning system are located; so also are amenities such as the electrical distribution system.

Example of a basement wine cellar
(Source: https://www.viator.com/en-MY/tours/San-Sebastian/Wine-cellar-visit-with-tasting-and-lunch/d5505-40961P18)

Example of a basement utility area
(Source: and https://www.standard-freeholder.com/opinion/columnists/handyman-hints-how-to-create-and-sustain-basement-life)

It is not possible to provide a comprehensive list of confined spaces. Some places may become confined spaces when work is carried out, or during their construction, fabrication or subsequent modification. Confined spaces can be below or above ground. Confined spaces can be found in almost any workplace. A confined space, despite its name, is not necessarily small. Examples of confined spaces include silos, vats, hoppers, utility vaults, tanks, sewers, pipes, access shafts, truck or rail tank cars, aircraft wings. Ditches and trenches may also be a confined space when access or egress is limited.

• HAZARDS THAT OCCUR IN CONFINED SPACES

In 2015 in the USA, 136 workers died due to confined space-related accidents. All hazards found in a regular workspace can also be found in a confined space. However, they can be even more hazardous in a confined space than in a regular worksite.

Hazards in confined spaces can include any or all of the following, namely:

(Source: https://safels.com/hazards-in-confined-space/)

- ✓ Poor air quality or toxic atmosphere: There may be an insufficient amount of oxygen for the worker to breathe. The atmosphere might contain a poisonous substance that could make the worker ill or even cause the worker to lose consciousness. Natural ventilation alone will often not be sufficient to maintain breathable quality air.
- ✓ Too much oxygen: this hazard is rare. But, once it is present, a fire or explosion could occur.
- ✓ Chemical exposures due to skin contact or ingestion as well as inhalation of 'bad' or toxic air.
- ✓ Fire Hazard: There may be an explosive/flammable atmosphere due to flammable liquids and gases and combustible dusts which if ignited would lead to fire or explosion.
- ✓ Process-related hazards such as residual chemicals, release of contents of a supply line.
- ✓ Noise.
- ✓ Safety hazards such as moving parts of equipment, structural hazards, entanglement, slips, falls.

- ✓ Radiation.
- ✓ Temperature extremes including atmospheric and surface.
- ✓ Shifting or collapse of bulk material.
- ✓ Barrier failure resulting in a flood or release of free-flowing solid.
- ✓ Uncontrolled energy including electrical shock.
- ✓ Visibility.
- ✓ Biological hazards from an unintended flow of contaminated solids or liquids

Exposure to hazards can arise in confined spaces because of, for example:

✓ **A lack of oxygen or too much oxygen**

The lack of oxygen is a leading cause of death among workers entering confined spaces. Low oxygen levels cannot be detected by sight or smell. The air must be tested for this hazardous condition. A very low level of oxygen can damage the brain and cause the heart to stop after a few minutes. It is possible:

- ✓ where there is a reaction between some soils and the oxygen in the atmosphere;
- ✓ following the action of groundwater on chalk and limestone which can produce carbon dioxide and displace normal air;
- ✓ in the holds of ships, freight containers, and trucks as a result of the cargo reacting with oxygen inside the space;
- ✓ inside steel tanks and vessels when rust forms;
- ✓ when it is used up during combustion such as during cutting or welding; and
- ✓ when it is used up by micro-organisms in sewer lines and fermentation vessels.

Too much oxygen is not as common a hazard as low oxygen, but it is also dangerous. Too much oxygen greatly increases the risk of fire or explosion in the confined space. Materials that would not normally catch fire or burn in normal air may do so extremely quickly and easily where there is a high level of oxygen.

The only way to know how much oxygen is present in a confined space is to use an oxygen monitor. The monitor must be in good working order and properly maintained and calibrated. The alarm must be set at the right level. Someone trained to use the monitor must test the air before anyone enters the confined space. Workers must not enter a confined space containing less than 19.5% oxygen without taking appropriate precautions, including the use of supplied-air respirators.

✓ **Poisonous gases, fumes or vapour**

These can:

- ✓ build-up in sewers and manholes and in pits connected to the system;
- ✓ enter tanks or vessels from connecting pipes;
- ✓ leak into trenches and pits in contaminated land, such as old refuse tips and old gas works

Contaminants in the air can result in an atmosphere that is toxic to workers and may result in injury or death as a result of work done inside the confined space.

In one-third of accidents involving harmful gases or lack of oxygen, the danger was not present in the confined space when the worker first entered it. Rather, the work in the confined space created the hazardous atmosphere.

Activities that may lead to the release of harmful substances into the air include grinding, descaling, insulation removal, metal spray applications, rubber lining, painting, fibre-glassing, cutting, welding, and the use of internal combustion engines. Cleaning agents may be toxic, may react with tank residues, or may release a deadly gas from porous surfaces inside the space. Cleaning a tank that contains dusts can cause the dust to become airborne and create a hazardous atmosphere.

✓ **Contamination from outside sources**

A confined space may be located next to a source of a hazardous contaminant. The contaminant could enter the confined space through porous walls, such as those that may be found in sewers or trenches, or through difficult-to-seal openings such as conduits. Normally, mechanical ventilation is set up to bring outside air into the confined space. If the intake hose is located beside a running vehicle or equipment with an internal combustion engine, the intake hose brings in exhaust fumes.

- ✓ **Liquids and solids**

Liquids and solids can suddenly fill the space, or release gases into it, when disturbed. Free flowing solids such as grain can also partially solidify or 'bridge' in silos causing blockages which can collapse unexpectedly.

Liquids may produce hazardous atmospheres if they evaporate — for example, liquid fuel in a tank producing vapours. Dangerous conditions can develop when pockets of gas in waste materials are disturbed during cleaning.

- ✓ For example, an organic material such as manure sludge can release the toxic gas hydrogen sulphide when manure pits are cleaned out.
- ✓ Rotting pulp in tanks also creates hydrogen sulphide.
- ✓ If grains in silos ferment, they use up oxygen and produce deadly gases.

- ✓ **Fire and explosions**

Three elements are necessary for a fire or explosion to occur: oxygen (air), flammable material (fuel), and an ignition source also known as the triangle of fire. Fires and explosions in confined spaces are often caused by gases or vapours igniting. Coal dust and grain dusts may explode when a certain level of dust in the air is reached. Containers of fuels such as gasoline and propane should not be taken into a confined space as fuel can easily burn or explode.

Some other common substances that can cause explosions or fires in confined spaces include, for example:

- ✓ Acetylene gas from leaking welding equipment
- ✓ Methane gas and hydrogen sulphide gas produced by rotting organic wastes in sewers or tanks
- ✓ Hydrogen gas produced by contact between aluminium or galvanized metals and corrosive liquids
- ✓ Grain dusts, coal dust
- ✓ Solvents such as acetone, ethanol, toluene, turpentine, and xylene, which may have been introduced into the space through spills or by improper use or disposal

A trained person must test the atmosphere for gases and vapours that will burn or explode. These dangerous gases and vapours cannot always be seen or smelt. If any measurable explosive atmosphere is detected, the air must be further evaluated by a qualified person to ensure that it is safe to enter the confined space.

Ignition sources include:

- ✓ Open flames
- ✓ Hot surfaces
- ✓ Discharge of static electricity
- ✓ Sparks from metal impact
- ✓ Welding arcs
- ✓ Arcing of electrical motors
- ✓ Lighting
- ✓ Chemical reaction

Many processes can generate static charge, including steam cleaning, purging, and ventilation procedures. To reduce the risks from these ignition sources, the use of non-sparking tools is recommended and all equipment must be bonded or grounded properly.

- **Residues**

Residues left in tanks, for example, or remaining on internal surfaces which can give off gas, fume or vapour present as hazards in confined spaces.

- ✓ **Dust**

The presence of dust in high concentrations, such as in flour silos also are particularly hazardous in confined spaces

- ✓ **Hot conditions**

Exposure to hot working conditions inevitably lead to a dangerous increase in body temperature. Special precautions are needed before workers enter equipment such as boilers and low-temperature systems. A qualified person must provide these procedures. Allow enough time for cooling of confined spaces that have been steam-cleaned.

✓ **Moving parts of equipment and machinery**

Mechanical equipment such as augers, mixers, or rotating tanks can be dangerous if activated or not secured. Residual energy, such as gravity or accumulated pressure, may also pose a risk unless the equipment is locked out and de-energized. This must be done by following a written lockout procedure that is specific for each piece of equipment and that states each place where a lock must be applied.

Even when the power is shut off and the equipment is locked out at control points, unsecured equipment can move, especially if it is out of balance.

Before doing any work in confined spaces:

- ✓ Shut the power off
- ✓ Ensure that the equipment is locked out at control points
- ✓ Test the lockout
- ✓ Secure any equipment that can move, even when it has been locked out

✓ **Electrical shock**

Electrical shock can result from defective extension cords, welding cables, or other electrical equipment. Work done in metal enclosures or in wet conditions can be particularly dangerous. Install ground fault circuit interrupters (GFCIs) or use assured grounding where there may be a danger of electrical shock. All electrical sources that pose a hazard to workers inside the space must be locked out following the written lockout procedure for the particular confined space.

✓ **Poor visibility**

Poor visibility increases the risk of accidents and makes it harder for a standby person to see a worker who may be in distress. If poor visibility results from inadequate lighting, the light levels should be increased although area lighting is not always required. If activities such as sandblasting or welding result in poor visibility, appropriate ventilation may be needed to reduce harmful substances in the air. If portable lighting is used where there may be an explosive atmosphere, the lighting must be "explosion-proof." Emergency lighting such as flashlights or battery-operated area units must be provided where necessary, so that workers can locate exits and escape.

✓ **Noise**

Noise produced in confined spaces can be particularly harmful because of reflection of sound off wall surfaces. Noise levels from a source inside a small, confined space can be up to 10 times greater than the same source placed outdoors. If the noise levels cannot be reduced, proper hearing protection must be worn where necessary.

✓ **Risk of drowning**

Confined spaces should be fully drained or dry when entered. Spaces that are not fully drained or dry may pose a risk of drowning. The risk of drowning in a vat or tank with a large amount of liquid is easily recognized. However, workers have drowned in small pools of liquid. For example, insufficient oxygen, the presence of a toxic gas, or a blow to the head can cause workers to lose consciousness. Workers who have fallen face-down into a small pool of water have drowned.

• CONFINED SPACE WORK IS MORE HAZARDOUS THAN OTHER WORKPLACES

As previously stated, some of these conditions may already be present in the confined space. However, some may arise through the work being carried out, or because of ineffective isolation of plant nearby, such as, for example, leakage from a pipe connected to the confined space.

The enclosure and working space may increase other dangers arising through the work being carried out, for example:

- ✓ machinery being used may require special precautions, such as provision of dust extraction for a portable grinder, or special precautions against electric shock;
- ✓ gases, fumes or vapours can arise from welding, or by use of volatile and often flammable solvents, and adhesives; and
- ✓ if access to the space is through a restricted entrance, such as a manhole, escape or rescue in an emergency will be more difficult to execute quickly and successfully.

Many factors need to be evaluated when identifying hazards in a confined space. There is smaller margin for error. An error in identifying or evaluating potential hazards can have more serious consequences. In some cases, the conditions in a confined space are always extremely hazardous. In other cases, conditions are life threatening under an unusual combination of circumstances. This variability and unpredictability is why HIRA is extremely important and must be taken very seriously each and every time one is done.

Some examples include:

- ✓ The entrance/exit of the confined space might not allow the worker to get out in time should there be a flood or collapse of free-flowing solid.
- ✓ Self-rescue by the worker is more difficult.
- ✓ Rescue of the victim is more difficult. The interior configuration of the confined space often does not allow easy movement of people or equipment within it.
- ✓ Natural ventilation alone will often not be sufficient to maintain breathable quality air. The interior configuration of the confined space does not allow easy movement of air within it.
- ✓ Conditions can change very quickly.
- ✓ The space outside the confined space can impact on the conditions inside the confined space and vice versa.
- ✓ Work activities may introduce hazards not present initially.

Working in confined spaces is challenging but can be done safely without threat to the health or the workers doing the work provided the proper precautions are taken after proper pre-planning of those work operations.

• REVIEW QUESTIONS

1. What is a confined space?
2. Give examples of confined spaces and the dangers associated with working in them.
3. What are some of the hazards of working in confined spaces?
4. What makes fires more hazardous in confined spaces?
5. Describe how you would ensure that activities in confined spaces are carried out healthily and safely.

• REFERENCES

Anonymous (nd). Loose notes on safety, health and environment management

Canadian Center for Occupational Health and Safety (n.d.). Confined Space Available at: https://www.ccohs.ca/oshanswers/hsprograms/confinedspace_intro.html?print.

Denise (2018). *The Destroy Depression™ System - Cure Depression Naturally*. [online] Destroy Depression. Available at: https://www.mydepressionrelieftools.com/another-word-for-depression-in-geography.

HSA (2017). [online] Confined Space. Available at: https://www.hsa.ie/eng/Publications_and_Forms/Publications/Codes_of_Practice/COP_Confined_Space.pdf.

HSE Books, 1999. Health risks management: A guide to working with solvents HSG188. ISBN0717616649 HSE Books, 2000. Latex and you INDG320

Krug, T.W., 2016, January. OSHA's Confined Spaces in Construction 2015 Standard. In *ASSE Professional Development Conference and Exposition*. American Society of Safety Engineers.

New Zealand Government, W. (n.d.). *Excavation safety*. [online] WorkSafe. Available at: https://worksafe.govt.nz/topic-and-industry/excavation/excavation-safety-gpg/.

Oregon OSHA. (nd). Pesticide Use and your personal protective equipment. https://nasdonline.org/7314/d002518/oregon-osha-pesticide-use-and-your-personal-protective.html

Pembina Trials (2017). *Pembina Trails School Division*. [online] Physical Hazards. Available at: https://www.pembinatrails.ca/WhatWeOffer/SafetyHealthandEnvironment/Safety/ConfinedSpaceEntry/Pages/Physical-Hazards.aspx.

Rangu,Sridhara., Sundaragiri, Suraj., and Rangu, Saritha. 2016. Suffocation due to irrespirable gases in confined spaces: accidental deaths of rescuers. International Journal of Research in Medical Sciences, 4(5), 1775-1777

Srihari, Vedartham., Sanjay, Das., Nimbalkar, Priya. and Amrut, Shipekar., 2017. Overview on Safety Guidelines for Confined Space Working Environment. *International Journal of Engineering and Management Research (IJEMR)*, *7*(6), pp.158-164.

Stefana, Elena., Marciano, Filippo., Cocca, Paola., Rossi, Diana. and Tomasoni, Giuseppe., 2019. Oxygen deficiency hazard in confined spaces in the steel industry: assessment through predictive models. *International journal of occupational safety and ergonomics*, pp.1-15.

Sulardi, Sulardi. and El-Ridho, Nuruddin.Kafy., 2019. Hazard Identification and prevention methods on work in confined spaces. *Identifikasi: Jurnal Ilmiah Keselamatan, Kesehatan Kerja Dan Lindungan Lingkungan*, *5*(2), pp.142-151.

Worksafe New Zealand. (n.d.) Excavation Safety. Available at: https://worksafe.govt.nz/topic-and-industry/excavation/excavation-safety-gpg/.

PART 3

MANAGEMENT OF OCCUPATIONAL HEALTH

This section comprises of five chapters that cover the management of occupational health in general industry in South Africa. In particular, the following are addressed, namely:

- ✓ The anticipation, identification, evaluation and control of occupational stressors
- ✓ Role of the occupational hygienist
- ✓ Functions of occupational health services
- ✓ Legislative requirements with regards to occupational health in the workplace
- ✓ Clinic management
- ✓ Fitness to work and disability evaluation
- ✓ Health monitoring and surveillance
- ✓ Absenteeism
- ✓ Hazard evaluation and control
- ✓ Drug abuse screening
- ✓ Worker assistance and counselling
- ✓ Health promotion
- ✓ Health risk assistance, and
- ✓ Primary health care.

Chapter 17 highlights the importance of effectively managing occupational health in the workplace while Chapter 18 deals comprehensively with the occupational health clinic and its function in the organization including dealing with COVID-19. Chapter 19 deals with several health challenges that could present in the workplace and how the impact of these could be mitigated. Chapters 20 and 21 cover the management of health hazards and the promotion of health in the workplace.

CHAPTER 17

INTRODUCTION TO OCCUPATIONAL HEALTH MANAGEMENT

Occupational illnesses and diseases tend to be chronic and are generally characterized by temporary or permanent physical dysfunction. Most of these symptoms are not generally visible. Managers are therefore more prone to look after the safety of workers and neglect the heath aspect giving credence to the phrase 'health is the second cousin to safety'. This misdirected focus by management has contributed to the historic poor management of the health of their workers with disastrous effects. The recent COVID-19 pandemic has enforced a shift in this focus, making employers more aware of their responsibility to fulfil their constitutional obligations to their workers of providing them with a working environment that is safe and does not present a threat to their health and wellbeing. At least in the future the envisaged partnership captured in the Act between employers and workers will be operationalized in the workplace post-COVID-19 with improved health and wellbeing of everyone in the organization. It is therefore imperative that occupational health in the workplace needs to be addressed systematically and become an organizational value rather than a priority, so that it can include all aspects of the interaction of workers with their work environment.

In order to accurately address occupational health issues in the workplace, several important aspects need attention which include the following, namely:

- The anticipation, identification, evaluation and control of occupational stressors
- Role of the occupational hygienist
- Functions of occupational health services
- Legislative requirements with regards to occupational health in the workplace
- Clinic management
- Fitness to work and disability evaluation
- Health monitoring and surveillance
- HIV and AIDS in the workplace
- Corona viruses
- Hazardous substances
- Absenteeism
- Hazard evaluation and control
- Drug abuse screening
- Disaster Management
- Worker assistance and counselling
- Health promotion
- Health risk assistance, and
- Primary health care.

These have to take cognisance that the workplace comprises of four basic elements all of which must be addressed, namely

- The worker
- The tool or equipment
- The process or work activity
- The work environment in which work activities are executed.

According to the World Health Organisation (WHO), occupational health is made up of three important legs, namely

- Occupational Hygiene
- Occupational Medicine
- Occupational Safety

On the other hand, In South Africa occupational health is seen slightly differently and includes:

- Occupational Hygiene
- Occupational Medicine
- Primary Health Care

Occupational medicine and primary health care are practiced by the occupational health medical practitioner and the occupational health nursing practitioner, while occupational hygiene is a specialized field practiced by qualified occupational hygienists. A total health care program can be developed and integrated when all these professionals in their various roles and functions work together to develop a holistic occupational health strategy placing importance on all the fields of the various organisations in an environment where occupational health is a value of the organization.

• IMPORTANT OCCUPATIONAL HEALTH RELATED DEFINITIONS AND CONCEPTS

Ergonomics is a multidisciplinary activity dealing with the interaction between workers and their total working environment and includes traditional environmental elements as atmosphere, heat, light, and sound as well as all tools and equipment used in the work place.

Occupational Health is the promotion and maintenance of the highest degree of physical, mental and social well-being of workers in a balanced synthesis of several disciplines including occupational hygiene, occupational medicine and biological monitoring.

Primary Health Care involves the immediate treatment of minor and serious injury or illness, chronic diseases and minor ailments occurring at the workplace, referral of the worker to the relevant doctor or hospital and follow-up treatment to ensure rapid recovery. It also covers aspects related to social welfare and health education and rehabilitation.

Occupational medicine is a specialist discipline within the context of occupational health. It primarily deals with the relationship between the worker and his/her environment that has the potential to result in impaired or ill-health. AII medical examinations pre-employment, post injury/illness, medical surveillance and biological monitoring fall into this category.

Occupational hygiene is the science and art devoted to the anticipation, recognition, evaluation and control of environmental factors or stresses, that arise from the place of work and cause illness, impaired health and well-being, or significant discomfort and inefficiency among workers or among the community at large.

Occupational health nursing concerns the prevention of illness and the promotion of health to assist the worker in maintaining the optimum state of physical and psychological well-being.

Occupational health medical practitioner is registered with the Health Professions Council of South Africa (HPCSA) with a post graduate qualification in occupational health and adequate experience in the field.

Biological monitoring is the study of the ability of chemical substances that could cause adverse effects in the worker through planned programs of periodic collection and analysis of body fluids. It involves the measurement of the worker's exposure and response to the substance.

Medical surveillance is a planned programme over an extended period of time incorporating various examinations under supervision of an occupational health medical practitioner.

Toxicology is the science that studies poisons and toxic substances, their mechanisms and effects on living organisms. It is the study of the adverse effects of chemicals on biologic systems, or the capacity of a substance to produce undesirable physiological effects when the chemical reaches a certain concentration in the body.

• RELATIONSHIP BETWEEN WORK AND HEALTH

The following diagram illustrates that work and the working environment will have an impact on worker health and wellbeing. The converse is also true, namely that the health of workers will have an impact on the work they do and eventually on the overall wellbeing of the organisation. Because of this relationship it is important for business owners not only to take interest in the health of their workers which is required in any case in terms of the Constitution, but also to ensure that their health is being taken care of as far a reasonably possible.

Importantly, it is not the current or present occupation of a worker that is the cause of the occupational disease, but all previous occupations should be considered since the disease or affliction may have been contracted earlier than current employment. For example, a person could have been exposed to asbestos at a previous place of work, and is not exposed to it currently, but only now presents with cancer of the lungs in the present workplace since the effects of the previous exposure may be latent for up to 50 years.

Apart from the previous occupations of workers, their hobbies or recreational activities could also be responsible for the contracting of an occupational health disease or risk. Workers who, for instance, work in their garden at home may be exposed to insecticides and pesticides without adequate protection. The effects of health at work, in converse, can typically be considered at three levels:

1. Workers who have no apparent health problems;
2. Workers whose work ability have been impaired by illness or injury; and
3. Workers whose health may impact the health and safety of fellow workers or their community.

An example of the last level could be a worker who was exposed to the *coronavirus* coming to work and passing the virus to fellow workers through contact resulting in the spread of the infection to their respective families and communities.

• OCCUPATIONAL HEALTH CHALLENGES

According to the World Health Organization, the most important occupational health challenges for the future include the following, namely:

- occupational health problems linked to new information technologies and automation;
- new chemical substances and physical energies;
- health hazards associated with new biotechnologies;
- transfer of hazardous technologies;
- aging working populations;
- special problems of vulnerable and underserved groups such as those who are chronically ill, handicapped and disabled, including migrants and the unemployed; and,
- problems related to growing mobility of worker populations and occurrence of new occupational diseases of various origins such as various coronaviruses in recent years.

• OCCUPATIONAL HYGIENE STRESSORS

The environmental stressors of occupational hygiene or causes of the health problems of workers are divided into five categories, namely:

- ✓ biological,
- ✓ chemical,
- ✓ physical,
- ✓ ergonomic, and
- ✓ psychosocial.

Occupational hygiene uses science and engineering to prevent ill health caused by the working environment. It helps both employers and workers to understand the risks to be able to improve working conditions and working practices.

Work has always involved hazards to health such as, for example

- ✓ Plumbers have been poisoned by the lead they used for pipes and joints when lead was the material of choice;
- ✓ Coal miners died from cancer caused by the inhalation and ingestion of coal dust; and

✓ Cutlery grinders died young from lung diseases caused by the silica in grinding wheels.

With good occupational hygiene practice, some historical risks have been eliminated and others brought under control. COVID-19 has elevated the need for and importance of both personal and occupational hygiene to prevent the spread and rate of infection in the workplace. Employers have additional responsibilities imposed on them as a consequence.

• OCCUPATIONAL HEALTH SERVICES

The International Labour Organization (ILO) Convention on Occupational Health Services (No. 161) and the ILO Recommendations on Occupational Health Services (No. 171) were adopted in 1985. In the Convention the following definition was given: - "The term 'occupational health services' means services entrusted with essentially preventive functions and responsible for advising the employer, the workers and their representatives in the undertaking, on:

- the requirements for establishing and maintaining a safe and healthy working environment which will facilitate optimal physical and mental health in relation to work;
- the adaptation of work to the capabilities of workers in the light of their state of physical and mental health".

This definition alone is not enough to describe the range of activities carried out by Occupational Health Services. There are a number of functions listed in the Convention, and these are summarised as follows, namely: -

- Identification and assessment of the risks from health hazards in the workplace. This involves surveillance of the factors in the working environment and working practices which may affect workers' health. It also requires a systematic approach to the analysis of occupational "accidents", and occupational diseases.
- Advising on planning and organisation of work and working practices, including the design of work-places, and on the evaluation, choice and maintenance of equipment and on substances used at work. In so doing, the adaptation of work to the worker is promoted.
- Providing advice, information, training and education, on occupational health, safety and hygiene and on ergonomics and protective equipment.
- Surveillance of workers' health in relation to work.
- Contributing to occupational rehabilitation and maintaining in employment people of working age, or assisting in the return to employment of those who are unemployed for reasons of ill health or disability.
- Organising first aid and emergency treatment.

Some of the challenges that could affect the provision of occupational health services include:

- Lack of awareness among workers, employers, health planners, policy makers, health professionals and the public at large.
- Lack of properly trained human resources.
- Inadequate, inaccessible, and inequitably distributed health service institutions
- Lack of multidisciplinary staff, absence of field-testing equipment for conducting environmental and biological monitoring of the work place and the health of the workers
- Insufficient budget for carrying out regular inspections and conducting research activities
- The profile of the workers who are generally from low income backgrounds, illiterate or poorly educated.
- The poor working environment.
- Absence of training institution on occupational health.
- Little or no collaboration or cooperation among stakeholders.
- Absence of integration of occupational health with general health service provision.
- Little or no collaboration or cooperation among stakeholders creating fragmentation and poor co-ordination.

• THE NEED FOR OCCUPATIONAL HEALTH

According to the estimates of the International Labour Organization (ILO), the overall economic losses resulting from work-related diseases and injuries were approximately 4-5 % of the world's Gross National Product, making improvement to occupational health important. The ILO also claims that 1.2 million working peoples die of work related accident and diseases every year

- More than 160 million workers fall ill each year due to workplace hazards
- UN estimates 10,000,000 occupational disease cases occur each year globally while severity and frequency is greatest in developing countries such as South Africa
- Women, children and migrant workers are least protected and most affected
- Micro and small enterprises account for over 90 % of enterprises where conditions are often very poor and the workers particularly in the informal sectors are often excluded from all legal protection

In many developing nations, death rates due to occupational accident among workers are five to six times higher than those in industrialized countries. Health-related matters impact organisations in many ways. These include, for example,

- **Legislation**: compliance with many acts such as the Occupational Health and Safety Act (OHSA), Labour Relations Act (LRA), Employment Equity Act (EEA) and Basic Conditions of Employment (BCEA). The organisational health department helps to interpret and provide guidance to the organisation to enable it to react appropriately with respect to these legislative and regulatory requirements.
- **Corporate governance**: The King IV report of 2016 in South Africa in terms of Principle 3 item 14 (a) and global reporting Initiative have a number of health-related issues which require attention from organisations.
- **Costs of healthcare continue to spiral**: South African corporate organisations are spending about 10% of their wage bill on medical aid subsidies. Organisational health helps to manage these costs, and promotes healthy living habits in order to reduce the cost impacts on medical aids.
- **HIV and AIDS**: the devastating HIV retrovirus and the AIDS syndrome need to be carefully managed to ensure an "AIDS-friendly" environment which will allow workers who are HIV^+ and/or with AIDS to be as productive in the workplace as possible for as long as possible.
- **COVID-19:** the global impact of the novel coronavirus with origins in Wuhan, China has affected South African industry like no other and will demand new and different approaches to workplace hygiene and health. Daily infections increase the total national number of infections by between 3,000 and 4,000 as the results of testing become available.

• OBJECTIVES OF OCCUPATIONAL HEALTH

The objectives of any occupational health service should be in line with the recommendations made by the ILO. These objectives are defined as:

- Protecting workers against health hazards which could arise from work or conditions in which work is carried out.
- Promotion of the health of workers by eliminating occupational factors and conditions hazardous to health and safety at work.
- Development of corrective measures to improve the health of workers by dealing with the various occupational hygiene stressors.
- Contributing to the physical and mental adjustment of workers, in particular by the adaptation of the work to the workers and placing them in jobs to which they are best suited.
- Enhancement of physical, mental and social well-being of workers and support for the development and maintenance of their working capacity, as well as professional and social development at work.
- Development and promotion of sustainable work environments and work organizations.

All guiding principles should ensure, as far as reasonably practical, a safe and healthy workplace, and the availability of adequate health care to all workers.

• ROLE OF THE OCCUPATIONAL HEALTH NURSE

In South Africa, the occupational health nurse (OHN) works alone most of the time, with the occupational doctor or medical practitioner available only on certain occasions. The OHN is therefore responsible in the main for the health and safety management portfolio of an organisation. The roles and responsibilities of the OHN are divided into FOUR main categories, namely:

- Administrative
- Personal
- Professional
- Welfare

All these roles and responsibilities overlap to some extent and one is not more important than another.

• HEALTH RISK ASSESSMENT

Before introducing occupational health to an organisation, it is important to conduct a thorough health risk assessment. A health risk assessment will help to determine to what extent occupational health services are needed in an organisation for the types of exposure workers are likely to have at work.

Health risk assessments include a comprehensive study of all raw materials, processes, chemicals and products used in the organization. On completion of the study the degree of risk exposure and the severity of the exposure should be determined and the subsequently developed medical surveillance programme should be implemented.

• THE OCCUPATIONAL HEALTH MODEL

The following occupational health model highlights the various functions that organisational health plays in South Africa

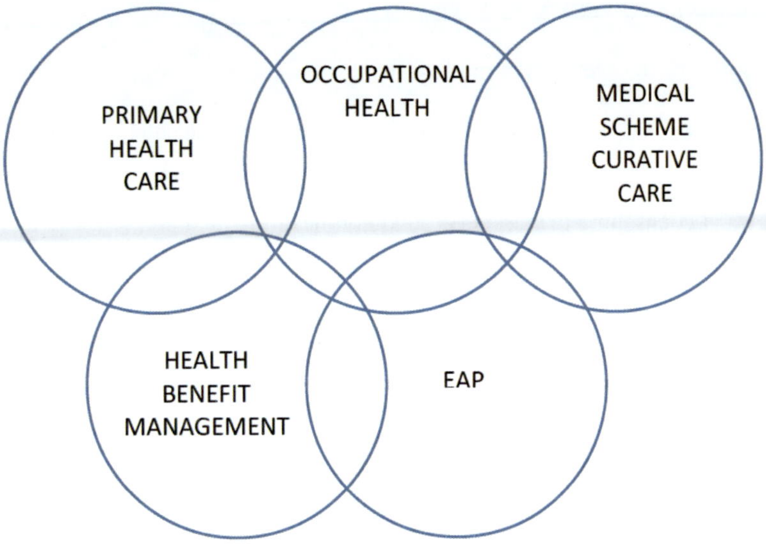

This particular model integrates the five key areas of health delivery in the workplace and was designed and implemented with great success within the Chamber of Mines in the late 1980s. Many industrial and commercial organisations have adapted the model to meet their own health-related needs. The integrated delivery of service allows organisations to meet the challenges of complex health issues such as HIV and AIDS, COVID-19, silicosis, asbestosis, cannabis use, incapacity management and the management of human capital and resources.

• PRIMARY HEALTH CARE

Primary health care includes all preventative and promotive health, health interventions, treatment of minor ailments and trauma, preventive and promotive health includes health education and screening, and, health interventions would include things like family planning for workers and immunisation.

In the workplace primary health care typically focuses on:

- Identification, treatment and monitoring of chronic diseases
- Risk screening that includes on site screening for possible COVID-19 infections
- Treatment of minor ailments
- Treatment and management of minor or serious injuries and diseases
- Completion of claims to the Compensation Commissioner
- Social services and counselling
- Rehabilitation after being away from work due to injury or illness
- Health education lifestyle management
- HIV and AIDS counselling
- Current health threats published by the WHO and NCID such as COVID-19.

The purpose of occupational health service is not to provide a facility where the sick can be treated, but primary health care as an emergency service that also assists workers to complete a shift or to identify a serious emergency which needs to be referred to the personal doctor of the worker.

The most important task of the occupational health staff is to link minor complaints to the worker's workplace and exposure and task, to identify how the workplace can affect the health of the worker, and to address the issues timeously.

• OCCUPATIONAL HEALTH (OH)

Occupational health is a specialist branch of medicine that focuses on the physical and mental wellbeing of workers in the workplace. It is a diverse science applied by occupational health professionals, engineers, environmental health practitioners, chemists, toxicologists, doctors, nurses, safety professionals and others who have an interest in the protection of the health of workers in the workplace.

The aim of occupational health is to prevent work-related illness, injury and fatalities by:

- encouraging healthy and safe working practices;
- ergonomics studying how work is executed and how work can be improved;
- monitoring the health of the workforce; and
- supporting the management of sickness absence.

An occupational health service in an organization might also:

- working with an employer to implement policies and ensure health and safety compliance;
- conducting pre-employment health assessments;
- supporting health promotion and education programmes;
- providing advice and counselling to workers around non-health-related problems;
- providing an employer with advice and guidance around making reasonable adjustments to existing working conditions.

Occupational health provision will depend on the size of the organisation. Occupational health can be provided by a nurse with occupational health training and a part-time doctor, or through a range of specialists, including:

- physiotherapists;
- hygienists;
- psychologists;
- ergonomic experts;
- occupational therapists;
- specialist occupational health nurses and doctors.

Occupational health is usually provided at the place of work, but if the employer does not have a dedicated service, the worker may need to travel to attend appointments with external providers.

Occupational health incorporates workplace hazard analysis, post exposure prophylaxis, universal precautions, training of first aiders, as well as the identification, evaluation and control of risk in the workplace to prevent exposure, the assistance with recruitment where appropriate ensuring fitness to work, screening for signs and symptoms of illness associated with exposure to hazards and the early identification of people at risk and appropriate placement. The occupational health leg of the South African occupational health model involves some of the following occupational health facets, namely:

- **Controlling identified hazards**: If and when occupational health services are involved in the control of various occupational hazards, the prevalence and severity of incidents and occupational diseases can be greatly reduced or eliminated completely.
- **Identification of unidentified hazards**: Sometimes management and other uninformed people might overlook hazards, where the occupational health staff has the duty to investigate all kinds of health hazards from the trivial which do no more than interfere with the comfort and efficiency to those endangering life. There are generally three ways to identify unidentified hazards through toxic testing, epidemiological enquiry and clinical observation.
- **Placing**: The pre-placement medical examination is a tool for the selection of workers, based on the inherent requirements of the job. The aim of using a medical examination prior to employment is identify whether a worker is physically and mentally fit enough for a certain job or position in an organisation. For example, a person cannot be placed in an environment where they will need to operate a tower crane if they have a fear of heights.
- **Retaining people in suitable work**: It is the responsibility of the occupational health services of the organisation to identify if a worker has been wrongly placed in a job that is not suited to him/her and to suggest positions that will be more suited to their health and circumstances.
- **Provision of treatment**: The main function of the medical services in South Africa is the provision of treatment. Treatment will include the treating of injuries, acute poisoning and minor ailments efficiently and speedily to prevent complications and to assist in the rehabilitation of the worker. It is important for physicians and nurses to know what a worker's job entails to ensure that they receive the right medication and treatment for their ailments.
- **Counselling**: This is the empathetic listening to and understanding of the problems of another person. Health practitioners that are involved with counselling do not only listen to health-related problems, but also social and work problems. The nursing practitioner plays an important role in counselling and will also assist the

worker in decision making. The counsellor should also assist a worker after making a decision with further information, referral and encouragement.

- **Supervision of vulnerable groups**: There are always vulnerable people in all groups of workers. When someone is absent for subsequent long periods of time they could be considered to most likely belong to a vulnerable category. The absences can be due to chronic illnesses. It is the role of the occupational health staff to supervise these vulnerable groups of people and to make recommendations suited to the needs of the individual.
- **Health Education and Safety Training**: Health education can be done both through formal training during working hours or through educational visits to the nurse or any of the occupational health workers. Training helps to prevent accidents and incidents and poor health. It is therefore important to ensure reduced medical costs and personnel suffering. Training and education give workers the power in their own hands to prevent becoming sick and responsibility for their own actions.

• EMPLOYEE ASSISTANCE PROGRAM (EAP)

An Employee Assistance Program (EAP) leg of the occupational health model in South Africa is a psycho-social support program for workers. EAP will include peer support and training, life skills, assisting with accessing benefits, referral system for stressed and/or troubled workers and trauma procedures.

• HEALTH BENEFIT MANAGEMENT

The health benefit management leg of the occupational health model in South Africa includes all different types of input and advice given to workers and management with regards to absenteeism, disability and incapacity management, COIDA interaction with medical aid and medical adjudication.

• MEDICAL SCHEME AND CURATIVE CARE

Finally, the medical scheme and curative care leg of the occupational health model in South Africa includes all the clinical services given to manage HIV, COVID-19, cannabis and other substance abuse for counselling and support and wellness management. This will ensure benefits such as being medically sound and giving support productively.

• THE ROLE OF THE OCCUPATIONAL HYGIENIST

Occupational hygienists analyse, identify, and measure workplace hazards or stresses that can cause sickness, impaired health, or significant discomfort in workers through chemical, physical, ergonomic, or biological exposures. They understand how these exposures may affect the health of the workforce, and in turn, the health of the business. Occupational hygienists specialise in controlling health risks in practical and cost-effective ways by assessing and resolving practical problems in a wide range of settings, including factories, hospitals, offices and building sites. They concentrate on the short and long-term effects on health arising from both acute and chronic exposure to hazards, and enable organisations to respond effectively to legislative requirements

Work activities which vary between specialist areas and employers, but may include all or any of the following, namely:

- undertaking workplace surveys and evaluating situations in the workplace;
- assessing risks to health arising from many different factors, such as chemicals, noise, and poor lighting or ventilation;
- accurately measuring and sampling levels of exposure, often through precise use of specialist equipment;
- accurately recording facts or details of procedures in the workplace;
- eliminating or significantly reducing risk by facilitating organisational change and by selecting and designing relevant facilities;
- finding cost-effective solutions to risks to health in workplaces of all types, large and small;
- recommending remedies or control methods;
- compiling data, writing reports, and presenting report findings to clients;
- liaising with a wide range of people, including employers and workers, in the process of evaluating workplaces;
- considering all options of control, such as ventilation, containment, and personal protective equipment;
- deciding on or devising the most appropriate solution for specific situations;
- providing clear and accurate information on complex health and safety issues;
- training organisation staff on health and safety issues;
- gaining the confidence and cooperation of the workforce;
- persuading management of the organisation to develop effective hazard controls when required;
- writing guidance information on health and safety;
- working as part of a team to meet health and safety objectives;
- keeping up to date with scientific and legal developments; and
- providing expert witness services.

• REVIEW QUESTIONS

1. What are the challenges for the development of occupational health and safety?
2. How can work affect health and health affect work? Give practical examples.
3. Why is a focus on occupational health necessary?
4. What are the five legs of the South African occupational health model?
5. What is the role of an occupational hygienist?

• REFERENCES

Amponsah-Tawiah, Kwesi. and Mensah, Justice., 2016. Occupational health and safety and organizational commitment: Evidence from the Ghanaian mining industry. *Safety and health at work*, *7*(3), pp.225-230.

Anonymous (n.d.). Loose notes on safety, health and environment management

Attridge, Mark., Farley, Thomas., Marion, David. and Cycenas, Joel., 2020. EAP and COVID-19: Employee Assistance Programs (EAP) in the COVID-19 Era: Support for Substance Abuse.

BOHS (n.d.). BOHS Publications - British Occupational Hygiene Society (BOHS). [online] British Occupational Hygiene Society (BOHS). Available at: http://www.bohs.org/about-us/bohs-publications/.

Burdorf, Alex., Porru, Fabio. and Rugulies, Reiner., 2020. The COVID-19 (Coronavirus) pandemic: consequences for occupational health. *Scandinavian Journal of Work, Environment & Health*, *46*(3), pp.229-230.

Ethiopia Public Health Training Initiative. 2006. Occupational Health and Safety. https://www.cartercenter.org/resources/pdfs/health/ephti/library/lecture_notes/env_occupational_health_students/ln_occ_health_safety_final.pdf

Goelzer, Berenice. (n.d.). Chapter 30 - Occupational Hygiene. [online] Ilocis.org. Available at: http://www.ilocis.org/documents/chpt30e.htm.

Graduate Education in Occupational Hygiene: A Rational Framework. (2005). The Annals of Occupational Hygiene, 49(8), pp.649-659.

Gul, Muhammet. and Ak, M. Faith., 2018. A comparative outline for quantifying risk ratings in occupational health and safety risk assessment. *Journal of cleaner production*, *196*, pp.653-664.

Hale, Andrew., 2019. From national to European frameworks for understanding the role of occupational health and safety (OHS) specialists. *Safety science*, *115*, pp.435-445.

Hallenbeck, William. H., 2018. *Quantitative risk assessment for environmental and occupational health*. CRC Press.

Haupt, Theodore., Smallwood, John., Kalindindi, S., and Varghese, K. (2010): The health and wellbeing of Indian construction workers: A comparison between older and younger, In Barrett, Amaratunga, Haigh, Keraminiyage and Pathirage (Eds) Building a Better World: Proceedings of CIB World Congress 2010, Manchester, U.K., May 10-13, 2010, paper 1634, ISBN 9781-905732-91-3

HSE (n.d.). Occupational Hygiene - HSL. [online] Hsl.gov.uk. Available at: https://www.hsl.gov.uk/what-we-do/occupational-hygiene.

Reese, C.D., 2018. *Occupational health and safety management: a practical approach*. CRC press.

ILO (1985). Convention C161 - Occupational Health Services Convention, 1985 (No. 161). [online] Ilo.org. Available at: http://www.ilo.org/dyn/normlex/en/f?p=NORMLEXPUB:12100:0::NO::P12100_INSTRUMENT_ID:312306.

ILO (1985). Recommendation R171 - Occupational Health Services Recommendation, 1985 (No. 171). [online] Ilo.org. Available at: https://www.ilo.org/dyn/normlex/en/f?p=NORMLEXPUB:12100:0::NO::P12100_ILO_CODE:R171.

Prall, Joshua. and Ross, Michael., 2019. The management of work-related musculoskeletal injuries in an occupational health setting: the role of the physical therapist. *Journal of exercise rehabilitation*, *15*(2), p.193.

Richmond, Melissa. K., Pampel, Fred. C., Wood, Randi. C. and Nunes, Ana. P., 2017. The impact of employee assistance services on workplace outcomes: Results of a prospective, quasi-experimental study. *Journal of occupational health psychology*, *22*(2), p.170.

WHO, (1986). Constitution: Basic documents. Geneva.

WHO, (1994). Global strategy on occupational health for all: The way to health at work. Geneva.

WHO, (2006). Declaration of workers health. WHO Collaborating Centres of Occupational Health: Stresa, Italy

Zarate, Pablo., Cuellar, Diego., Velazquez, Luis. and Cura, Leila., 2018. 1017 Occupational health nurses working as worksite health promotion agents.

CHAPTER 18

THE OCCUPATIONAL HEALTH CLINIC AND ITS FUNCTIONS

Occupational health services are often times established only long after a factory, workshop or office was built, implying that the occupational health nursing practitioner has to make do with the space that she has been given. If however a new facility is built, the nursing practitioner or hygiene practitioner should be involved in the design of the clinic. Building the factory, or establishing the clinic, means that a thorough health risk assessment for the type of industry should be done to determine the need for special equipment or facilities that mighty be needed. The recent experiences with COVID-19 impose additional requirements for occupational health clinics and facilities at work to protect the workforce against infection in the first instance and spread of the virus in the workplace in the second instance.

• PHYSICAL ASPECTS OF THE CLINIC

When designing a clinic or occupational health facility the following should be kept in mind, namely:

- Adequate ventilation in all areas of the clinic
- Large enough to allow for 1.5m spatial separation between workers being examined and treated in the clinic
- Location close to the factory floor or production area yet in a relatively quiet environment
- Clear sign-posting
- Hand sanitizing or hand washing facility at entrance to the clinic
- Reception area with restricted number of chairs in accordance with applicable regulations to limit the number of patients waiting for attention while at the same time taking into account the need for 1.5m social distancing and educational posters
- Reception desk that has a transparent screen to protect the receptionist and other healthcare staff from possible infection through transmission of viruses from workers reporting at the facility
- Dedicated reception area for symptomatic patients
- An office or offices
- At least two examination rooms
- Isolation rooms with adequate ventilation of not less than 60 L/s per worker or hospital standard air-conditioning
- Lockable medicine cupboard
- One major treatment room for minor procedures, with a separate clean area for stitching and small surgical procedures
- One special examination room for special testing such as audiometry
- Doors wide enough for transportation of patients
- Accessibility to emergency vehicles with a ramp at the entrance
- Toilets and ablutions
- Storeroom

• BASIC AND SPECIALIZED EQUIPMENT NEEDED FOR A CLINIC

The basic equipment needed in all occupational clinics includes the following:

- Thermometer
- Ear, nose and throat (ENT) set
- Glucometer
- Haemoglobinometer
- Cholesterol meter
- Baumanometer
- Stethoscope

Specialised equipment needed in occupational clinics includes the following:

- Audiometer
- Vision Testing Equipment
- Lung Function Testing

- Electrocardiograph (ECG) machine
- Defibrillator
- 0_2 Treatment

• RECORDKEEPING IN THE OCCUPATIONAL HEALTH CENTER

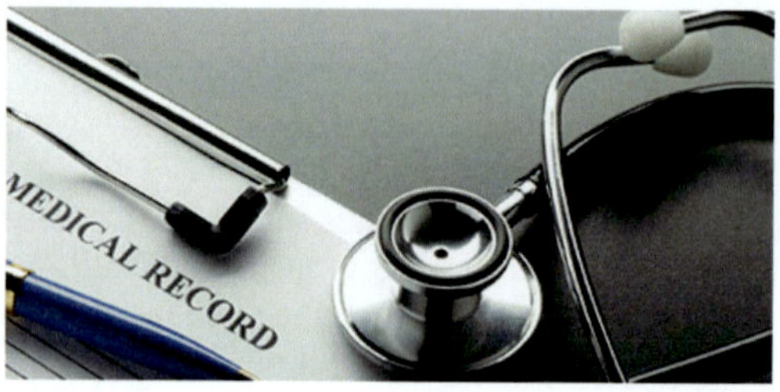

It is vitally important to keep accurate record of the occupational health nurse's daily routine. Patient records are legal documentation and failure to keep them up to date could have legal implications should they be needed in a court case. There are three basic types of records namely administrative records, individual health records and miscellaneous records:

Administrative Records includes the following records and information on file:

- Incident register
- Medicines Register
- Daily attendance sheet/book
- Management report
- List of departments, workers and transfers.

Individual Health Records consist of a personal medical file for every worker. Individual health records should be opened when the worker joins the firm and should be kept up to date with each medical visit that the worker pays to the clinic. On these records it should also be indicated when the worker receives any health and safety education and training and when he/she is due for more training. These records are highly confidential and only the occupational health staff is allowed to access these files. Should anyone require information on these files, be it managers or owners of the organisation, it could only be given them with the written consent of the worker.

Miscellaneous Records are records other than administrative and individual health records. These can include records of all equipment in the clinic and records of any maintenance or calibration done to these instruments.

• HEALTH MONITORING AND SURVEILLANCE

Occupational health surveillance is "the tracking of occupational injuries, illnesses, hazards, and exposures

The purpose of the Occupational Health Surveillance Program is to help assure the health of workers who:

- ✓ Have workplace exposure to particular health hazards such as, for example, high noise levels, high levels of dust or animal allergens known to pose risks for a potentially serious health condition, illness, or injury; OR
- ✓ Perform specific work tasks such as, for example, respirator use or driving commercial vehicles that require a certain degree of health and fitness to assure worker and/or public health and safety.

• OCCUPATIONAL HEALTH SURVEILLANCE

Workers typically undergo a medical work history and in some cases, medical examination by a licensed medical practitioner. These screenings are used to establish an initial baseline of the health of the workers and are then used to monitor their future health as they pertain to potential occupational exposures to hazardous agents.

Medical examinations and consultations, either as part of screening programmes or on an as-needed basis, serve five main purposes, namely

- evaluation of the effectiveness of control measures in the workplace;
- detection of pre-clinical and clinical abnormalities at a point when intervention is beneficial to individuals' health;
- prevention of further deterioration in workers' health;
- reinforcement of healthy and safe methods of work and of health maintenance; and
- assessment of fitness for a particular type of work, the present concern being the adaptation of the workplace to the worker.

During medical examinations and consultations, it may be appropriate for the occupational health practitioner to:

- inform workers of potential injuries and diseases and control measures necessary for prevention;
- inform workers of potential diseases and conditions of work and exposures which are medically contra-indicated, and to advise them where they can get help in the treatment or correction of their condition;
- inform workers and their employers of the effectiveness or otherwise of control measures;
- help the employer to place workers in occupations that take into account their capacity for particular work;
- draw the attention of young persons to their physical and mental aptitudes, in order to facilitate appropriate vocational guidance; and
- prevent the total exclusion of any worker from employment, and provide for the employment of each worker, despite any contra-indications, in work which he or she is capable of performing, taking into account the respective employment opportunities available.

These medical examinations can be for different reasons or to determine various diseases. Some of these examinations include:

- Pre-placement/pre-employment medical examination
- Fitness to work and disability evaluation
- Periodic medical examination
- Executive medical examination
- Exit medical examinations

• PRE-PLACEMENT/PRE-EMPLOYMENT MEDICAL EXAMINATION

When a pre-placement medical examination is conducted, the medical practitioner strives to determine from the health status of a worker whether if he/she is healthy enough to do the required task that he/she has been recruited for. Doing a pre-employment medical examination will protect the organization against any future compensation or legal claims, by ensuring that the worker will not be harmed, be a danger to others or that any pre-existing defect will be aggravated as a result of his/her work. A thorough pre-employment medical examination will provide the organisation with an important medical baseline for any future treatment of the worker. The scope of the examination will depend on the type of industry, the hazards involved and the type of work that needs to be done in the prospective job of the worker. When the medical examination is conducted, it is important to note the name and number of the worker's previous occupations, as well as the tasks done at the previous job. Also note the hobbies and pastimes at home of the worker. It is important to conduct a baseline hearing test, if the worker is applying for work in a noise zone.

Once the worker has been examined, they should be placed in categories according to their abilities to work for example:

- Worker without limitations: People placed in this category is fit and healthy enough to work in any part of the factory
- Employable with permanent restriction: A person placed in this category can work in certain areas only
- Rejected for employment: This person is not fit enough to work anywhere in the organisation.

• FITNESS-TO-WORK AND DISABILITY EVALUATION

A "fit to work" or "fitness to work" medical assessment is typically done when an employer wishes to be sure a worker can safely perform a specific job or task. The purpose of the assessment is to determine whether medically the worker is able or not to perform the job, work activity or task under the working conditions. Fit to work assessments are usually done to determine medical fitness after an illness or injury, but are sometimes done after employment has been offered, as requested by the employer, or as a condition of a job transfer.

The worker will visit a medical practitioner at the clinic or other facility who will determine if the worker is able to do that particular job. The medical practitioner may consider, for example, physical or mental capacities, sensory acuity, and level of skill. The medical practitioner will typically only report one of three following conditions back to the employer, namely

- fit,
- unfit, or
- fit subject to work modifications.

A fit-to-work assessment may be done, for instance, for the following reasons, namely:

- ✓ there has been a significant change in the working conditions,
- ✓ the job at work has been modified and the returning worker is still going for physiotherapy, rehabilitation, or both,

- ✓ there has been a change in the health of a worker such as, for example, returning to work after recovery from a serious illness or injury,
- ✓ a medical condition may limit, reduce or prevent the person from performing a new or current job effectively such as, for example, musculoskeletal conditions that limit mobility,
- ✓ a medical condition is likely to make it unsafe to do the work activity such as, for example, a worker may unpredictably become unconscious in a hazardous situation,
- ✓ a medical condition is likely to make it unsafe both for the worker, co-workers or the public such as, for example, where driving is essential to the job but the person is subject to unpredictable and sudden unconsciousness, or
- ✓ the medical condition may be made worse by the job such as, for example, excessive physical exertion by a worker with a heart or lung disorder.

• DISABILITY

For the evaluation documentation has to be sent out such as medical reports, and declarations that the worker, the line manager, and the doctor/s need to complete. These documents are generally forwarded to the worker because they are responsible for the disability claims process. However, remember that the worker is forced to deal with the stress of adjusting to living with their disability and that the organisation should be able to take over some of the documentation processing to prevent them from having to deal with small administration duties, such as forwarding documents and obtaining signatures when the disability they have to deal with may be crippling and render them incapable of normal functioning. Obtaining medical reports and medical insurance records from the appropriate professionals and people in such situations is pivotal and will speed up the process for both the benefit of the employer and the worker.

• INCAPACITY IN TERMS OF THE LABOUR RELATIONS ACT, 1995 AS AMENDED (LRA)

The LRA sets out a process relating to the incapacity of a worker due to ill health or injury. An employer, therefore, has the following obligations, namely:

1. An employer has to determine whether a worker is temporarily or permanently unable to work.
2. If the worker is temporarily unable to work, the employer should investigate the extent of the incapacity and alternative solutions to accommodate the worker at work.
3. If the incapacity is permanent, the employer should ascertain the possibility of securing alternative employment or adapting the duties or work circumstances of the worker to accommodate the disability.
4. In any investigations related to incapacity, the worker should be allowed to state his/her case in the course of the investigation and to be assisted by a trade union representative or a fellow worker.
5. The degree of incapacity is relevant to the fairness of any dismissal, whether for temporary or permanent incapacity.
6. At all times during assessments, an employer should consider:
   - i) Whether the worker is capable of performing the work;
   - ii) If the worker is not capable, the extent to which the worker is able to perform the work;
   - iii) The extent to which the work circumstances of the worker can be adapted to accommodate the disability or, where this is not possible, the extent to which the duties of the worker can be adapted;
   - iv) The availability of any suitable alternative employment.

• EXPEDITING THE PROCESS

Should the worker become aware that they have become incapable of performing the majority of their duties, or has been absent from work for longer than three weeks, it is necessary for the immediate supervisor, line manager or HR to contact the worker and make an appointment with a doctor to provide an assessment of their ability to perform their duties.

The claims process must begin as early as possible. Most corporate health insurers will have a waiting period where the employer will have to pay the full salary of the affected worker from the date on which the worker becomes disabled until the disability claim is approved or rejected. The employer can then claim for these funds to be reimbursed if the claim is approved.

The LRA requires corporate health insurers to evaluate the work activities of the worker and how serious the disability is by using the various medical reports supplied by doctors, hospitals and/or other institutions where the worker was treated.

• DISABILITY CLAIM APPROVAL

Once an application is approved, disability benefits are generally paid monthly. The employer is no longer obliged to pay the salary or wages of the worker. Group schemes usually pay workers between 60%-75% of their pre-disability salary.

Becoming disabled or temporary impaired is a life-changing event for any worker. Submitting the disability claim may seem like a tedious, bureaucratic and endless process but the process is necessary to prevent fraud and abuse of the disability benefit which is a not uncommon practice in South Africa. It is beneficial to the organization to assist a worker during this difficult time of adjustment and guide them through the laborious and tedious process of claiming an income after becoming disabled.

• PERIODIC MEDICAL EXAMINATION

Some periodic medical examinations are required by the law. In terms of Section 8(2) of the Act states that any health hazard should be monitored periodically. Statutory periodic examinations that need to be specifically monitored according to the law include the following categories of workers, namely:

- Pilots
- Public vehicle drivers
- Apprentices and miners
- Commercial drivers and listed work

Specified hazards that are to be examined are:

- Noise
- Lead
- Hazardous substances
- Heat/cold

Organisations conduct annual screening of their workers for conditions that are work related even though it is not required by the law. These annual screenings are intended to identify high risk conditions such as high cholesterol levels and levels of hypertension in workers early. Screenings are normally done to certain target areas where target organs and systems can be affected. The aim of such screenings is to identify potential problems at an early stage, before damage is too advanced. Being preventative like this can save huge medical expenses for your workers and also high costs due to absenteeism and worsened medical conditions in the future.

• EXECUTIVE MEDICAL EXAMINATIONS

Executive medical examinations are designed to improve the general health and fitness of the participants to reduce the mortality and morbidity due to lifestyle-related diseases. Workers and managers are encouraged to voluntarily participate in executive health programmes where they are offered and where their performance and condition will be carefully monitored. Participation in these programs are usually entirely by choice or by organization policy. The Executive Health Programme should be monitored by a biokinetic or trained occupational health practitioner to ensure risk factors are minimised and controlled as well as to provide ongoing motivation and encouragement.

• EXIT MEDICAL EXAMINATION

The exit medical examination examines workers before they leave the employ of the organization or their current job position and go to another position either within the organisation or to a new organisation. The importance of this examination is often misunderstood. An exit medical examination safeguards the organization against any future legal claims if they can prove that when the worker left the organisation no damage had been done to the worker due to employment in a certain department at the organization. It is evident that an environment in which good two-way communication between the human resources department and the occupational health centre creates positive health outcomes for both the organization and the workers.

• REVIEW QUESTIONS

1. What are some of the key considerations when deciding to establish a clinic at work?
2. How would an existing facility have to be adapted given the recent experiences with COVID-19?
3. In what ways can an organization assist a worker who has been disabled due to an accident at work?
4. What is the purpose of a fitness-to-work medical examination?
5. Why is an exit medical examination important?

• REFERENCES:

Anonymous (n.d). Loose notes on safety, health and environment management

Attridge, Mark., 2019. A global perspective on promoting workplace mental health and the role of employee assistance programs.

Boudreau, Yvonne. (2010). *State-based Occupational Safety and Health Surveillance | | Blogs | CDC*. [online] Blogs.cdc.gov. Available at: https://blogs.cdc.gov/niosh-science-blog/2010/08/30/surveillance/.

CCOHS (n.d.). *Fit to Work*. [online] Ccohs.ca. Available at: https://www.ccohs.ca/oshanswers/psychosocial/fit_to_work.html.

Chennells, Colleen. (2015). *Paralegal Manual*. [online] Labourguide.co.za. Available at: https://www.labourguide.co.za/workshop/1364-paralegal-manual.

Deacon. Clair., Smallwood, John. and Haupt, Theodore. Conrad. 2005. "The Health and Well Being of Older Construction Workers," Assessment and Promotion of Work Ability, Health and Well-Being of Ageing Workers, International Conference Series, Pp. 172-177, ISBN 0-444-51989-0

Department of Health (2003). *OH services for health care workers in the National Health Service of South Africa*. [online] Kznhealth.gov.za. Available at: http://www.kznhealth.gov.za/occhealth/OHmanual.pdf.

Department of Labour (1993). Occupational Health and Safety Act No: 85 of 1993.

Haupt, Theodore. Conrad., Deacon, Claire. and Smallwood, John. 2005. "The Health Status of South African Construction Workers." Paper presented at Third International Conference on Construction in the 21st Century, Advancing Engineering, Management and Technology, September 15-17, Athens, Greece, pp. 263-277, ISBN 960-254-655-7

Haupt, Theodore. Conrad., Deacon, Claire. and Smallwood, John 2005. Importance of Healthy Older Construction Workers, *Acta Structilia*, 12 (1), 1-19, ISSN 10230564

Kalia, Nimisha., Alfriend, Amy. S., Minor, Sebastian. A., Bernacki, Edward. J., Lavin, Robert. A., Leung, Nina., Yuspeh, Larry., Tsourmas, Nicholas. F. and Tao, Xuguang. Grant., 2020. Assessing Workplace Clinic Utilization and Performance: Lessons and Implications for Care Delivery. *Journal of Occupational and Environmental Medicine*, *62*(8), pp.e407-e413.

Labour Relations Act 66 of 1995 (1995). [online] Gov.za. Available at: https://www.gov.za/sites/default/files/gcis_document/201409/act66-1995labourrelations.pdf.

LaPenna, A.Michael., 2017. *Workplace clinics and employer managed healthcare: A catalyst for cost savings and improved productivity*. CRC Press.

Mabuza, Mbuso., 2018. Impact of an onsite occupational health clinic on organisational performance and employee wellbeing at a southern African maritime port. *Eurasian Journal of Medicine and Oncology*, *2*(3), pp.152-164.

Minicozzi, Salvatore. J. and Russell, Brent. S., 2017. On-Site Chiropractic Care as an Employee Benefit: A Single-Location Case Study. *Journal of Chiropractic Medicine*, *16*(3), pp.183-188.

Parkinson, Michael. D., 2018. The Healthy Health Care Workplace: a Competitive Advantage. *Current cardiology reports*, *20*(10), p.98.

Rantanen, Jorma. and Fedotov, Igor. (1995). *Standards, Principles and Approaches in Occupational Health Services*. [online] Ilo.org. Available at: http://www.ilo.org/safework/info/publications/WCMS_110439/lang--en/index.htm.

Sharan, Deepak., Rajkumar, Joshua. Samuel. and Jose, Jerrish. A., 2018, August. Effectiveness of Onsite Occupational Health Clinics in Management of Work Related Musculoskeletal Disorders in 12000 Information Technology Professionals. In *Congress of the International Ergonomics Association* (pp. 642-646). Springer, Cham.

WHO (2001). *Occupational health: A manual for primary health care workers*. [online] World Health Organization. Available at: https://www.who.int/occupational_health/publications/emhealthcarework/en/.

WHO (2001). *The Role of the Occupational Health Nurse in Workplace Health Management*. [online] Who.int. Available at: https://www.who.int/occupational_health/regions/en/oeheurnursing.pdf?ua=1.

CHAPTER 19

HEALTH CHALLENGES IN THE WORKPLACE

• MUSCULOSKELETAL AND SENSORY EFFECTS

Repeated bodily movements while executing work-related activities and working continuously in static positions such as, for example, kneeling for extended periods of time are major contributors to repetitive movement disorders that include meniscus lesions, bursitis, tendonitis, ganglion cyst, trigger finger and osteoarthritis of the knees. Other contributory movements include repetitive and heavy lifting, bending and twisting, frequent repetition of a movement, uncomfortable working position, exerting more force that necessary, and working for long periods without any breaks. Carpal tunnel syndrome (CTS) manifests as tingling, numbness, weakness or pain in the fingers or hand due to pressure on the median nerve in the wrist. Contributory causes include repetitively making the same hand and/or wrist movements as well as wrist injuries and spurs.

Lower back pain is caused by some of the following causes, namely:

- Lumbar strain where ligaments, tendons and/or muscles of the lower back are stretched;
- Nerve irritation of the lumbar spine;
- Lumbar radiculopathy caused by damage to the discs between the vertebrae
- Bone encroachment where the space for the adjacent spinal cord and nerves is reduced; and
- Bone and joint conditions that arise from injury such as fractures.

Slipped or herniated discs, where the tissue that separates the vertebral bones of the spinal column has been ruptured, manifest as a result of repetitive body movements during construction activities. Workers experience pressure against one or more spinal nerves causing a shooting pain known as sciatica, weakness or numbness in the area served by these nerves. Treatment is in the form of anti-inflammatory and muscle-relaxant medications, exercises, and physical therapy. Surgery may be required in severe cases.

• MIGRANT WORKERS

Since South Africa has opened its economy post-1994 it has become an attractive destination for migrants in the main from countries that are members of the South African Development Community (SADC). Both internal migrants within the borders of the country and international migrants face daily stresses associated with the challenges of moving to a new area or region, seeking work, struggling to access safe housing and a secure livelihood, feeling alone and without social support. Maternal, new-born and child health; HIV, COVID-19 infections and TB; non-communicable diseases and violence and injury all affect migrants who enter workplaces in South Africa. It is hard to track workers who migrate as part of survival. Follow up medical appointments are easily missed and prescribed medication courses are not completed. As a consequence, migrants bring adverse health conditions into the workplace. Some diseases require lengthy periods on treatment which are interrupted when workers migrate to another location and workplace in that location. Very often these workers do not have their medical history and records available for their new employer creating a risk of having pre-existing health conditions introduced into the workplace. It is therefore necessary for the health system at work to recognize these challenges and develop a migration-aware health system.

• CORONAVIRUS INFECTIONS

Human coronavirus infections are again receiving global attention following the emergence at a market in Wu Han, China in December 2019 of a novel coronavirus (COVID-19). The animal reservoirs, zoonotic transmission, high mutation rates, and demonstrated human-to-human spread of these ribonucleic acid (RNA) viruses have made them of particular public health concern. The current potential for global dissemination of COVID-19, in addition to the growing number of reported cases, have increased urgency across the entire world to understand this outbreak which has caused unprecedented numbers of infections and deaths everywhere.

• BACKGROUND ON HUMAN CORONAVIRUSES

Coronaviruses (CoV) are a large family of viruses that cause illnesses ranging from the common cold to more severe diseases. Based on genome sequencing, all known human coronaviruses have emerged from animal reservoirs. These RNA viruses have high mutation rates that allow them to adapt to varied hosts, increasing their potential for rapid human-to-human spread once a spill over event has occurred. The COVID-19 is the seventh identified human coronavirus, and appears to have notable similarities to two other highly pathogenic human respiratory coronaviruses, namely severe acute respiratory syndrome coronavirus (SARS-CoV) and Middle East respiratory syndrome coronavirus (MERS-CoV), both of which have generated large-scale public health responses in the last 20 years.

The COVID-19, SARS-CoV and MERS-CoV belong to the family of betacoronoviruses, and likely share a common reservoir in bats. Intermediate hosts for zoonotic transmission to humans proposed for each of these three pathogenic strains include civets (SARS-CoV), dromedary camels (MERS-CoV), and an unconfirmed but likely mammalian source (COVID-19). These betacoronaviruses typically produce respiratory and gastroenteritis symptoms in human and animal hosts, respectively. The remaining identified human coronaviruses (HCoV-229E, HKU1, NL63, OC43) are limited in their severity of disease and often fail to produce symptoms greater than the common cold in immunocompetent hosts.

Coronaviruses are zoonotic and therefore transmitted between animals and people. For example, SARS-CoV was transmitted from civet cats to humans and MERS-CoV from dromedary camels to humans. Several known coronaviruses are circulating in animals that have not yet infected humans. The novel coronavirus (nCoV) is a new strain that has not been previously identified in humans which was previously named 2019-nCoV, now designated COVID-19. This virus can be spread by human-to-human transmission via respiratory droplets or direct contact with an infected individual.

The challenge with the novel coronavirus (2019-nCoV) is its detection and spread of the emerging respiratory pathogen and the uncertainty over its key epidemiological, clinical and virological characteristics of the novel pathogen. Of great concern is its ability to spread in the human population and its virulence or case-severity. This snake-associated virus has never been seen before and as far as is known humans have no internal defence against it. Currently, a case fatality rate (CFR) of slightly >3% has been reported, with some reports as high as 5%. Comparatively, MERS and SARS exhibited CFRs of 35% and 15%, respectively. Initially, mortality among COVID-19 victims was highest among the elderly and those with multiple comorbidities. The World Health Organization (WHO) model, formulated by Ferguson and colleagues, suggests a basic reproductive number, R_0 which is the number of new cases that can develop from 1 confirmed case of 2.6–3.1, twice higher than for seasonal Influenza.

Tremendous worldwide efforts are being made to understand the molecular and clinical virology of this virus. Unbelievable molecular knowledge about the genomics, structure, and virulence of this virus has already been achieved. However, there will be NO drugs or vaccines available in the near future to protect against or limit the infection of people across the globe. Only symptomatic support is available. To date, the average incubation period is estimated to be a median of 5 days with a range between 2 and 14 days although isolation and quarantine periods can be as long as 21 days.

Common signs of COVID-19 infection include respiratory symptoms, fever, cough, shortness of breath and breathing difficulties. Severe cases result in infections such as pneumonia, severe acute respiratory syndrome, kidney failure and even death. In China, most patients had presented with fever (83%–98%) or cough (76%–82%) and roughly one-third had shortness of breath. Less common symptoms included myalgias (11%), rhinorrhea (10%), headache (8%), chest pain (2%), and gastrointestinal symptoms (3%). Additionally, the majority of patients presented with more than one symptom and were mostly males with a median age of 59 years.

Standard recommendations to prevent infection spread include regular hand washing, covering mouth and nose when coughing and sneezing, thoroughly cooking of meat and eggs, and avoiding close contact with anyone showing symptoms of respiratory illness such as coughing and sneezing.

Strategies to prevent or limit transmission in healthcare settings such as a clinic within the workplace

The following strategies are recommended by the World Health Organization (WHO), namely:

1. Ensuring triage, early recognition, and source control

Clinical triage includes a system for assessing all workers to the clinic at admission allowing early recognition of possible COVID-19- infection and immediate isolation of workers with suspected nCoV infection in an area separate from other workers (source control). To facilitate the early identification of cases of suspected nCoV infection, clinics at work should:

- encourage occupational nurses to have a high level of clinical suspicion;
- establish a well-equipped triage station at the entrance of the clinic, supported by trained staff;
- institute the use of screening questionnaires according to the updated case definition (https://www.who.int/publications-detail/global-surveillance-for-human-infection-with-novel-coronavirus-(2019-ncov) and
- post signs in public areas reminding symptomatic patients to alert occupational nurses.

The promotion of hand hygiene and respiratory hygiene are essential preventive measures. The following flowchart was adapted from the one recommended by the CDC:

A.	**IDENTIFY** If in the first 14 days of the onset of symptoms a history of either		
	Travel to infection 'hot spot'	← **OR** →	**Close contact** with a person known to have COVID-19 illness
B.	**AND** the person has		
	Fever or symptoms of lower respiratory illness (Such as for example a cough or shortness of breath)		
If both symptoms and illness are present ↓			
1.	**ISOLATE**		
	• Place a face mask on the patient • Isolate the patient in a private room or separate area • Wear appropriate personal protective clothing and equipment (PPE)		
2.	**ASSESS CLINICAL STATUS**		
	Examination	Is fever present? ☐ Subjective? ☐ Measured? _____ ^{0}C	Is respiratory illness present? ☐ Cough? ☐ Shortness of breath?
3.	**INFORM**		
	• Contact health department or similar to report at-risk patients and their clinical status • Assess need to collect specimens to test for COVID-19 • Decide disposition		
If discharged to home ↓			
	Instruct patient As needed depending on severity of illness and health department consultation		
	• Home care guidance • Home isolation guidance		
	Advise patient If the patient develops new or worsening fever or respiratory illness		
	• Call clinic to determine if re-evaluation is needed • If re-evaluation is needed call ahead and wear a face mask		

2. Applying standard precautions for all workers being treated

Standard precautions include hand and respiratory hygiene, the use of appropriate personal protective equipment (PPE) according to risk assessment, injection safety practices, safe waste management, proper linens, environmental cleaning and sterilization of patient-care equipment.

It is important to ensure that the following respiratory hygiene measures are used, namely:

- ensure that all workers being treated cover their nose and mouth with a tissue or elbow when coughing or sneezing;
- offer a medical mask to workers with suspected COVID-19 infection while they are in the waiting or public areas of the clinic or in cohorting rooms; and
- perform hand hygiene after contact with respiratory secretions.

Occupational nurses should apply the WHO's My 5 Moments for Hand Hygiene approach before touching a worker, before any clean or aseptic procedure is performed, after exposure to body fluid, after touching a worker, and after touching the surroundings of a worker. These include

- hand hygiene that includes either cleansing hands with an alcohol-based hand rub (ABHR) or with soap and water;
- alcohol-based hand rubs that are preferred if hands are not visibly soiled; and
- washing hands with soap and water when they are visibly soiled.

The rational, correct, and consistent use of PPE also helps to reduce the spread of pathogens. The use of PPE effectiveness strongly depends on adequate and regular supplies, adequate staff training, appropriate hand hygiene and specifically appropriate human behaviour.

It is important to ensure that environmental cleaning and disinfection procedures are followed consistently and correctly. Thoroughly deep cleaning environmental surfaces with water and detergent and applying commonly used hospital-level disinfectants such as sodium hypochlorite are effective and sufficient procedures. Bleach has also been found to be effective as a cleaning and disinfectant agent. Medical devices and equipment and medical waste should be managed in accordance with healthy and safe routine procedures.

3. Implementing empiric additional precautions

3.1 Contact and droplet precautions

- In addition to using standard precautions, all workers and occupational nurses should use contact and droplet precautions before entering the room where suspected or confirmed nCoV infected workers are admitted;
- Infected workers should be placed in adequately ventilated single rooms where adequate ventilation is considered to be 60 L/s per worker or hospital-level air-conditioning;
- when single rooms are not available which is typically the case in a clinic at work, workers suspected of being infected with nCoV should be grouped together until moved to a specialist facility or hospital;
- While in the clinic the beds of all workers held in the clinic should be placed at least 1.5 m apart regardless of whether they are suspected to have nCov infection or not;
- Where possible, a team of occupational nurses should be designated to care exclusively for suspected or confirmed cases to reduce the risk of transmission;
- Occupational nurses should:
  - use a medical mask, wear eye protection or goggles or facial protection in the form of a face shield to avoid contamination of mucous membranes;
  - wear a clean, non-sterile, long-sleeved gown; and
  - use gloves

Note that the use of boots, coveralls and aprons is not required during routine care.

- After patient care, appropriate doffing and disposal of all PPE and hand hygiene should be carried out;
- a new set of PPE is needed, when care is given to a different worker;
- equipment should be either single-use and disposable or dedicated equipment such as, for example, stethoscopes, blood pressure cuffs and thermometers.
- if equipment needs to be shared among workers, they should be cleaned and disinfected between use for each individual person such as, for example, by using ethyl alcohol 70%;
- occupational nurses should refrain from touching eyes, nose or mouth with potentially contaminated gloved or bare hands;

- avoid moving and transporting workers out of their space or area unless medically necessary;
- use designated portable X-ray equipment and/or other designated diagnostic equipment;
- where transport is required, predetermined transport routes should be used to minimize exposure for other workers and have the infected worker being moved use a medical mask;
- ensure that occupational nurses who are transporting workers perform hand hygiene and wear appropriate PPE as previously described;
- notify the facility receiving the worker of any necessary precautions as early as possible before the arrival of the worker at the facility;
- routinely clean and disinfect surfaces which the worker was in contact with; and
- maintain a record of all persons entering the worker's space.

3.2 Airborne precautions for aerosol-generating procedures

Some aerosol-generating procedures have been associated with an increased risk of transmission of coronaviruses (SARS-CoV and MERS-CoV), such as tracheal intubation, non-invasive ventilation, tracheotomy, cardiopulmonary resuscitation, manual ventilation before intubation, and bronchoscopy.

Ensure that occupational nurses performing aerosol-generating procedures in accordance with the WHO strategies for COVID-19 infection prevention and control during health care that include:

- perform procedures in an adequately ventilated room – that is, natural ventilation with air flow of at least 160 L/s per patient or in negative pressure rooms with at least 12 air changes per hour and controlled direction of air flow when using mechanical ventilation;
- use a particulate respirator at least as protective as a US National Institute for Occupational Safety and Health (NIOSH)-certified N95, European Union (EU) standard FFP2, or equivalent.
- when occupational nurses put on a disposable particulate respirator, they must always perform the seal check. Note that if the wearer has facial hair such as, for example, a beard it may prevent a proper respirator fit;
- use eye protection such as goggles or a face shield;
- wear a clean, non-sterile, long-sleeved gown and gloves. If gowns are not fluid resistant, occupational nurses should use a waterproof apron for procedures expected to have high volumes of fluid that might penetrate the gown; and
- Isolation rooms and care processes should be configured to minimize the number of health care personnel exposed to the worker as well as the duration of contact.
- If a worker needs to be hospitalized and an airborne infection isolation room is not available, the worker should be transferred to a facility with appropriate capabilities.
- Although workers with mild symptoms can be discharged, there are reports of clinical decline in the second week of illness with over 50% of workers having dyspnea around the eighth day. Therefore, discharge may be inappropriate for workers who are unable to adhere to recommendations, return precautions, self-monitoring, or those with inadequate housing.
- workers with evidence of severe illness such as hypotension, tachycardia, hypoxia, or other signs of shock or those with comorbidities in the form of older age, immunocompromised, chronic conditions such as diabetes, cardiovascular, or chronic lung disease will require admission to isolation units and/or the ICU of a specialist facility.

4. Implementing administrative controls

4.1. Administrative measures related to occupational nurses

Administrative controls and policies for the prevention and control of transmission of COVID-19 infections within the clinic include, but may not be limited to:

- establishing sustainable Infection Prevention and Control (IPC) infrastructures and activities;
- educating workers' caregivers;
- developing policies on the early recognition of acute respiratory infection potentially caused by COVID-19;
- ensuring access to prompt laboratory testing for identification of the etiologic agent;
- preventing overcrowding, especially in the treatment area;
- providing dedicated waiting areas for symptomatic patients;
- appropriately isolating workers who are to be hospitalized;
- ensuring adequate supplies of PPE;
- ensure the adherence of IPC policies and procedures for all facets of health care;
- provision of adequate training for occupational nurses;
- ensuring an adequate worker-to-clinic staff ratio;

- establishing a surveillance process for acute respiratory infections potentially caused by nCoV among occupational nurses;
- ensuring that occupational nurses and the workforce understand the importance of promptly seeking medical care; and
- monitoring occupational nurse compliance with standard precautions by a medical doctor and providing mechanisms for improvement as needed.

5. Using environmental and engineering controls

These controls address the basic infrastructure of the clinic. These controls aim to ensure there is adequate ventilation in all areas of the clinic, as well as adequate environmental cleaning. Additionally, spatial separation of at least 1.5 meters should be maintained between all workers being examined and/or treated in the clinic. Both spatial separation and adequate ventilation can help reduce the spread of many pathogens in the clinic.

Cleaning and disinfection procedures must be followed consistently and correctly. Cleaning environmental surfaces with water and detergent and applying commonly used hospital disinfectants such as sodium hypochlorite is an effective and sufficient procedure. Medical waste must be managed in accordance with healthy and safe routine procedures.

Duration of contact and droplet precautions for patients with nCoV infection

Standard precautions should be applied at all times. Additional contact and droplet precautions should continue until the patient is asymptomatic. More comprehensive information about the mode of COVID-19 infection transmission is required to define the duration of additional precautions.

Collecting and handling laboratory specimens from workers with suspected 2019-nCoV infection

All specimens collected for laboratory investigations from workers should be regarded as potentially infectious. Occupational nurses who collect, handle or transport any clinical specimens should adhere rigorously to the following standard precaution measures and biosafety practices to minimize the possibility of exposure to pathogens, namely:

- ensure that occupational nurses who collect specimens use appropriate PPE such as eye protection, a medical mask, a long-sleeved gown, and gloves. If the specimen is collected with an aerosol-generating procedure, personnel should wear a particulate respirator at least as protective as a NIOSH-certified N95, an EU standard FFP2, or the equivalent;
- ensure that all personnel who transport specimens are trained in safe handling practices and spill decontamination procedures;
- place specimens for transport in leak-proof specimen bags such as secondary containers that have a separate sealable pocket for the specimen such as a plastic biohazard specimen bag, with the particular worker's label on the specimen container, namely the primary container, and a clearly written laboratory request form;
- ensure that laboratories adhere to appropriate biosafety practices and transport requirements, according to the type of organism being handled;
- deliver all specimens by hand whenever possible. Pneumatic-tube systems should not be used to transport specimens;
- document clearly each worker's full name, date of birth and suspected nCoV of potential concern on the laboratory request form. Notify the laboratory as soon as possible that the specimen is being transported.

Recommendation for offsite precautions

The basic principles of IPC and standard precautions should be applied for primary care and offsite. For COVID-19 infection, the following measures should be adopted:

- triage and early recognition;
- emphasize hand hygiene, respiratory hygiene and medical masks to be used by workers with respiratory symptoms;
- appropriate use of contact and droplet precautions for all suspected cases; and
- educate families about the early recognition of symptoms, basic precautions to be used and which health care facility they should refer to.

Actions that should be taken by employers

- Adapt and apply action steps through ongoing two-way communication and engagement between employer and workers that regularly addresses misunderstandings, misinformation, rumours and frequently asked questions.
- Encourage everyone to adopt protective behaviours.

- Manage expectations and communicate uncertainties.
- Coordinate and encourage collaboration among response partners.
- Assess the initial perception of risk among affected and at-risk populations.
- Provide information and guidance.
- Develop and test risk communication systems.
- Activate the spokespersons identified and appointed for the emergency.
- Draw up timelines for communication activities and products.
- Monitor processes that delay the release of information and create confusion.
- Assign responsibilities for internal communication within and between each response agency and external communication to the public.
- Coordinate message preparation, consistency and dissemination.
- Create a co-operative, harmonious and humane working environment with respect to open two-way communication with workers to understand and respond to their concerns, attitudes, beliefs, and barriers to following health guidance and flexibility of working arrangements to accommodate threats of COVID-19 infections, self-isolation and quarantine without fear or panic.
- Liaise with employer-worker health and safety forums, unions and other worker representative bodies to ensure informed collective decision-making about appropriate steps and actions to take to deal with COVID-19 effectively.
- Encourage workers not to smoke and to avoid alcohol and explain the reasons for this.
- Particular attention must be given to workers with reduced immune systems, including those who are HIV positive, diabetics, cancer, heart or lung complaints and other chronic diseases.
- Implement non-punitive "emergency sick leave" policies.
- Cross-train workers to perform essential functions so the workplace can operate even if key workers are absent.
- Discourage workers from using the phones, desks, offices, or other work tools and equipment of other workers. Or if necessary, clean and disinfect them before and after use.
- Implement flexible work hours, staggered shifts or working at home to enable additional work space for social distancing.
- Increase physical space between employees and customers by introducing drive throughs and partitions including see-through separators and shields.
- Deliver services remotely through phone, video, or web.
- Deliver products through curbside pick-up or delivery.
- Avoid gatherings during lunch and tea breaks. Workers must spread out and not eat at the same tables in the cafeteria or dining area. Create a roster with flexible lunch times to avoid crowding.
- Increase welfare and hygiene facilities and ensure that they are properly stocked with soap and sanitizers and regularly deep cleaned.
- Provide PPE to all workers in the form of masks or transparent face shields, soap and water or 60%-70% alcohol-based sanitizer and gloves as needed.
- Set up clear communication channels to all suppliers on contracts, orders and projects and provide regular updates.
- Liaise with clients and all other stakeholders, for example, in the supply chain on contingency arrangements.
- Ensure payments to supply chains are up to date.
- Review all contractual arrangements to understand options now and in the future.
- Clarify the position over suspension of any contracts, orders and projects.
- Liaise with funders and banks to secure sufficient credit to sustain business activities.
- Set up project-specific communication networks with supply chains.
- Review health and safety critical elements of projects and workplaces.
- Review production and delivery programmes and schedules.
- Identify if any activities can continue remotely or be brought forward.
- Ensure remote access to critical business and operational information is in place.
- Provide the means to operate from remote locations by providing workers who need them with internet and Wi-Fi access at the cost of the employer.
- Offer those with the ability to productively work from home the opportunity to do so without fear of reprisals taking into account the most vulnerable groups.
- Train all workers in communication platforms such as ZOOM, MS Teams and Google Meet to enable communication with and by them.
- Protect business critical functions such as payroll and accounts departments.
- Split work teams up to avoid everyone becoming ill at the same time.
- Arrange work rosters with registers, screening of everyone upon entry to the workplace, ensure protocols are strictly enforced and all workers who do not pass the initial screening are to be denied access to the work premises and advised to either self-isolate or be referred for testing.
- Reduce contact between leadership teams to ensure operational continuity.
- Cancel or move meetings and overseas travel to conference calls or online meetings where possible.

- Ensure that production, projects and deliveries are ready to recommence as soon as possible after the crisis has been overcome.
- Obtain all required permits after developing the required plan to demonstrate clearly that the workplace is 'COVID-19 ready'.
- Develop safe operating procedures or working procedures for the recommencement of operations at the workplace
- Ensure that the workplace and work surfaces are regularly disinfected and sanitized.
- Place doormats sprayed with disinfectant or household bleach at all doorways and entrances and make sure that everyone entering the premises walks over them. Shoes can be contaminated from infected droplets that fall on the ground.
- Doors should be automated where possible, both at the entrance and throughout the workplace.
- Actively force social distancing of at least 1.5 meters in all work areas by creating spaces between workstations and placing markers on floors.

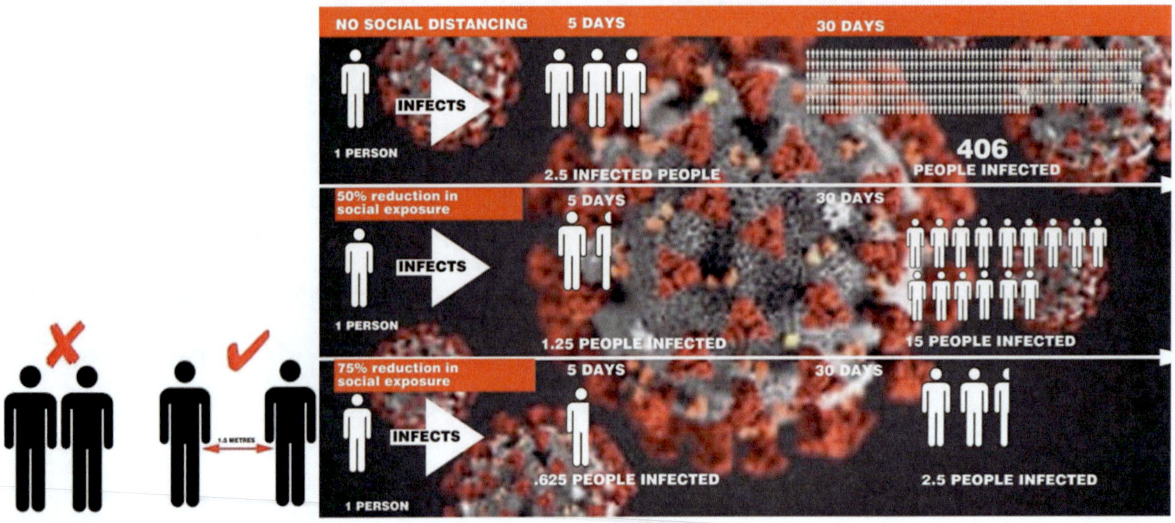

(Source: https://globalnews.ca/news/6709071/coronavirus-social-distancing-math/)

- Assist workers in terms of special UIF provisions to prevent undue loss of income by workers.

Communication in the workplace

- Should the need arise, announce the first nCoV case early to workers and the public when appropriate to reduce unnecessary panic, and update information after a risk assessment at the workplace and an analysis of risk perception have been done based on existing formal and informal information
- Provide any relevant information as soon as it is received, even if it is not complete, and openly explain the degree to which information is uncertain. Manage uncertainty - communicate what is known and what is not known: explain the degree to which uncertainty exists;
- Place visible appropriate signage as widely as possible in the workplace to reinforce correct messaging about coping with COVID-19;
- Provide the workforce and public with regular channels through which they can get updated information such as, for example, hotlines or a website.
- Produce and test messages to ensure accuracy and relevance, including messages about public health advice and manage misinformation and fake news.
- Ensure messages are consistent across all sectors and levels of the work environment.
- Use trusted and effective communication channels that the target audiences regularly use.
- Engage, train and activate trusted occupational nurses and other official spokespersons.
- Monitor possible barriers to the uptake of protective behaviours including developing and using a lexicon to ensure unambiguous shared understanding of behaviour-related messages.
- Translate information material into relevant languages and adapt them to appropriate literacy levels.
- Develop short multimedia pieces that present key information such as, for example, explain the disease etiology, symptoms, transmission, how to protect oneself and what to do if someone gets sick and that can be shared widely on multiple platforms.

Steps workers should be encouraged by employers to take to protect themselves include the following:

1) **NO HANDSHAKING!** Use a fist bump, slight bow, elbow bump, hand wave or other appropriate signs of acknowledgement or greeting.

2) Use **ONLY** knuckles to touch light switches, elevator buttons, and door handles, for example. Lift any necessary dispensers with a paper towel or use a disposable glove.
3) Open doors with a closed fist or hip - do not grasp the handle with the hand, unless there is no other way to open the door. Especially important on bathroom and office or commercial doors. Try to minimise usage of door handles. Keep doors open or semi-open when possible.
4) Use disinfectant wipes at the stores when they are available, including wiping the handle and child seat in grocery carts.
5) Wash hands with soap for 10-20 seconds and/or use a greater than 60% alcohol-based hand sanitizer whenever workers return home from ANY activity that involves locations where other people have been even if their hands are not visibly dirty.
   Why? Washing hands with soap and water or using an alcohol-based hand rub eliminates the virus if it is on the hands.
6) Keep a bottle of sanitizer available at each entrance at home AND in cars for use after refuelling or touching other contaminated objects when hands cannot immediately be washed. Keep a bottle at the workstation.
7) Always wear a mask or face shield in the workplace.
8) Avoid touching eyes, nose and mouth
   Why? Hands touch many surfaces which can be contaminated with the virus. If eyes, nose or mouth are touched with contaminated hands, the virus can be transferred from the surface to a person.
9) Maintain social distancing – maintain at least 1.5 metre distance between workers, particularly those who are coughing, sneezing and have a fever.
   Why? When a worker who is infected with a respiratory disease, like COVID-19, coughs or sneezes they project small droplets containing the virus. If one is too close, one can breathe in the virus
10) If possible, cough or sneeze into a disposable tissue and discard safely in a suitable dispenser – not in the road, pathway or pavement. Use an elbow only if necessary. The clothing on the elbow will contain infectious virus that can be passed on for up to a week or more – remove and wash properly as soon as possible!
11) If a worker has fever, cough and difficulty breathing, seek medical care early. Tell immediate supervisor if there has been close contact with someone who has been infected or tested positive for COVID-19 and has respiratory symptoms.
    Why? Whenever a worker has fever, cough and difficulty breathing it is important to seek medical attention promptly as this may be due to a respiratory infection or other serious condition.
12) If there are mild respiratory symptoms carefully practice basic respiratory and hand hygiene and stay home until recovered, if possible.
13) Try to keep open windows when possible in offices and work areas.
14) If possible, stop smoking or reduce the number of cigarettes being smoked per day as COVID-19 attacks the lungs and respiratory system which are weakened by regular exposure to tobacco smoke inhalation.

The following should be readily available, namely:

1) Latex or nitrile latex disposable gloves for use when going shopping and all other outside activities when in contact with contaminated areas.

   Note:
   This virus is spread in large droplets by coughing and sneezing. This means that the air will not infect a person! BUT all the surfaces where these droplets land are infectious for about a week on average - everything that is associated with infected people will be contaminated and potentially infectious. The virus is on surfaces and someone **will not** be infected unless their unprotected face is directly coughed or sneezed upon. This virus only has cell receptors for lung cells meaning that it only infects the lungs. The only way for the virus to infect anyone is through the nose or mouth via hands or an infected cough or sneeze onto or into the nose or mouth.
2) Stock up with either disposable or cloth masks and use them to prevent from touching the nose and/or mouth – a person touches the nose/mouth about 90 times per day without knowing it. This is the only way this virus can infect someone - it is lung-specific. The mask will not prevent the virus in a direct sneeze from getting into the nose or mouth - it is only to keep a person from touching their nose or mouth.
3) Stock up with hand sanitizers and latex/nitrile gloves - get the appropriate sizes for the family. The hand sanitizers must be alcohol-based and greater than 60% alcohol to be effective.
4) Stock up with zinc lozenges. These lozenges have been proven to be effective in blocking coronavirus and most other viruses from multiplying in the throat and nasopharynx. Use as directed several times each day when one begins to feel the start of ANY "cold-like" symptoms. It is best to lie down and let the lozenge dissolve in the back of the throat and nasopharynx.

General Advice on wearing masks

Wearing a medical mask is one of the prevention measures to limit spread of certain respiratory diseases, including 2019-nCoV, in affected areas. However, the use of a mask **alone** is insufficient to provide the adequate level of protection and other equally relevant measures. Given that there is likely to be a universal shortage of medical and surgical masks, the use of these special masks should be discouraged by the general public and workers so that these may be reserved for use in the first instance by frontline medical practitioners and health workers. Rather encourage the use of cloth masks which workers can make for themselves.

Masks management

- If medical masks are worn, appropriate use and disposal is essential to ensure they are effective and to avoid any increase in risk of transmission associated with the incorrect use and disposal of masks. These masks are recommended to be reserved for use by medical practitioners and health workers since they will likely be in short supply and they have a greater need for them.

- Place mask carefully to cover mouth and nose and tie securely to minimise any gaps between the face and the mask.
- While in use, avoid touching the mask.
- Remove the mask by using an appropriate technique namely, do not touch the front of the mask but remove the lace from behind the head.
- After removal or whenever a used mask is inadvertently touched, clean hands by using an alcohol-based hand rub or soap and water if visibly soiled.
- Replace masks with a new clean, dry mask as soon as they become damp/humid.
- Do not re-use single-use masks.
- Discard single-use masks after each use and dispose of them immediately upon removal.
- Cloth such as cotton or gauze masks are usually not recommended but can be considered so that specialist types of masks can be made available to frontline medical practitioners and health workers given that these will likely be in short supply.

• VACCINES AND VACCINATION

Given the currency of the information on vaccines and vaccinations the following online sources of information were consulted and used to provide the latest information and details, namely
https://www.who.int/emergencies/diseases/novel-coronavirus-2019/covid-19-vaccines/html. and
https://sacoronavirus.co.za/vaccine-updates/html

The aim of vaccines and vaccination is to prevent morbidity and mortality. Vaccines work by training and preparing the body's natural defences – the immune system – to recognize and fight off the viruses and bacteria they target. After vaccination, if the body is later exposed to those disease-causing germs, the body is immediately ready to destroy them, preventing illness. As of 18 February 2021, at least seven different vaccines across three platforms have been rolled out in countries. At the same time as part of a global exercise, more than 200 additional vaccine candidates are in development, of which more than 60 are in clinical development. COVAX is part of the ACT Accelerator, which WHO launched with partners in 2020 and aims to end the acute phase of the COVID-19 pandemic by:

- speeding up the development of safe and effective vaccines against COVID-19;
- supporting the building of manufacturing capabilities; and
- working with governments and manufacturers to ensure fair and equitable allocation of the vaccines for all countries – the only global initiative to do so.

Vaccines are a critical new tool in the battle against COVID-19. Working as quickly as they can, scientists from across the world are collaborating and innovating that will produce tests, treatments and vaccines that will collectively save lives and end the current COVID-19 pandemic.

COVID-19 vaccines go through a rigorous, multi-stage testing process, including large trials that involve tens of thousands of participants. These trials, which include people at high risk for COVID-19, are specifically designed to identify any common side effects or other safety concerns. Once a clinical trial shows that a COVID-19 vaccine is safe and effective, a series of independent reviews of the efficacy and safety evidence is required, including regulatory review and approval in the country where the vaccine is manufactured, before the WHO considers a vaccine product for prequalification. An external panel of experts convened by the WHO analyses the results from clinical trials, along with evidence on the disease, age groups affected, risk factors for disease, and other information. The panel recommends whether and how the vaccines should be used.

Available vaccines details (as at February 18, 2021 based on data released by the WHO and NCID in South Africa are:

Pfizer /BioNTech Vaccine

- Regulatory: Emergency Use Authorizations by US Food and Drug Administration including WHO PREQUALIFICATION Programme.
- Efficacy: > 90% protection – 2 dose vaccine
- Rollout has happened in a few countries.
- Storage: minus 70 deg C (limitation for SA as we have limited commercial ultra-low cold chain storage)
- Effective against the 501Y.V2 variant

AstraZeneca/University of Oxford Vaccine

- Regulatory: Approved as Emergency Use Authorizations by Medicines and Healthcare products Regulatory Agency and Drugs Controller General of India.
- Efficacy: 70% efficacy – 2 dose vaccine
- AZ has outsourced the production of the vaccine to various sites globally including the largest vaccine producer globally – the Serum institute of India (SII).
- This vaccine likely to be widely used globally due to temperature stability and volumes.
- Storage: 2 – 8 deg C

Johnson & Johnson

- Regulatory: Emergency Use Authorisation by FDA
- Single dose product.
- Vaccine has shown to be 66% effective.
- Product will also be manufactured at the Aspen facility in South Africa.
- Refrigerator storage
- Effective against the 501Y.V2 variant

Moderna

- Regulatory: Emergency Use Authorizations by FDA
- Two dose vaccine
- Storage: minus 20 deg C
- Effective against the 501Y.V2 variant

Sputnik V

- In Phase 3 clinical trials in the UAE, India, Venezuela and Belarus.
- Sputnik V is already registered in 17 countries
- Sputnik V is a two dose vaccine
- Efficacy of over 90%.
- Storage: The lyophilized vaccine can be stored at a temperature of +2 to +8 degrees Celsius

CoronaVac (Sinovac)

- In phase three trials in various countries.
- Interim data from trials in Turkey and Indonesia show 91.25% and 65.3% effective respectively
- Storage: Refrigerator at 2-8 degrees Celsius

Herd immunity

When many people in a community are vaccinated, the pathogen has a difficulty circulating because most of the people it encounters are immune. This is called herd immunity. But no single vaccine provides 100 per cent protection, and herd immunity does not provide full protection to those who cannot safely be vaccinated. But with herd immunity, these people will have substantial protection, thanks to those around them being vaccinated. Vaccinating not only protects those vaccinated, but also protects those in the community who are unable to be vaccinated because of their health condition or age, for example.

South Africa is rolling out its national COVID-19 vaccine programme, which aims to vaccinate 40 million South Africans. The South African government is working closely with the South African Health Products Regulatory Authority (SAHPRA) to ensure there is no delay approving the vaccine for use. There are a number of variants of COVID-19 that have arisen around the world. No single vaccine will be effective against all the variants. South Africa's vaccination campaign is guided by science and this means the country may need to change the choice of vaccine it uses. This was demonstrated in the case with the AstraZeneca vaccine, which was less effective against the 501Y.V2 variant while the Johnson & Johnson vaccine has proved effective against the COVID-19 501Y.V2 variant. While the Johnson & Johnson vaccine may have a lower efficacy rate than other alternatives, the facts that it does not require the rigorous storage requirements and is a single dose vaccine makes it the most viable for South Africa.

The vaccine will be administered free of charge at various points of service across all parts of the country. The country's vaccination campaign draws on the principles of universal health coverage where all adults living in South Africa have access to the vaccine. This is the largest vaccination campaign undertaken in the history of the country and stretches across 52 districts and 280 wards to reach 40 million people. The programme entails procurement, distribution, actual vaccination, monitoring, communication and mobilisation.

The COVID-19 vaccine will be rolled out in three phases. It is anticipated that by the end of the final phase, 40,350,000 citizens would have been immunised. While the SA government will lead the vaccine rollout initiative, it requires a multi sectoral collaboration to ensure that the vaccine drive is effective. The rollout will proceed in the form of an implementation study with the partnership of the Medical Research Council and the National Department of Health vaccination sites across the country. This will provide valuable information about the pandemic in the post-vaccination community and ensure early identification of breakthrough infections should they occur amongst vaccinated health workers.

The rollout of the vaccine will take a three-phase approach that begins with the most vulnerable in the SA population and target 67 per cent of the population, which will allow the country to achieve herd immunity.

Phase 1 will focus on frontline healthcare workers, with a target of 1,250,000 people.

Phase 2 will include;

- Essential workers. Target population: 2,500,000
- Persons in congregate settings. Target population: 1,100,000
- Persons >60 years. Target population: 5,000,000
- Persons >18 years with co-morbidities. Target population: 8,000,000

Phase 3 will focus on persons older than 18 years, targeting 22,500,000 of the population.

The vaccination system will be based on a pre-vaccination registration and appointment system at a specific vaccination site. The system will help the government plan on the number of doses needed at any point in time. All South Africans who are vaccinated will be placed on a national register and provided with a vaccination card.

After targeted groups receive the vaccine, mass vaccinations will take place in urban centres at pharmacies, health facilities, community halls and schools. These sites will have to be registered and must comply with several requirements to secure and safeguard the vaccination process.

An electronic vaccination data system (EVDS) will assist with the rollout of COVID-19 vaccines across the country. EVDS is an online self-enrolment portal where South Africans can register via a digital device for an appointment. Those who qualify will be sent a notification through SMS informing them of the time and place that the vaccine will be available. They will have to provide their ID, a contact number and unique code that is sent to them when at the vaccination site. Those residents who do not have access to the internet can approach healthcare facilities to assist them with assisted registration on the EVDS.

Safe and effective vaccines will be a gamechanger: but for the foreseeable future everyone must continue wearing masks, physically distance themselves and avoid crowds. Being vaccinated does not mean that caution can be thrown to the wind because it is still not clear the degree to which the vaccines can protect not only against disease but also against infection and transmission.

Mutations and variants

The virus continues to mutate which presents serious challenges to containing the spread of the virus. Each new wave of infections has been found to be associated with a certain variant such as the Delta variant which a highly contagious strain originating in India.

According to WHO, mutations of the virus are more likely when it is widely circulating in a population and causing many infections. However, depending on where the mutations are in the genetic material of the virus, they may affect transmission where the virus may spread more or less easily or they may affect severity where the virus may cause more or less severe disease.

It is important to note that vaccines are expected to provide at least a degree of protection against any new virus variants. The vaccines cause a broad immune response involving a range of antibodies and cells and therefore the vaccines do not become ineffective. However, in cases where the vaccines are found to be less effective against a certain variant, their composition can be changed to provide protection. This is ongoing as a result of continuous monitoring and study.

It is also likely that vaccines may need to incorporate more than one strain of the virus during their development process. Booster shots or doses may be necessary in the future to ensure protection. Ongoing studies of the impact of vaccines as they are deployed are also essential in order to understand their impact.

Finally, monitor information-providing sources such as WHO, Department of Health, and NICD for more current information and trends on dealing with COVID-19. Also consult the provisions of the Disaster Management Act of 2002 as amended from time to time in terms of Section 27(2). COVID-19 will not be the last pandemic or epidemic considering that it has been the 5th such large-scale health challenge to the world in the last century. Therefore, the lessons learnt from dealing with COVID-19 will be invaluable when faced with similar health challenges in the workplace in the future.

• BIOLOGICAL HAZARDS

Many biological hazards such as exposure to viruses, bacteria, parasites, fungi, moulds, organic dusts or any living organism that can cause disease to human beings have been found to occur in occupational exposures. In the industrialized countries around 15 % of workers may be at risk of viral or bacterial infection, allergies and respiratory diseases. In many developing countries including South Africa biological agents present as a major occupational challenge. For example, Hepatitis B and C viruses and other blood borne pathogens, tuberculosis infections, asthmas and chronic parasitic infections, are the most common occupational diseases that result from such exposures.

Biological hazards can be transmitted to a worker through:

- Inhalation
- Injection
- Ingestion
- Contact with the skin

Knowing the biohazard and their groupings is important to decide on what to do to safeguard the workers from exposure to the hazards. It is important to remember that:

- Any accident involving biohazard material can result in infection.
- When working with biological agents or materials for which epidemiology and etiology is not known or not completely understood, it must be assumed that the materials present a biological hazard.

• VIBRATION

Noise and vibration in industry go hand in hand. Vibration refers to back and forth movement. Severe vibration exposure occurs in the use of chain saws, jackhammers, percussion drills and percussive tools are the tools that cause the most vibration damage. Transportation systems such as tractors, trucks or earth moving machinery produce tremendous vibration when operated. Less severe vibration can be found at places such as textile mills with looms and sewing machines or foundries with mould shake-outs. Vibration causes vascular disorders of the arms and bony changes in the small bones of the wrist. Vascular changes can be detected by X-ray examination of the wrist. The most common findings are rarefaction of the lunate bone.

Whole-body and hand-arm vibrations (HAVS) occur during the use of hydraulic and power tools, and mechanical plant and equipment. HAVS is associated with an increased risk of lower back pain, sciatic pain and degenerative changes in the spinal system including lumbar intervertebral disc disorders. Vascular disorders relative to damage of blood vessels manifest with symptoms such as whiteness of the fingers with numbness and 'pins and needles' and loss of finger dexterity. Neurological disorders resulting from damage to the nervous system manifest in numbness, tingling sensations and loss of hand-grip strength. Other symptoms include pain or stiffness in the hands and lower arms, and a decline in strength and dexterity.

Whole Body Vibration	Consequences	•Damage to bones and joints •Morphological changes in spine •Digestive system problems •Female reproductive damage •Visual impairment. •Defects in vestibular system of ear •Variations in blood pressure which may lead to heart problems
	Recommendations for prevention	•Do not remain on a vibratory surface longer than necessary •Maintain vibratory machinery regularly to prevent the development of excess vibration. •Keep machine controls remotely located from vibratory surface if possible •Try to have the source of vibration or the station where workers are stationed vibration-isolated to reduce exposure
Segmental Vibration	Consequences	•Get a condition Vibration White Finger or Raynaud' s syndrome •Nerve degeneration that result in the loss of senses of touch and heat •Muscle atrophy and tenosynovitis •Carpal Tunnel Syndrome
	Recommendations for prevention	•Do not smoke while using the vibratory hand tool (nicotine reduces the blood supply to the hand and fingers) •Let the tool do the work, while you just hold it lightly •Only use the tool when necessary and operate less than full throttle if possible •If you feel signs of tingling, numbness and white or blue fingers, consult a doctor. •Take regular implement rest breaks to avoid continuous exposure over long periods of time.

Vibration can be classified as either (a) the whole body vibration that is transmitted through the feed, back or buttocks, or, (b) segmental vibration such as hand-arm transmitted through the hands and arms when working with hand tools. The table summarises the consequences when exposed to the two different types of vibration, and recommendations for treating each of these. Vibration can be controlled in various ways, and should be done to the best possible optimal ways to protect workers from extreme vibration:

- Isolating the vibrations by using special mountings that adjust the centre of gravity as low as possible
- Restrict the duration and magnitude of exposure
- Consideration of working methods that eliminate or reduce exposure such as minimising the transport of goods or materials
- Selection of ergonomically designed work environment
- Damping of vibrations, such as wearing padded gloves, or designing tool with damping material between the tool and the handle
- By maintaining properly sharpened cutting tools
- Using sandwich structures.
- Designing an operating platform remotely from the vibrating object
- Warming the operators' hands or handles of the vibratory tools in cold conditions.
- Training workers to protect their hands visually from injury and to inspect their hands regularly for injury.

• ALCOHOL AND DRUGS IN THE WORKPLACE

A worker may not work when under the influence of alcohol or drugs such as dagga (Cannabis) or mandrax. Any worker who uses alcohol or drugs while working will face severe disciplinary action such as dismissal or criminal charges. When a worker works while being intoxicated he will put himself and his fellow workers at risk, which in itself is very dangerous and could cause an accident.

(Source: https://www.military.com/benefits/veterans-health-care/facts-of-drug-addiction.html)

Medication with a high alcohol content, like cough mixtures, can also cause intoxication and drowsiness. This medication will also cause delayed reaction and reduced concentration. It is very important that the chemist be asked if the medication that is being used will have these side effects or the pamphlet supplied with the medication should be read. If it does, ask for medication that does not have these side effects. It is very important to inform the supervisor and health and safety representative of the medication that is being taken and of its side effects.

(Source: https://www.alcoholproblemsandsolutions.org/alcohol-in-cough-syrup-and-other-over-the-counter-medicines/)

• CANNABIS IN THE WORKPLACE

Cannabis refers to the tobacco-like greenish or brownish material made up of the dried leaves, flowers, stems and seeds from the Cannabi sativa or indica (hemp) plants. It is also referred to by the World Health Organization as a natural existing drug. Cannabis resin or "hash" is the dried black or brown secretion of the flowering tops of the cannabis plant, which is made into a powder or pressed into slabs or cakes. Cannabis oil or "hash oil" is a liquid extracted from either the dried plant material or the resin.

(Source: https://time.com/4107603/dea-medical-marijuana-joke-2/)

• LEGALIZATION OF CANNABIS

The legalization of the personal use of cannabis also known as weed or marijuana in South Africa by the Constitutional Court on September 18, 2018 presents a challenge for the construction industry where its use has been covert or clandestine and in many cases seen to be synonymous with high levels of productivity on sites. The ruling legitimizes the possession, purchase and cultivation of cannabis, which is a psychotropic drug for personal use by an adult in a private dwelling. In terms of this ruling the maximum amount allowed for private use at home per person is around 1 kg and the number of plants that a person may grow is 8 with the maximum per household of 2 adults being 16 plants. The penalty for exceeding these amounts is severe and 15 years has been cited as a sentence if found guilty by a court of law.

Several challenges arise from the court ruling in South Africa. These include what constitutes private use. Technically, if someone in possession of cannabis steps outside of their home, retains the substance in their pocket and it is for personal use they have not broken the law. Possession in itself would no longer carry the previous legal censure. Further, allowing people to purchase marijuana would amount to the court sanctioning dealing in the substance. Should the user want to grow their own cannabis they would have to purchase the seeds or small plants from another party who would be deemed to be a dealer in marijuana which is still an illegal practice. The purchaser would be an accomplice to dealing in cannabis. A decision by the South African government would need to be made about what quantities are allowed per person strictly for personal use.

• COMMON TERMS USED FOR CANNABIS

In order to deal with cannabis in the workplace it is necessary to know the terms that are frequently used by workers when they are referring to the use of the substance or discussing its use among themselves. There are more than 1,200 words used for cannabis around the world. Some of the more common names used in workplaces include:

- Marijuana;
- Weed;
- Pot;
- Grass;
- Dagga;
- Ganga; and
- Herb;
- Joint;
- Reefer;
- Pill;
- Boom; and
- Zol.

• HEALTH EFFECTS OF CANNABIS

Cannabis is a psychotropic drug that affects and alters brain activities associated with mental processes, perception and behaviour. Psychotropic drugs are also referred to as psychotherapeutic or psychoactive. More particularly, cannabis is a psychoactive drug. As such, it is a chemical substance that acts primarily upon the central nervous system where it alters brain function, resulting in temporary changes in perception, mood, consciousness and behaviour. Cannabis also acts as a psycho-stimulant where it elevates the mood, produces feelings of excitement, alertness, attention, energy and euphoria. This elevation is usually followed by a break down.

What makes cannabis so appealing and of interest is the fact that there are 483 known compounds in the plant including at least 65 other cannabinoids demands. These cannabinoids form one of a class of diverse chemical compounds that acts on cannabinoid receptors in cells that alter neurotransmitter release in the brain. In simple terms, cannabinoids slow down communication between cells in the body and the brain.

Delta-9-tetrahydrocannabinol (THC) is the chemical that is responsible for the manner in which the brain and body of a user react to cannabis. It is argued that there are potential therapeutic benefits like feeling relaxed and happy and health risks in the form of unpleasant, unwanted or negative effects on the brain and body that arise from the use of cannabis.

• IMPACT OF CANNABIS ON THE BRAIN

Some short-term effects on the brain have been cited as:

- Confusion;
- Changes in mood;

- Fatigue or sleeplessness;
- Impaired ability to remember, concentrate or pay attention;
- Impaired body movement;
- Anxiety, fear or panic; and
- Reduced ability to react quickly.

The use of cannabis is reported to possibly involve psychotic episodes. These generally are severe mental disorders that cause abnormal thinking and unrealistic perceptions. Consequently, people with psychoses lose touch with reality. When cannabis is taken in high doses, these psychotic episodes are characterized by, inter alia:

- Temporary paranoia;
- Delusions;
- Worsening schizophrenia;
- Disorganized thinking; and
- Temporary hallucinations.

Cannabidiol (CBD) is one of the over 100 phytocannabinoids identified in Cannabis sativa and constitutes up to 40% of the plant's extract, being the second most abundant component. It is possible for cannabidiol which is a chemical in cannabis itself to reduce some of the psychoactive effects of its use such as disturbances in mood, perception and psychotic symptoms. Drugs taken together are known to have a cumulative or synergistic effect, which increases the overall psychoactive experience. There is evidence that when cannabis which has 50% more tar than high tar cigarettes is used in combination with other substances such as tobacco in blunts and spliffs/mulled cigarettes or alcohol the severity of some psychoactive effects and the risk of mental health outcomes may increase. The combined use of cannabis and tobacco presents a significant potential for nicotine exposure, and may lead to excessive tobacco use patterns, nicotine addiction, and compounded health effects.

• IMPACT OF CANNABIS ON THE BODY

The short-term effects on the body include:

- Damaged blood vessels caused by smoke;
- Decreased blood pressure resulting in fainting or passing out;
- Intense nausea and vomiting;
- Chest pain;
- Increased heart rate leading to increased risk of heart attack especially in people with heart conditions; and
- Headaches.

In terms of impairment, Delta-9-tetrahydrocannabinol (THC) affects:

- Co-ordination;
- Reaction time;
- Change in sense of time;
- Ability to pay attention;
- Decision-making abilities; and
- Ability to judge distances.

Impairment can last for more than 24 hours after the use of cannabis which is long after the other effects have faded. One of the consequences in regular users could be the difficulty with skills needed to drive safely for weeks after their last use because cannabis smoking increases motor vehicle accident risk.

Long-term effects of the frequent use of cannabis that continues over periods of weeks, months or years develop gradually over time. Of concern is that these effects are found to last beyond when the use of cannabis has stopped and may not be fully reversible. Long-term effects on the brain that affect brain development include:

- increased addiction,
- damaged memory,
- inability to concentrate,
- reduced intelligence quotient (IQ) up to 8 IQ points between ages 13 and 38; and
- reduced ability to think and make decisions.

Long-term effects on the body have been found to include:

- risk to lung function,
- increased frequency of lung infections,
- bronchitis,
- chronic long-term cough,
- increased mucus build up in the throat,
- cardiac arrest and strokes,
- seizures, and
- hyperactivity.

The frequent use of cannabis has also been linked to an increased risk of suicide, depression and anxiety disorders. Cannabis addiction has been found to lead to, inter alia:

- Absenteeism and failing to execute major tasks and duties at work;
- Giving up important activities because of cannabis use;
- Unintentional increased frequency of use in larger doses; and
- Inability to reduce or control the use of cannabis.

• WAYS OF USING CANNABIS

The main ways of using cannabis are by means of smoking (inhalation) or eating (ingestion). In the case of smoking which includes vaping, cannabis begins to work fastest as THC is carried to the brain in the bloodstream and a user may start to feel 'high' within seconds or minutes. The amount of THC peaks in about 30 minutes and then fades after one to four hours. There are several ways in which cannabis can be smoked that include:

- Hand rolled into a cigarette known as a joint;
- In a pipe or water pipe referred to as a bong;
- In a cigar that has been hollowed out and refilled with cannabis called a blunt; and
- In the form of sticky resins drawn from the cannabis plant which have a higher concentration of THC.

When cannabis is consumed or ingested the effects are slower than if smoked. Typical periods before the onset of the 'high' feeling are between 30 minutes and two hours. The after-effects last up to eight hours. Cannabis can be mixed, inter alia:

- In food such as brownies, cookies and candy;
- In teas in which it is brewed;
- In hash oil or honey oil which is a sticky glue-like liquid;
- In wax or butter form which is a soft texture-like lip balm;
- In vaporizer in liquid form;
- Shatter which is a hard amber-coloured solid, or
- In tonics and tinctures.

• REPORTED THERAPEUTIC BENEFITS OF USING CANNABIS

Several therapeutic benefits for the use of cannabis for medicinal reasons have been reported, for example:

- Feeling of relaxed well-being and heightened senses making colours appear brighter;
- Increased sociability;
- Muscle relaxant in the case of stiff muscles or muscle spasms from multiple sclerosis;
- Analgesic effect to relieve ongoing pain which is most common use for medical cannabis;
- Help with sleep problems in persons suffering from fibromyalgia and sleep apnea;
- Appetite stimulation in cases where persons with AIDS suffer with loss of appetite and weight loss;
- Antiemetic effect which prevents vomiting such as from chemotherapy;
- Anticonvulsant effect; and
- Lower intraocular pressure, namely pressure in the eye.

• REASONS FOR USING CANNABIS

Several reasons have been posited for the use of cannabis as a drug on construction sites. These reasons directly or indirectly relate to the working conditions on job sites or the workers themselves. Directly related reasons or coping strategies include the need to:

- Increase concentration;
- Increase self-confidence;

- Have more physical strength; and
- Reduce anxiety and tension.
- Other non-work related reasons include
- Tradition and customs;
- Personal lifestyles; and
- Peer pressure.

• SYMPTOMATIC SIGNS OF HAVING USED CANNABIS

In order to control the use of cannabis it is vital that management and workplace security are able to detect the symptoms of use. There are many tell-tale signs that could alert the use of cannabis by workers and them still being 'high.' However, these signs also might be indicative of the use of other substances that include alcohol or strong medication. These signs include the worker having or demonstrating, *inter alia*:

- Glassy red eyes;
- Poor muscle and limb co-ordination;
- Delayed reaction times;
- Increased appetite;
- Sudden mood shifts; and
- Abrupt symptoms of anxiety, panic or hallucinations.
- The most distinctive sign of cannabis use is its skunk-like smell or odour which generally permeates the clothes and body of the worker.

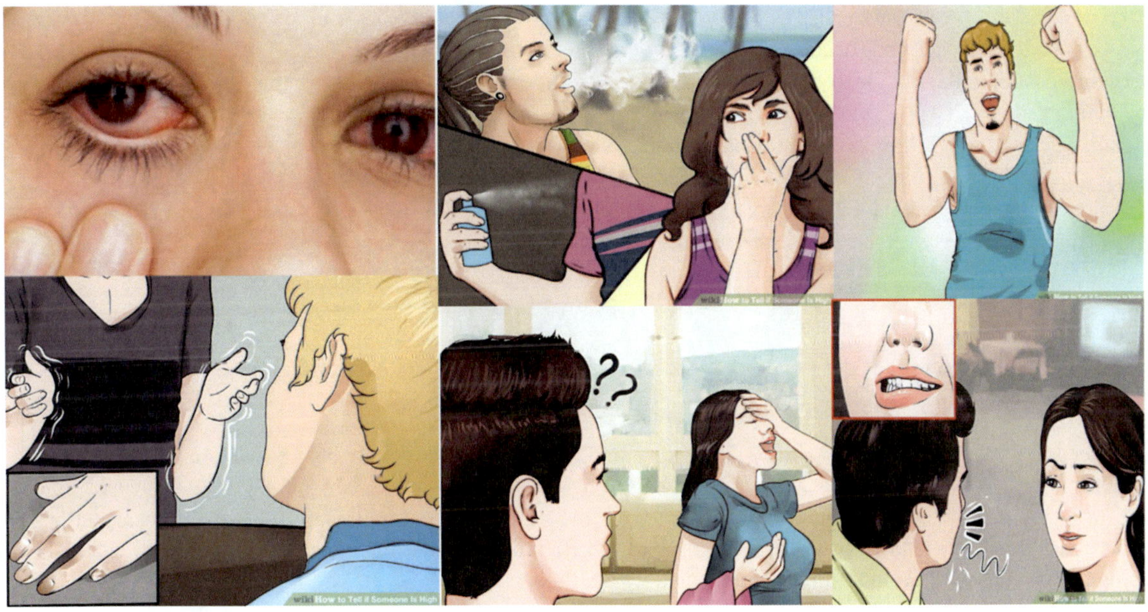

(Source: Adapted from https://www.wikihow.com/Tell-if-Someone-Is-High)

• PARAPHERNALIA AND HIDEAWAY PLACES

Cannabis users will be 'inventive' to conceal their cannabis-related activities in the workplace. For example, the following paraphernalia, *inter alia*, could be found at work, namely:

- Rolling papers;
- Pipes such as glass top of bottles (darkened from burn residue);
- Cigar papers with their content emptied; and
- Edibles with green hue.

Examples of places where cannabis users will hide cannabis at work include the following, namely:

- In cavities and crevices;
- Carved out spaces at the top of doors;
- Over-the-counter medication packages to avoid detection;

- Soda or cool drink cans;
- In drop or suspended ceiling spaces;
- In water bottles;
- In their clothing;
- In toolboxes;
- In hammer handles; and
- In spirit levels.

The effect of cannabis is felt within seconds of smoking and can last up to six hours while on the other hand after eating or ingestion the effects are felt within 30 minutes and can last up to 12 hours afterward. Given that the after effects of the use of cannabis last for many hours, it is likely that workers who have used cannabis at home or somewhere else could come to work feeling high. Consequently, they do not only place themselves at risk at work but also their fellow workers. In South Africa the Constitution which is the overarching piece of legislation in the country stipulates that workers are entitled to a working environment that does not present a threat to their health and safety. This entitlement is captured in the Occupational Health and Safety Act of 1993 as amended, in terms of which employers must ensure a working environment for all their workers that does not present a threat to their health and safety. The organization should have a 'zero tolerance' policy for the use of cannabis and similar substances at work with clear censures for non-compliance with its provisions.

• IMPACT OF CANNABIS ON WORKPLACES

The use and abuse of cannabis by workers given the likelihood that its use will be more visible and brazenly open particularly in South Africa after the court ruling will lead to several impacts on workplaces. Industry will also be affected unless its use at work or after effects from use at home or off site are controlled in the workplace. These include:

- High rates of absenteeism by workers who use cannabis and have after effects from use before coming to work;
- Loss of productivity at work because of the lasting effects of use off site;
- Violent and unpredictable behaviour that could even include petty crimes such as theft and pilfering to fund the cannabis habit;
- Steadily decreasing work quality resulting in unnecessary rework to fix poor workmanship and errors;
- Increasing inability to pay attention and concentrate for any length of time;
- Needless risk taking threatening workplace safety and the safety of fellow workers at work; and
- High labour turnover with the associated recruitment costs.

• RECOMMENDATIONS FOR CONTROL IN THE WORKPLACE

Suggestions to improve the control of the use of cannabis itself onsite or to reduce the likelihood of workers coming on to work with effects of cannabis use off site include:

- The need for management including supervisors and security personnel to be knowledgeable, aware and vigilant;
- Create behavioural change in the workplace through the combined use of the traditional extrinsic pathway governed by systems and rules with rewards and punishments and the intrinsic pathway which establishes voluntary compliance via individual commitment to health and safety;
- Development of intensive awareness programs presented using multiple media as part of wellness programs;
- Implementation of peer interventions and support initiatives since workers are more likely to respond to their peers than their supervisors;
- Development and communication with the involvement of trade unions and health and safety representatives of a 'no or zero tolerance' substance abuse policy that includes regular random testing;
- Disciplinary action where the substance abuse policy has been breached;
- Improving access and egress control of workplaces that may include random physical body searches and restriction of worker exits during working hours;
- Conducting frequent routine inspections of workplaces and ablution facilities;
- Setting up an anonymous helpline for workers to report use and also to seek counselling if addicted to cannabis;
- Consideration for a strict no smoking policy or provision of designated smoking areas that are properly monitored;
- Reducing workplace stressors that include adjusting workloads and improving workplace culture and controls through holistic wellness programs and
- Provision of lockers on site for workers in properly monitored areas.

- ERGONOMICS

The Ergonomics Regulations were gazetted on the 6th of December 2019 by the Department of Employment and Labour under Section 43 of the Occupational Health and Safety Act 85 of 1993. Ergonomics, also known as human engineering or human factors engineering, is the multi-disciplinary science dealing with the interactions between people, and machines and their work environment. The Ergonomics Regulations provide guidelines for the management of ergonomics within organisations and also present an opportunity for organisations to benefit from the dual objective of ergonomics which is to improve worker health and safety while also optimising work output.

The official International Ergonomics Association (IEA) definition of ergonomics is as follows:

"*Ergonomics is the scientific discipline concerned with the understanding of interactions among humans and other elements of a system, and the profession that applies theory, principles, data and methods to design in order to optimise human well- being and overall system performance (www.iea.cc).*

It seeks to improve the performance of systems by improving human machine interaction. It involves the science of designing machines, products, and systems to maximize the safety, comfort, and efficiency of the people who use them. Ergonomists view workers and the objects they use in the workplace as one unit, and consequently ergonomic design blends the best abilities of people and machines. An ergonomically designed system provides optimum performance because it takes advantage of the strengths and weaknesses of both its human and machine components.

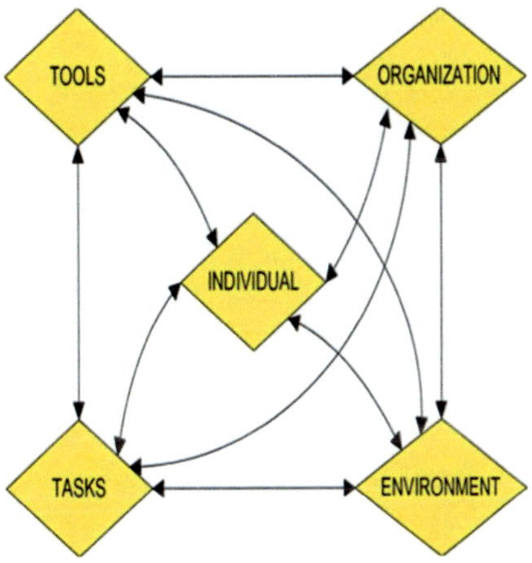

(Source: https://www.researchgate.net/figure/Work-System-Model-Carayon-Smith-2000-Smith-Carayon-Sainfort-1989_fig1_228995864)

Ergonomists draw on the principles of industrial engineering, psychology, anthropometry or the science of human measurement, and biomechanics which is the study of muscular activity to adapt the design of products and workplaces to the sizes and shapes of workers and their physical strengths and limitations. They also consider the speed with which workers react and how they process information, and their capacities for dealing with psychological factors, such as stress or isolation.

In general, ergonomics deals with the interaction between humans and such additional environmental elements such as heat, light, sound, atmospheric contaminants and all tools and equipment pertaining to the work place.

The advantages of ergonomics or the proper designing of work systems based on human factors include:

- more efficient operations
- fewer accidents
- reduced training time
- lower costs of production
- more effective use of workers

Key questions to be asked are:

- What is the task to be performed (task variable)?
- Would the person be able to do the task (human variable)?
- What is the type of work environment (environmental variable)?

Systems can be improved through the

- Design of user-interface to make it more compatible with the task and the user, making the system easier to use and more resistant to errors that workers are known to make.
- Changing the task to make it more compatible with the user characteristics
- Changing the way work is organised to accommodate the psychological and social needs of workers, or
- Changing the work environment to make it healthier and safer and more appropriate for the task.

In the field of ergonomics, attention should be given to:

- Thermal exposure and the design of the physical environment
- Vision, light and lighting
- Hearing sound noise and vibration
- Anatomy, posture and body mechanics,
- Anthropometric principles in workspace and equipment design
- Static work: Design for standing and seated workers
- Design of repetitive tasks
- Design of manual handling tasks
- Work capacity, stress and fatigue,
- Physiology applied in industry
- Human information processing, skill and performance
- Displays, controls and virtual environments
- Human computer interaction, memory and language
- Human machine interaction, human error and health and safety
- Systems design: organisational and social aspects

• ERGONOMIC HAZARDS

It is claimed that between 10% and 30% of the workforce in industrial countries and between 50% and 70% in developing countries such as South Africa may be exposed to heavy physical workload or to ergonomic working conditions such as lifting and moving of heavy items or repetitive manual tasks. Repetitive tasks and static muscular load are found in many industrial and service occupations. In many countries musculoskeletal disorders are the main cause of both short-term and permanent work disability.

Postural Stress is the term used to denote the mechanical load on the body by virtue of its posture. Postural stress can cause pain. People who work with the spine flexed forward about 60° for more than 5% of the day or 30° for more than 10% of the day are prone to suffer back pain.

A stable posture can only be kept if the different parts of the body are supported and maintained in an appropriate relation to their feet. For example: when a standing person leans forward to touch their toes, the pelvis moves backwards to compensate for the forward displacement of the centre of gravity of the upper body. In order to reduce postural stress, it is important to provide sufficient space around standing operators and plenty of room for the feet if losses of balance are to be avoided.

By doing the manual work in the wrong way, unnecessary stress and damage will be done to the soft tissue in the back of a worker, and can cause permanent back damage. It is extremely important to train workers who do hard manual work how to handle their bodies to avoid damage. Training should be done annually to refresh the memories of workers, and managers and supervisors should always be on the look-out for people· doing harm to their bodies.

• ERGONOMIC CONSIDERATIONS

Matching workers with their work

It is important to match worker capabilities with the related requirements of a given job or work activity or task. If the job demands are equal to the capabilities of the worker or if they exceed them, the worker will be under much strain and may not be able to perform the job, activity or task.

Work classification

Generally, the work demands are classified from light work to extremely heavy work in terms of energy expenditures per minute and the relative heart rate in beats per minute. The energy requirement for light work is 2.5 Kcal/minute and the heart rate is 90 beats rate per minute, while it for extremely heavy work the energy requirement is 15 Kcal/minute or the equivalent of 6 times higher and heart beat is 160/minute or the equivalent of 1.75 times faster.

Workstation design

The workstation is the immediate area where the worker performs his/her work activities. The goal of designing a workstation is to promote ease and efficiency of the performance of the worker. Productivity will be compromised if, for example, the operator is uncomfortable and the workstation is poorly designed.

Workplace design

Given that the workplace is where the worker performs his/her duties, the most basic requirement is that it must accommodate the worker working in it. Specifically, this means that:

1. The workspace for the hands should be between hip and chest height in front of the body.
2. Lower location is preferred for heavy manual work.
3. Higher locations are preferred for tasks that require close visual observations.

The workplace should be designed relating the physical characteristics and capabilities of the worker to the design of equipment and to the layout of the work place. When this match is achieved, there is arguably:

- an increase in worker efficiency
- a decrease in human error
- consequent reduction in accident frequency

Workspace dimension

The dimension of the workspace can be grouped into three basic categories, namely minimal, maximal, and adjustable dimensions. Minimal workspace provides clearance for ingress and egress in walkways and doors. Maximal workspace dimensions permit smaller workers to see the equipment by selecting a workspace dimension over which a small person can reach or by establishing control forces that are small enough so that even a weak person can operate the equipment. Adjustable dimensions permit the operator to modify the work environment and equipment so that it conforms to those workers on a particular set of anthropometric characteristics.

- REVIEW QUESTIONS

1. What threats do migrant workers present in the workplace?
2. Why should exposure to biological hazards be taken seriously?
3. What actions should an employer take to deal with the threat of COVID-19 in the workplace?
4. What are the consequences of avoiding a focus on vibration as a result of work activities?
5. What steps should be taken to prevent the use of cannabis and its effects on workers at work?
6. How would ergonomics contribute to improving worker productivity and efficiency?
7. Select ONE of the following work environments or workplaces and develop a safe work or operating procedure for it to ensure that it is 'COVID-19 ready' taking into account all information and regulations issued by the World Health Organization, Department of Health, Department of Labour and National Institute for Communicable Diseases:
    - Bank
    - Construction firm
    - Public transport company
    - Education and training institution
    - Municipal office
    - Airline
    - Restaurant
    - Beauty salon

    Illustrate your SOP or SWP as necessary with pictures, diagrams or graphics. Prepare a short presentation of no more than 5 slides which you will present to the management of your organization.

- REFERENCES:

Bovenzi, Massimo. and Hulshof, Carel. (1999). An updated review of -epidemiologic studies on the relationship between exposure to whole-body vibration and low back pain (1986–1997). International Archives of Occupational and Environmental Health, 2(6), 351–365.

Canada. (n.d.). Health Effects of Cannabis. Available at: https://www.canada.ca/content/dam/hc-sc/documents/services/campaigns/27-16-1808-Factsheet-Health-Effects-eng-web.pdf

Carayon, Pascale., Alvarado, Carla. and Hundt, Ann. 2003. Reducing workload and increasing patient safety through work and workspace design. Paper commissioned by the Institute of Medicine Committee on the Work Environment for Nurses and Patient Safety, Center for Quality and Productivity Improvement, University of Wisconsin-Madison, CQPI Technical Report No. 185 http://www.engr.wisc.edu/centers/cqpi/reports/pdfs/r185.pdf. Accessed August 26, 2020

CBS News (2018). South Africarules marijuana legal in private for adults https://www.cbsnews.com/news/south-africa-rules-marijuana-legal-in-private-for-adults/

Department of Labour (1997). Basic Conditions of Employment Act 75 Of 1997. Department of Labour.

Department of Labour (1998). Employment equity act 55 of 1998.

Department of Labour (1995). Labour Relations Act 66 Of 1995. Department of Labour.

Department of Labour (1993). Compensation for Occupational Injuries and Diseases Act no of 103 of 1993.

Department of Labour (1993). Occupational Health and Safety Act No: 85 of 1993.

DHSSPS (Department of Health, Social Services and Public Safety) (2014). Workplace Drugs and Alcohol Policies. [Available at http://www.dhsspsni.gov.uk/stats-workplace] Accessed 16 August 2014.

EU (2020). *Q & A on COVID-19*. [online] European Centre for Disease Prevention and Control. Available at: https://www.ecdc.europa.eu/en/novel-coronavirus-china/questions-answers.

Haupt, Theodore. (2019). "An appraisal of the use of cannabis on construction sites " *Acta Structilia*, 26(1), pp.148-166.

Haupt, Theodore. Conrad. 2019. The use of cannabis on construction sites: A Review. Journal of Construction, 11(3): 29-35.

Haupt, Theodore. Conrad., Akinlolu, Mariam. and Raliile, Mohlomi. Terah. 2019. The use and effect of cannabis among construction workers in South Africa: A pilot study, The 10th West Africa Built Environment Research Conference (WABER), Accra, Ghana, pp. 1126-1137, ISBN978 – 9988 – 2– 6010 –1

Health and Safety Executive (HSE) (2014). *Musculoskeletal Disorders in Great Britain.* [Available at http://www.hse.gov.uk/statistics/causdis/musculoskeletal/msd.pdf]. Accessed 11 December 2014.

McAleenan, Ciaran. and Oloke, David. (2015). *ICE manual of health and safety in construction.* 2nd ed. One Great George Street, Westminster, London SW1P 3AA: ICE Publishing, 2015.

NIOH. 2020. Guidelines on routine and deep cleaning in the workplace. https://www.nioh.ac.za/wp-content/uploads/2020/07/Guidance-on-routine-deep-cleaning_1-July-Fin.pdf

OSHA, Safety and Health Topics Page, Welding, Cutting, and Brazing 2014,

Samuels, W., Haupt, Theodore. Conrad., and Shakantu, WMW (2007): An Exploratory Examination of Poor Construction Practices: Opportunities for Ergonomic Interventions in the Western Cape, *Occupational Health Southern Africa*, 13 (2), pp. 19-25 ISSN 1024-6274

Smallwood, John. J., and Haupt, Theodore. Conrad. (2007): Construction Ergonomics: An Indian and South African Comparison, *Ergonomics SA - Journal of the Ergonomics Society of South Africa*, vol. 19, no. 1, pp. 30-43, ISSN 10 10 2728

Smallwood, John. and Haupt, Theodore. (2009): Construction ergonomics: perspectives of female and male production workers, In Dainty A. (ed.) Proceedings of 25th Annual Conference of Association of Researchers in Construction Management (ARCOM), Nottingham, U.K., September 7-9, pp. 1263-1272 ISBN 13: 978-0-9552390-1-4

Smallwood, John. J. and Haupt, Theodore. Conrad. (2009): Construction Ergonomics: Perspectives of Female and Male Production Workers, Paper 1BU0021, Proceedings of 17th World Congress on Ergonomics, Beijing, China, August 9-14, (CD-ROM)

Tyrer, John. (2015). A new means of quantifying laser safety hazards; The laser micro-mort. *International Laser Safety Conference.*

WHO (2020). *Coronavirus*. [online] Who.int. Available at: https://www.who.int/health-topics/coronavirus.

CHAPTER 20

ADDITIONAL HEALTH CHALLENGES IN THE WORKPLACE

• HIV AND AIDS IN THE WORKPLACE

According to the Southern African Development Community (SADC), Southern Africa, with a combined population of only 3.5% of the world's population, accounts for 35% of the people living with HIV and AIDS and about half of all infections in Africa. The high number of the productive adults in the labour force dying from or living with the disease is alarming. Further, UNAIDS estimates that over one million people died of AIDS in Southern Africa in 2001, and cumulatively the total number of deaths in the region since the start of the epidemic is over 20 million.

HIV and AIDS threatens productivity, profitability and the welfare of workers and their families. Note that HIV and Aids are not the same thing. HIV refers to the retrovirus with which workers can become infected. Only testing will confirm the serostatus of any worker. On the other hand, AIDS is the illness which manifests in HIV positive workers after several years for which there is as yet no known cure. It is possible for someone who is HIV positive or sick with AIDS to live a reasonably normal life for many years by using anti-retroviral drugs and treatments. It is only when the C4 white blood corpuscle count reaches very low levels such as around 200 that the worker will become extremely ill. Companion afflictions such as TB exacerbate the state of illness. Workplace HIV and AIDS policies and programmes can play a vital role in raising awareness around HIV, preventing HIV infection and caring for people living with HIV.

An HIV and AIDS policy:

- ✓ Defines the position of an organisation on HIV and AIDS and sets out clear guidelines on how HIV and AIDS will be managed within the workplace.
- ✓ Aligns the workplace response to the broader legal framework.
- ✓ Ensures fairness.
- ✓ Identifies and protects the rights and responsibilities of employers and workers in the context of HIV and AIDS.
- ✓ Sets standards of behaviour expected of all employers and workers.
- ✓ Establishes consistency within the organisation.
- ✓ Sets the standard for communication about HIV and AIDS.
- ✓ Provides a good foundation upon which to build an HIV and AIDS workplace programme.
- ✓ Informs workers about assistance that is available.
- ✓ Indicates commitment to dealing with HIV and AIDS.
- ✓ Ensures consistency with national and international practices.

Worker rights

- ✓ No worker, or applicant for employment, may be required by their employer to undergo an HIV test. HIV testing by or on behalf of an employer may only take place where the Labour Court has declared such testing to be justifiable in accordance with Section 7(2) of the Employment Equity Act.
- ✓ All persons with HIV or AIDS have a right to privacy, including privacy concerning their HIV or AIDS status. There is no legal duty on a worker to disclose their HIV status to their employer or to other workers.
- ✓ An employer cannot demand to know if the cause of an illness is HIV infection.
- ✓ A doctor or health care worker who informs an employer about the HIV status of a worker without their consent is acting against the law and failing to recognize the right of the worker to confidentiality.
- ✓ A worker with HIV and AIDS may not be dismissed because he or she is HIV positive or has AIDS in terms of Section 187(1)(f) of the Labour Relations Act, No. 66 of 1995. However, where there are valid reasons related to their ability to continue working and fair procedures have been followed, their services may be terminated in accordance with Section 188(1)(a)(i).
- ✓ An employer is obliged to provide, as far as is reasonably practicable, a safe and healthy workplace according to Section 8(1) of the Act which may include ensuring that the risk of occupational exposure to HIV is minimised.
- ✓ A worker who is infected with HIV as a result of an occupational exposure to infected blood or bodily fluids, may apply for benefits in terms of Section 22(1) of the Compensation for Occupational Injuries and Diseases Act, No. 130 of 1993.
- ✓ In accordance with the Basic Conditions of Employment Act, all workers must receive certain basic standards of employment, including a minimum number of days' sick leave.

- ✓ A registered medical aid scheme may not unfairly discriminate directly or indirectly against its members on the basis of their "state of health" according to Section 24(2)(e) of the Medical Schemes Act, No 131 of 1998.

• WELDING

Exposure to welding fumes

Many welders become ill from inhaling welding fumes. Some may be ill for only a short time while others may become permanently ill with, for example, asthma. There is no easy way to know which worker it could be. Welders stop welding altogether and find a new career after being ill for too long due to exposure to welding fumes.

Pneumonia

Welders are likely to have a lung infection that can lead to severe and frequently fatal pneumonia. Modern antibiotics usually stop the infection. However, in severe cases the worker could be hospitalised. It can affect both young as well as older welders. However, exposure to welding fumes in the past does not increase the chances of the worker contracting pneumonia later.

Occupational asthma

Welders need to protect themselves and follow the safe and healthy way of working provided by their employer in the form of SWPs or SOPs. Stainless steel fume contains chromium oxide (CrO3) and Nickel Oxide both of which can cause asthma. For this reason, stainless steel welding fume is considered more harmful than mild steel fume.

Cancer

Welding fume is internationally classified as possibly carcinogenic to humans under IARC classification group 1. Although primarily associated with stainless steel welding, this classification is not limited to stainless steel fumes as it covers all welding fumes.

Metal fume fever

It is not unusual for welders to experience and report flu-like symptoms after welding. Usually, these effects are worse at the start of the working week. Metal fume fever is usually linked to welding or hot work on galvanised metals. High exposures to mild steel weld fumes can also cause this illness. Metal fume fever does not usually have any lasting ill effects. Further, as is commonly held, drinking milk before welding does not prevent getting metal fume fever.

Irritation of throat and lungs

The effects of the gases and fine particles in welding fumes include experiencing dryness of the throat, tickling, coughing or a tight chest. These effects do not last long and rather tend to be short lived. Ozone is a particular cause of these effects when welding stainless steels and aluminium. High exposures to nitrous oxides generated during most arc welding operations can also cause these effects. Extreme exposure to ozone can cause pulmonary oedema or fluid on the lungs.

Temporary reduced lung function

Overall lung capacity and the ease at which one can breathe out or peak flow are affected by prolonged exposure to welding fumes. The effects tend to get worse through the working week but gradually improve when not exposed such as over the weekend or breaks from work.

• MANAGING ASBESTOS IN BUILDINGS

Managing asbestos is necessary to help prevent risk to workers or other users of buildings. Asbestos can be found in all pre-2008 constructed non-domestic buildings whatever the type of business.

Common areas include:

- ✓ Halls
- ✓ Stairways
- ✓ Lift shafts
- ✓ Roof spaces

Inhalation or ingestion of air containing asbestos fibres can lead to asbestos-related diseases such as cancers of the lungs and chest lining. Asbestos is only a risk when asbestos fibres are released into the air and inhaled. People who are mostly at risk of asbestos fibres are workers doing building maintenance and repairs. There is a long delay between first exposure and the onset of disease and this can vary between 15 and 60 years.

• ASBESTOS-RELATED DISEASES

Exposure to asbestos causes diseases such as non-malignant pleural disease, asbestosis, lung cancer and mesothelioma.

Non-malignant pleural disease is the development of diffuse pleural thickening or pleural plaques which may reduce lung volumes.

Asbestosis is a fibrotic interstitial lung infection resulting in shortness of breath on exertion. Dry cough is associated with the later stages of disease. Lung function may become impaired.

Lung cancer is caused by asbestos exposure. Smoking increases the risk of lung cancer. However, this cancer takes many years to develop.

Mesothelioma's early symptoms include weight loss, fever and night sweating. Chest pain, breathlessness on exertion, and/or pleural effusion are present. May also result in abdominal discomfort, change in bowel habit and weight loss. Long latency period of about 30 years

• TYPES OF ASBESTOS IN BUILDINGS

There are three main types of asbestos found in buildings. And all these types of asbestos are considered to be carcinogens, namely:

- ✓ Blue asbestos – crocidolite
- ✓ Brown asbestos – amosite
- ✓ White asbestos – chrysotile

Blue and brown asbestos are more dangerous than white forms of asbestos. However blue asbestos also known as Crocidolite is the most dangerous because its mineral fibres are finer and sharper.

• MOST AT RISK WORKERS TO ASBESTOS EXPOSURE

Occupiers of the building may be at risk of exposure to asbestos. However, the most at risk people include construction workers doing demolition work or maintenance and repairs, roofers, electricians, painters and decorators, joiners, plumbers, gas line installers, plasterers, shop fitters, heating and ventilation engineers and surveyors. Moreover, anyone dealing with electronics such as telephone services, IT engineers and alarm installers may well be at risk.

• COMMON PLACES WHERE ASBESTOS CAN BE FOUND IN BUILDINGS

Asbestos can be found within many old buildings as it was a commonly used building material for a very long time. The common places to find asbestos being used include:

- ✓ Sprayed asbestos (limpet) (blue or brown) has been used in fire protection ducts and on structural steel members.
- ✓ Lagging (blue or brown): > 85% content can be found in thermal insulation of pipes and hot water cylinders
- ✓ Asbestos insulating boards (AIB) (blue or brown): > 85% content has been used in fire protection, thermal insulation, wall partitions, ducts, soffits, ceiling and wall panels
- ✓ Textured coatings containing asbestos have been used in decorative plasters and paints
- ✓ Asbestos cement products – flat and corrugated sheets (± 10-15% content) have been used in roofing and wall cladding, gutters, rainwater pipes, water tanks.
- ✓ Bitumen or vinyl materials containing asbestos have been used in roofing felts, floor and ceiling tiles.

• DUTY TO MANAGE ASBESTOS

It is important and critical by maintenance and facilities managers to manage asbestos in non-domestic buildings. It is also the responsibility of every maintenance manager to ensure the health and safety of occupants or users within the affected buildings. Anyone who is aware of the presence of asbestos within the building must report that to the maintenance manager to ensure the health and safety of everyone who might be or have been exposed. The duties to manage asbestos are listed as follows:

- ✓ Finding out if there is asbestos present within the premises, its location and its condition
- ✓ Maintaining up-to-date records of location and condition of asbestos in premises
- ✓ Assessing the risk from asbestos material

- ✓ Preparing a plan which sets out in detail how the risk is going to be managed
- ✓ Implement plan of action
- ✓ Reviewing and monitoring the plan and arrangements made to put into action
- ✓ Providing information on location and condition to those who might have to work on or disturb it.

• STEPS TO CONSIDER WHEN ASSESSING RISK OF ASBESTOS EXPOSURE

The following steps must be considered when assessing potential risks caused by exposure to asbestos in the building:

Step 1: Finding out if asbestos is present within the building

The use of asbestos was prohibited in 2008 in South Africa. In assessing the presence of asbestos within a building, it may be assumed that asbestos is present in all buildings built or refurbished before 2008 unless there is a strong evidence against its presence. It may also be necessary to:

- ✓ examine building plans and any other relevant information, such as builders' invoices, which may state if and where asbestos was used in the construction or refurbishment of the premises
- ✓ carry out a thorough inspection of the premises both inside and out to identify materials that are, or may be, asbestos
- ✓ consult others, such as the architects, workers or safety representatives, who may be able to provide more information and who have a duty of co-operation to make this available.

Step 2: Assessing the condition of Asbestos Containing Materials

In assessing the condition of asbestos-containing materials, certain conditions must be considered:

- ✓ whether the surfaces of the materials containing asbestos are damaged, frayed or scratched
- ✓ whether the surface sealants are peeling or breaking off
- ✓ If the materials are becoming detached from their base (particularly within pipes and boiler lagging and sprayed coatings)
- ✓ Identify if the protective coverings designed to protect the material are missing or damaged
- ✓ whether there is asbestos dust or debris from damage near the material

Step 3: Survey and Sample for asbestos

The survey of the premises and identification of asbestos in materials must be conducted by a suitably trained person with the necessary certification, experience and must provide evidence of liability insurance.

Step 4: Keeping written records

Clear and simple records must be kept and always readily available. The records on asbestos must indicate known or presumed areas where asbestos may be present, type of asbestos, form and condition.

Step 5: Acting on findings

In order to act on the findings, a priority plan must be drawn. Materials with damage must be given a higher priority to ensure they are repaired, sealed or enclosed by a trained personnel. The following factors must be considered when assessing the likelihood of disturbances of materials containing asbestos:

- ✓ location, amount and condition of the materials containing asbestos
- ✓ position where materials are likely to be disturbed
- ✓ quantity of asbestos
- ✓ accessibility
- ✓ if asbestos is closer to normal working areas
- ✓ the numbers of people who use the area where asbestos is present
- ✓ if maintenance work, refurbishment is likely to be carried out where there is asbestos

Step 6: Keeping records up to date

Regular inspection of the condition of materials containing asbestos must be carried out regularly, at least every 6 to 12 months and updated in the records. The action plan must also be regularly reviewed and monitored.

• CHECKLIST ON MANAGING RISK OF ASBESTOS EXPOSURE

The checklist sets out the necessary steps for managing risk of asbestos:

✓ Find – check for presence of asbestos
✓ Condition – check condition of material
✓ Presume – material contains asbestos unless strong evidence it does not
✓ Identify – arrange for material to be sampled and identified by a specialist
✓ Record – location and condition on a plan or drawing
✓ Assess – decide if material likely to be disturbed
✓ Plan – prepare and implement plan to manage risks

Minor damage	**Good condition**
• Material should be repaired and/or encapsulated • Condition of material should be monitored regularly and if possible labelled • Inform contractor and any worker likely to work on or disturb material	• Condition of material should be monitored regularly • Where possible material should be labelled • Inform contractor and any worker likely to work on or disturb material
Poor condition	**Asbestos disturbed**
• Asbestos in poor condition should be removed	• Asbestos likely to be disturbed should be removed

• ASBESTOS-RELATED RECORD KEEPING

Health records of every worker exposed to asbestos must be kept for 40 years from date of last entry. Clinical records must also be kept that include copy of completed respiratory symptom questionnaire, details of examination findings and any other tests. The statistical records must be made available when requested.

• SKIN DISEASE

Occupational contact dermatitis is caused by exposure to nickel, chromium, cobalt and their compounds, latex and rubber chemicals, epoxy resins, and various paint and wood preservatives. Symptoms are red, sore, itchy, scaly and blistered skin. In severe cases the skin can crack and bleed and it could spread all over the body. Hands are the most frequent location of occupational skin disease followed by the face and legs. Allergic contact dermatitis occurs more frequently than irritant contact dermatitis. Allergens include epoxy resin, cobalt chloride, thiuram mix, nickel sulphide, p-phenylenediadiamine and colophony.

Interventions include the application of the hierarchy of control and substitution of materials containing the substances that cause skin disorders. Material Safety Data Sheets (MSDSs) have to be carefully scrutinized to determine the composition of materials. Where exposure cannot be avoided, workers need to be trained in the proper use of the materials and provided with the necessary protective clothing and equipment. Given that the prolonged use of wet gloves aggravates and/or exacerbates skin disorders gloves should be replaced when they become wet.

• LEAD EXPOSURE

Exposure to lead and lead-based paint presents similar challenges to exposure to asbestos. Effects of lead exposure include lung infections, high blood pressure, digestive problems, nerve disorders, memory and concentration problems, and muscle and joint pain.

Workers may also be required to wear special respirators and protective clothing to protect them from exposure. Lead paint debris has to be disposed of taking special precautions. Controls include the elimination of the use and production of lead in the workplace, changing the way work activities are executed, modifying processes, and use of an alternative lead-free substitute.

• SLEEP DEPRIVATION AND FATIGUE

Long hours of working and shifts result in sleep deprivation and mental fatigue. Lack of sleep or sleep deprivation can have potentially serious health effects on workers in the form of physical and mental impairments. These impairments include reduction of their ability to think lucidly and concentrate on the task at hand, handle stress, maintain healthy immune systems and moderate their emotions. Without adequate rest or recovery window, the ability of the brain to function properly quickly deteriorates. Consequently, the brain has to work harder to counteract the effects of sleep deprivation. Other typical effects of sleep deprivation include:

- Depression;
- Heart disease;
- Hypertension;
- Irritability;
- Slower reaction times;
- Slurred speech; and
- Tremors.

It is likely after a lengthy shift that workers will experience weakness and feelings of tiredness accompanied by sore muscles. Consequently, due to overwork and lack of sleep, workers may become fatigued when they have feelings of tiredness, exhaustion and lack of energy. The overall consequences of lack of sleep and mental fatigue can be catastrophic in the workplace and those workers who operate plant and equipment or drive are particularly susceptible. Exposure to consequences of overtime and shift work can be avoided by proper planning and consideration of workers.

• RADIATION

When workers are exposed to radioactive materials and radiation in the work environment without proper protection they will suffer damage to their bodies as well as ionizing radiation. Together with exposure from external sources, radiation exposure can occur internally from ingesting, inhaling, injecting or absorbing radioactive materials. Both external and internal sources may irradiate the whole body or a portion of the body.

The most common forms of ionising radiation are alpha and beta particles or gamma and X-rays. Malignancy is a likely consequence of high levels of exposure as also is genetic change. Genetic changes result in change in the genes or chromosomes of reproductive cells. Of concern is that these changes are handed down to descendants of the affected worker in the form of gene mutations, chromosome aberrations and changes in the number of chromosomes themselves. The health effects may be stochastic which results in cancer and changes in DNA or non-stochastic that produce burns and radiation sickness resulting in premature aging or even death.

On the other hand, non-ionizing radiation is relatively low-energy radiation with little energy to ionize atoms or molecules. Sources include electric power lines, microwaves, radio waves, infrared radiation, visible light and lasers. Overexposure can lead to health issues. Determinants of the effects are strength of the low frequency at source, distance and duration of exposure. Pacemakers, hearing aids and defibrillators are affected. Large amounts of radio frequency can damage skin or eyes and raise body temperature. Exposure to UV radiation can lead to sunburn or skin cancer, cataracts and a compromised immune system. Sources besides the sun include black light and welding tools.

The initial signs and symptoms of treatable radiation sickness are usually nausea and vomiting. The period of time between exposure and the onset of these symptoms is a reliable indicator of how much radiation has been absorbed. After the initial onset of these symptoms the worker then experiences a brief period of a few days without any apparent illness. This period is followed by nausea, weakness, hair loss, vomiting blood, bloody stools, diarrhoea, high fever, skin burns or diminished organ function.

Prevention or mitigation of the effects of exposure include reducing the time of exposure to radiation sources, increasing the distance from the sources and using shielding to reduce the magnitude of the exposure. Shielding could be heavy clothing and the use of lead.

• ELECTROMAGNETIC FIELDS (EMF'S)

EMFs are areas of energy that surround any electrical device which can emit magnetic fields, and electrical wiring. The strength of a magnetic field reduces rapidly with increased distance from the source. EMFs are not shielded by most common materials and pass easily through them. Workers most likely to be exposed include electricians, telephone linemen, electric power workers and welders.

Studies suggest that exposure to EMFs may contribute to Alzheimer's disease and is carcinogenic. Electromagnetic hypersensitivity is a condition reported by workers who have been exposed to EMFs.

Protection is in the form of restricting access where electric fields are very large. Where magnetic fields are very strong the only practical protective method available is to limit the number of workers exposed and the duration of that exposure.

• STRESS

According to the Health and Safety Executive (HSE) in the UK stress is the adverse reaction people have to excessive pressures or other types of demand placed on them. There is a clear distinction between pressure, which can create a 'buzz' and be a motivating factor, and stress, which can occur when this pressure becomes excessive. Occupational stress can lead to poor health which may result in injury.

About 1 in 5 people say that they find their work either very or extremely stressful. Over half a million people report experiencing work-related stress at a level they believe has actually made them ill. Each case of stress-related ill health leads to an average of 29 working days lost. The concept of occupational stress is often confused with job challenge, although these concepts are not the same.

Causes of occupational stress

Stress sets off an alarm in the brain, and the brain responds by preparing the body for a defensive action. This biological response is sometimes called fight or flight response and is important because it helps the body cope with threatening situations. The nervous system is aroused by this action and releases hormones to heighten the senses, increase heart rate, deepen respiration, and tense the muscles.

Occupational stress results from the workers' interaction with the working environment. However, differences in individual characteristics such as personality and coping style have an impact on predicting whether certain job demands will result in stress, for example, what is stressful for one worker may not necessarily be problematic to another worker. Therefore, it is generally accepted that working conditions play a vital role in causing job stress. Nevertheless, it is evident that stress causes several types of chronic ills such as cardiovascular disease, musculoskeletal disorders, and psychological disorders. Furthermore, stress related syndromes often manifest as psychosomatic symptoms.

Early symptoms of occupational stress include, for example:

- Headache
- Insomnia
- Short temper
- Upset stomach
- Job dissatisfaction
- Low morale
- Low energy levels
- Weak immune system

NIOSH Model of Job Stress

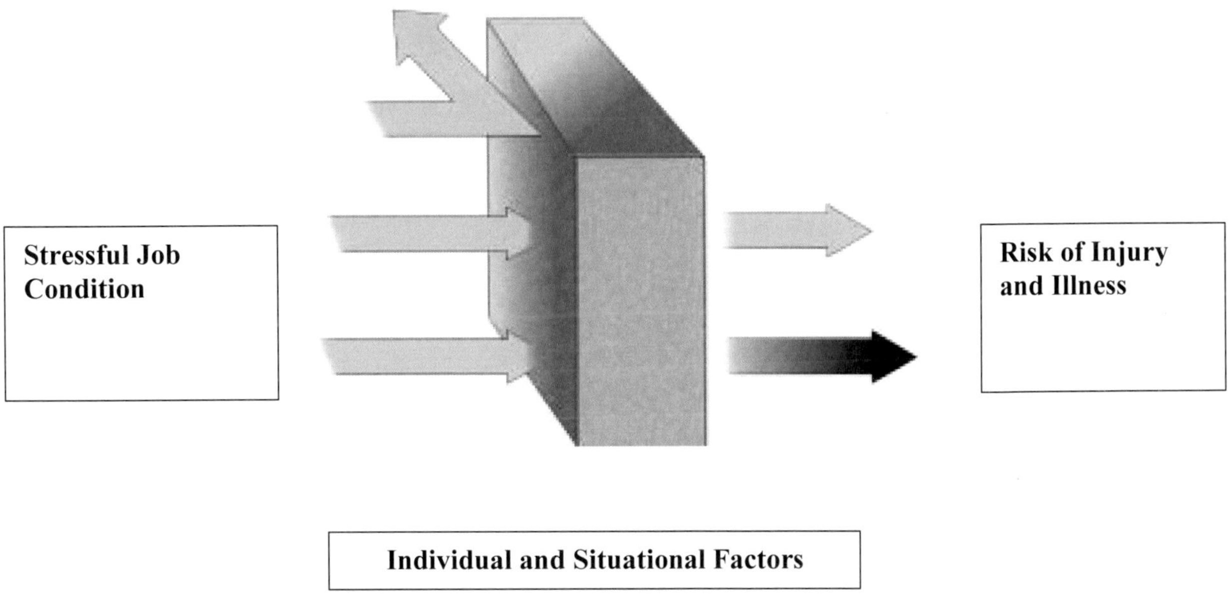

(Source: https://www.researchgate.net/figure/NIOSH-Model-of-Job-Stress_fig2_325618275)

Management of stress in the workplace

The Management Standards propagated by the HSE in the UK as a tool to manage stress in the workplace cover six key areas of work design that, if not properly managed, are associated with poor health and well-being, lower productivity and increased sickness absence. In other words, the six Management Standards cover the primary sources of stress at work. These are:

1. **Demands** – such as workload, work patterns and the work environment.
2. **Control** – such as how much say the worker has in the way they do their work.
3. **Support** – such as the encouragement, sponsorship and resources provided by the organisation, line management and colleagues.
4. **Relationships** – such as promoting positive working to avoid conflict and dealing with unacceptable behaviour.
5. **Role** – such as whether workers understand their role within the organisation and whether the organisation ensures that they do not have conflicting roles.
6. **Change** – such as how organisational change (large or small) is managed and communicated in the organisation.

Employers have a duty to ensure that risks arising from work activity are properly controlled. The Management Standards approach helps employers work with their workers and representatives to undertake risk assessments for stress. The balance between work and family or personal life could further alleviate stress and workers could further seek a support network of friends and co-workers and approach work in a relaxed and positive manner.

• REVIEW QUESTIONS

1. What rights do workers have pertaining to HIV and AIDS?
2. What are some of the health challenges presented by welding activities?
3. What steps must be considered when assessing potential risks caused by exposure to asbestos in the workplace?
4. How may the effects of exposure to ionizing radiation be prevented or mitigated?
5. What are key areas need to be addressed in terms of The Management Standards to manage stress in the workplace?

• REFERENCES

Bock, M., Schmidt, A., Bruckner, T. and Diepgen, T. (2003). Occupational skin disease in the construction industry. *British Journal of Dermatology*, 149, 1165–1171.

British Lung Foundation (2008). Construction tradespeople, Asbestos awareness survey. [Available at http://a-a-s-c.org.uk/wp-content/uploads/2012/05/Tradespeople-Asbestos-Awareness-Survey-Feb-2008.pdf] Accessed 17 August 2014.

Campbell, Fiona. (2006). Occupational Stress in the Construction Industry. London: The Chartered Institute of Building. [Available at http://www.ciob.org/filegrab/stress.pdf].

Darnton, Andrew., McElvenny, Damien. and Hodgson, John. (2006). Estimating the number of asbestos-related lung cancer deaths in Great Britain from 1980 to 2000. *Annals of Occupational Hygiene*, 50(1), 29–38.

Department of Labour (1997). Basic Conditions of Employment Act 75 Of 1997. Department of Labour.

Department of Labour (1998). Employment equity act 55 of 1998.

Department of Labour (1995). Labour Relations Act 66 Of 1995. Department of Labour.

Department of Labour (1993). Compensation for Occupational Injuries and Diseases Act no of 103 of 1993.

Department of Labour (1993). Occupational Health and Safety Act No: 85 of 1993.

Eichholz, Geoggrey. G. (1977). Environmental Aspects of Nuclear Power. Ann Arbor, MI, USA: Ann Arbor Science.

Harinarain, Nishani. and Haupt, Theodore. (2010): Impact of workplace HIV and AIDS polices on stigma and discrimination, In Haupt, T.C. (Ed) The Built Environment 5, Proceedings of the 5th Built Environment Conference, Durban, 18-20 July, pp. 453-465, ISBN 978-0-620-46703-2

Harinarain, Nishani. and Haupt, Theodore. Conrad. (2011): HIV and AIDS: The KZN Construction Industry Response, Journal of Construction, 4 (2), pp. 13-17, ISSN 1994-7402

Harinarain, Nishani. and Haupt, Theodore. (2011), Implications of ignoring HIV and AIDS by the construction industry: The South African experience, CIB W99 Conference: Prevention: The Means to the End of Construction Injuries, Illnesses and Fatalities, Washington, D.C., August 24-26

Harinarain, Nishani. and Haupt, Theodore. (2012). Threats to effective HIV and AIDS management in construction: Lessons from literature. International Cost Engineering Council - 8th ICEC World Congress "Quest For Quality: Professionalism In Practice". International Convention Centre: Durban, South Africa: 23 - 27 June 2012

Harinarain, Nishani., and Haupt, Theodore. (2014), The lack of management commitment to HIV and AIDS in South African construction, Proceeding of Association of Schools of Construction of Southern Africa (ASOCSA) 8th Built

Environment Conference, 28th-29th July, University of Kwa Zulu Natal, Durban, South Africa, pp. 289-300, ISBN 978-0-620-60356-0

Harinarain, Nishani. and Haupt, Theodore. (2014). The Vulnerability of the Construction Industry to HIV and AIDS, Journal of Construction, 7 (2), 61-66, ISSN 1994-7402

Harinarain, Nishani., and Haupt, Theodore. Conrad., (2014), Drivers for the effective management of HIV and AIDS in the South African construction industry— a Delphi study, African Journal of AIDS Research, vol. 13 (3) 291-303, ISSN 1608-5906

Haslam, Cheryl. and Mallon, Krissie. (2003). Post-traumatic stress symptoms among firefighters. Work & Stress, 17, 277–285.

Haupt, Theodore. Conrad., Deacon, Claire. and Smallwood, John. J. 2005. "Respiratory and Skin Infections in Older Construction Workers," *Occupational Health Southern Africa*, 10 (6), 4-9, ISSN 1024-6274

Health and Safety Authority. (n.d.). New Asbestos Guidelines. Available at: https://www.hsa.ie/eng/Your_Industry/Chemicals/Legislation_Enforcement/Asbestos/newasbestosguidelines.pdf

Health and Safety Executive (HSE): Asbestos – health and safety in the workplace http://www.hse.gov.uk/asbestos Google Scholar

Health and Safety Executive (HSE) (2002). Respirable Crystalline Silica, Phase 1. [Available at http://www.hse.gov.uk/pubns/priced/eh75-4.pdf] Accessed 17 September 2014.

Health and Safety Executive (HSE) (2014a). Statistics, Mesothelioma. [Available at http://www.hse.gov.uk/Statistics/causdis/mesothelioma/index.htm] Accessed 16 August 2014.

Health and Safety Executive (n.d.). Work-related stress and how to tackle it. Available at: https://www.hse.gov.uk/stress/what-to-do.htm

Health and Safety Executive for Northern Ireland (2009). Construction workers: advice on cancer prevention. [Available at http://www.hseni.gov.uk/advice_on_cancer-2.pdf] Accessed 17 August 2014.

ILO, 2013. The prevention of occupational diseases [Homepage of International Labor Organization], 4 [viewed 2015, April 8]. Available at http://www.ilo.org/wcmsp5/groups/public/---ed_protect/---protrav/---safework/documents/publication/wcms_208226.pdf.

McAleenan, Ciaran. and Oloke, David. (2015). *ICE manual of health and safety in construction.* 2nd ed. One Great George Street, Westminster, London SW1P 3AA: ICE Publishing, 2015.

National Institute for Occupational Safety and Health. (1999). STRESS…At Work (NIOSH Publication No. 99-101). Retrieved from http://www.cdc.gov/niosh/docs/99-101/

Peto, Julian., Rake, Christine., Gilham, Clare. and Hatch, Jane. (2009). Occupational, domestic and environmental mesothelioma risks in Britain. [Available at http://www.hse.gov.uk/research/rrpdf/rr696.pdf] Accessed 16 August 2014.

Rom, W., Hammar, S., Rusch,V., Dodson, R. and Hoffman, S. (2001). Malignant mesothelioma from neighbourhood exposure to anthrophylite asbestos. *American Journal of Industrial Medicine*, 40, 211–214.

Ronca, Debra. (2014). "How Radiation Works" 23 July 2008. HowStuffWorks.com. [Available at http://science.howstuffworks.com/radiation.htm]. Accessed 20 August 2014.

Ugale, C. 2009. Relationship between Occupational Stress & Job Satisfaction: A Case Study of InfoTekNetAlia, Unpublished MBA dissertation, University of Wales

WHO (World Health Organization) (2010). Healthy Workplaces: a model for action. Geneva: World Health Organization.

WHO (2020). *HIV/AIDS.* [online] Who.int. Available at: https://www.who.int/news-room/fact-sheets/detail/hiv-aids.

CHAPTER 21

HEALTH PROMOTION

• EMPLOYEE ASSISTANCE PROGRAMS

An Employee Assistance Program (EAP) is commonly the most overlooked, yet valuable, benefits offered by both large and small organizations. An (EAP) can give workers a way to cope with personal issues that can have a negative impact on their job performance, health and wellbeing such as work-related stress or problems outside of the workplace. Many workers deal with issues, including substance abuse, depression, marital strife and grief over the loss of a loved one. An EAP provides an outlet to help them resolve these life-stressors and issues, or at least learn to cope with them so that they will not adversely impact job performance. An EAP can, therefore, help workers stay focused on their duties at the workplace, keep productivity and performance high, and genuinely improve the well-being of the workers in the process.

EAPs generally include short-term counselling and referral services for workers and their household members. EAP counsellors typically provide assessment, support, and referrals to additional resources such as counsellors for a limited number of program-paid counselling sessions. The issues for which EAPs provide support vary, but examples include:

- ✓ Substance abuse
- ✓ Occupational stress
- ✓ Emotional distress
- ✓ Major life events, including births, accidents and deaths
- ✓ Health care concerns
- ✓ Financial or non-work-related legal concerns
- ✓ Family/personal relationship issues
- ✓ Work relationship issues
- ✓ Concerns about ageing parents
- ✓ Conflict Management,
- ✓ Assertiveness in the Workplace,
- ✓ Interpersonal Skills,
- ✓ Time Management,
- ✓ HIV and AIDS programmes

The services provided by EAP are generally free to the worker and the members of their families, having been pre-paid by the employer. Typically, an employer contracts with an outside organization to manage its EAP. These organizations rely upon other vendors or contracted parties for specialized services to supplement their own services, such as, for example, financial advisors, attorneys, travel agents, and geriatric or child care specialists.

Confidentiality must be maintained in accordance with privacy laws and ethical standards. With an EAP, the employer is not notified when a worker uses the service. With the EAP, the worker can seek help during the early stages of a problem, which can prevent it from turning into turning into something more serious.

An EAP can also provide benefits to employers. Because workers have access to a resource that can remedy a potentially debilitating problem, it can reduce absenteeism or even turnover. As a result, the workplace is arguably likely to be a happier and more productive one. Employers may also be able to minimize the cost of health insurance plans and medical aid schemes, because workers can use the EAP to prevent stress-related illnesses, meaning fewer visits to the doctor or medical practitioner.

Some of these programmes could be done in-house, but many of these assistance programmes will be out-sourced. It is important to find and appoint the best possible supplier for these programmes.

Benefits for employers include:

- ✓ lower medical costs,
- ✓ reduced turnover and absenteeism,
- ✓ and higher worker productivity

The standards of the Employee Assistance Professionals Association (EAPA) in the USA include the following, namely:

- Standard crisis intervention services are available 24 hours a day, seven days a week and 365 days per year.
- EAP expertise and procedures for determining when and how to refer workers to EAP services.
- Training for the managers and supervisors of the employer so they learn to recognize issues and can intervene by referring workers to their EAP.
- The EAP must staff professionals who maintain professional competence and knowledge.
- The ability to meet changing needs by adding services that are needed as trends change.

Other components of a comprehensive EAP program include:

- Multiple communication methods of service delivery implying that phone counselling should not be the only form of service available.
- Legal and regulatory compliance support.
- Expert-driven policy development in templates that employers can manipulate and wordsmith with ease.
- Worker-focused communication and training.
- Comprehensive websites and promotional materials customized to multiple languages and cultures.
- Customized data reporting on industry trends.
- Quality local services that are sensitive to language, culture, faith, particularly during emergencies, when unique guidance and assistance may be required.
- Quality of services and ease of access are prioritized over the cost of the program.
- Website and online materials are provided in multiple languages and are able to relate to EAP customers in those languages using common phrases, and plain word use for ease of understanding.

• ABSENTEEISM

Absenteeism has been defined as the intentional or habitual absence from work of a worker. It is therefore a deliberate act on the part of a worker. It is the term given when a worker is habitually and frequently absent from work and excludes paid leave and occasions where an employer has granted the worker time off. While employers expect workers to miss a certain number of workdays each year, excessive absences can equate to decreased productivity and can have a major effect on organization finances, morale and other factors.

• CAUSES OF ABSENTEEISM

Workers stay away from work for many reasons which may or may not be legitimate. Some of the common causes of absenteeism include the following according to High Speed Training (https://www.highspeedtraining.co.uk/hub/reducing-absenteeism-in-the-workplace – accessed May 26, 2020), namely:

- ✓ Bullying and harassment – Workers who are bullied or harassed by co-workers and/or bosses are more likely to call in sick to avoid the situation
- ✓ Burnout, stress and low morale – Heavy workloads, stressful meetings or presentations and feelings of being unappreciated can cause workers to avoid going into work. Personal stress outside of work can also lead to absenteeism.
- ✓ Childcare and eldercare – Workers may be forced to miss work in order to stay home and take care of a child or an elder when normal arrangements have fallen through such as, for example, a sick caregiver or if a child or elder is sick.
- ✓ Depression – Depression can lead to substance abuse if people turn to drugs or alcohol to self-medicate their pain or anxiety.
- ✓ Disengagement – Workers who are not committed to their jobs, co-workers and/or the organization are more likely to miss work simply because they have no motivation to go.
- ✓ Illness – Injuries, illness and medical appointments are the most commonly reported reasons for missing work although not always the actual reason. Not surprisingly, each year during the cold and flu season, there is a dramatic spike in absenteeism rates for both full-time and part-time workers. Workers suffering from AIDS are also likely to miss work more frequently as the illness progresses.
- ✓ Injuries – Accidents can occur on the job or outside of work, resulting in absences. In addition to acute injuries, chronic injuries such as back and neck problems are a common cause of absenteeism.
- ✓ Job hunting – Workers may call in sick to attend a job interview
- ✓ Partial shifts – Arriving late, leaving early and taking longer breaks than allowed are considered forms of absenteeism and can affect productivity and workplace morale.

• CONSEQUENCES OF ABSENTEEISM

Absenteeism has consequences for both the employer and the worker. Some of the consequences that affect employers include:

- Reduced productivity levels.
- High administration costs.
- Increased labour costs if replacement temporary workers have to be hired.
- Understaffing which could lead to poor customer service.
- Poor morale among colleagues. This will be particularly prevalent if workers constantly have to fill in for absent workers and if there are no retributions for being absent.

On the other hand, the consequences for the absent worker include:

- Loss of pay for their time off.
- Reduced productivity on their return to work due to a need to 'catch up'.
- If absenteeism is frequent and unexplained, the worker could even face dismissal.

Clearly, absenteeism is costly for all of the workforce, and it is important to know the causes, and strategies to reduce it, for the benefit of the organization.

• STRATEGIES TO REDUCE ABSENTEEISM

After understanding the causes for absenteeism, the following strategies might be useful:

Create a clear attendance policy and set expectations

This policy should explain how to report absences, the procedures in place surrounding absences, and the policy for recording absences. It should also contain information about how repeated absences will be follow up on and what the repercussions are for excessive absenteeism. The policy should be discussed with all workers and be included as part of induction or orientation of new workers and be signed by them to show that they understand it. The policy will act as a resource to workers about the expectations of the organization. The policy should be applied consistently and followed in all possible situations.

Provide support

If a worker is absent due to personal issues, like bereavement or mental health problems, they should be provided with support both when they are absent and, on their return, to work. Support will likely make them feel happier to return to work earlier and prevent repeated absenteeism.

Reduce workplace stress

Always try to rectify factors at work that may cause stress for workers. For example, if a piece of equipment is broken and workers are stressing, it should be fixed as soon as possible.

Provide feedback

Provide workers with frequent and effective feedback, especially when they have performed well. If feedback is never provided the workforce may feel that their hard efforts are not appreciated and their absence from work will not be noticed.

The following is an article taken from https://www.skillsportal.co.za/content/absenteeism-workplace (Accessed May 26, 2020)

Absenteeism at the workplace
Wed, 07 Mar 2012 10:28
By Bruno Bruniquel
Do not jump to conclusions

According to a recent report, police are abusing sick leave with staff at one police station alone taking an average of 77.2 days' sick each. Abuse of sick leave has apparently been identified as the 'number one risk' to the police, and a 'health risk manager' has now been appointed to refer serial absentees to 'neutral' doctors to find out if they are really too sick to go to work.

"There are people in the department who, according to the sick leave register, are supposed to have died, because there isn't any part of their body that has not been ill. However, they are strong and kicking. They are not dead," said the chairwoman of Parliament's police oversight committee, Sindi Chikunga.

The typical 'knee-jerk reaction to being presented with figures which show high absenteeism and sick leave abuse is get out the 'big stick' and to discipline – preferably fire the malingerers. Do so at your peril.

For example, in a long drawn out legal battle stemming back to 2007 (Booysen v Minister of Safety & Security & Others (Labour Court Case Nos. C60/08 and 307/09 dated 25 January 2012) the Labour Court ruled that the chairperson was unfair to proceed with a disciplinary enquiry due to the worker being unfit for work. Expert witnesses had testified to the effect that Booysen's condition impaired his ability to participate in a disciplinary enquiry due to post traumatic stress syndrome (PTSS). One wonders how much tax payer's money was wasted on legal fees in this case.

Barometer

Absenteeism is a good barometer of staff morale. People who are motivated and committed to their work and employer have to be really ill before they book off sick. On the other hand, if absenteeism and sick leave are not monitored and action is not taken against absentees, sick leave soon turns into another form of 'annual leave' to be taken whether or not one is really ill.

Attendance / 'wellness' bonus

Some employers have reacted to this by introducing an attendance or 'wellness' bonus. If a worker does not take sick leave he is paid a bonus. This is not recommended. Why not? Well, in the first place, under common law and our labour law, a worker has a fundamental duty to be at work! Sick leave is for when the worker is sick.

If a worker is injured or genuinely sick, it is in neither the employer's, nor the worker's best interests for the worker to come to work. The worker might well infect others or be injured if working around moving machinery. Furthermore, it is unfair and demotivating to take away a bonus from someone who has been genuinely ill and who cannot attend work through no fault of his or her own.

Job satisfaction

If workers are not aligned with their organisation's goals and are demotivated at work, it should be expected that they are going to absent themselves. That is where leadership and performance management come into the equation.

Reasons for absence

People absent themselves from work for a variety of reasons and then claim sick leave in order to get paid. This is in effect fraud and it is a dismissible offence. The employer needs however, to investigate the reasons and of course, establish whether or not the worker is genuinely ill.

The first step is to examine the worker's attendance record and establish whether there is a trend or pattern of absenteeism. For example, absence usually occurs before and after weekends and public holidays, after paydays or on a particular day of the week.

A point of caution, a worker's absenteeism is usually accompanied by sick leave abuse, poor performance, poor timekeeping, unsafe acts, unsociable behaviour at work, family problems and financial difficulties, including the serving of garnishee orders on the employer. Look for a pattern.

Confront the absentee

If a pattern becomes apparent, the next step is to confront the worker to establish why he or she has taken to absenting themselves from work. The best way to do this is to show concern and to suggest that it would appear from the sick leave taken that the worker is heading for a serious ill health problem. In order to get to the bottom of this, suggest that the correct course of action will be for the worker to undergo a medical evaluation to establish his/her fitness to work.

From my experience, the mere act of confronting workers is often enough to put an end to a lot of Monday/Friday absence. For example, in one instance, a worker who had regularly taken his 10 days' sick leave per annum did not take another day's sick leave for the next two years after being confronted! Medical evaluation

If the worker agrees to undergo a medical, the employer should pay for this, conditional upon the worker agreeing for the results to be made known to the employer. If this is agreed, the worker should be sent to an occupational health doctor for examination.

The doctor should be supplied with a job description and a description of the physical conditions and pressures of the job. The results of the medical will reveal whether or not the worker is genuinely ill. If the worker is ill, the early detection of illness might well be a life saver. It is also likely to reduce medical and sick leave costs in the longer term.

Malingerers

If the results of the medical reveal that the worker is not genuinely ill, then one needs to establish what else could be the cause. For example, a worker who is in debt may absent himself after payday in order to avoid loan sharks from whom he has borrowed money! Likewise, a worker with a substance abuse problem may absent himself from work because he is suffering a hangover. A wife who is being abused will absent herself to avoid questions over injuries.

The best way to deal with personal problems giving rise to absenteeism or unsatisfactory performance is to offer the worker professional counselling. If the worker agrees, the real problem will usually emerge and can be dealt with by way of rehabilitative treatment or counselling (e.g. for alcoholism).

If a worker has a personal or substance abuse problem, only the worker can deal with it. If the worker chooses to face the problem, it can be overcome. If not, the pattern of absenteeism and unsatisfactory performance and minor misconduct will continue, regardless of disciplinary action (and deduction of bonuses). The employer needs to put the worker on terms.

All embracing final warning

This is best done by way of an all embracing final warning, setting out the aspects of the worker's performance or behaviour which are not acceptable and confirming steps taken by the employer to address these.

It should also note that as the worker has refused to undergo a medical, counselling or rehabilitative treatment, any further shortcomings in the areas listed could result in the termination of the worker's services.

Limit sick leave

It is important to note that in terms of the Basic Conditions of Employment Act, if a worker is absent for more than two consecutive days (not working days), or on more than two occasions in an 8-week period, the employer is entitled to require the worker to bring a certificate for every day's absence. This means that if a worker works a 5-day week, then absence on a Monday or Friday require a medical certificate. If the worker fails to produce a medical certificate for absences of more than two consecutive days or when the worker has been off for more than two days in an 8-week period, then the employer may treat it as absence without leave and may take appropriate disciplinary action against the worker.

If the worker has exceeded his sick leave entitlement he should be advised that he will not be paid sick leave, even if a sick certificate is produced. He should also be advised that his services could be terminated on the grounds of incapacity. If absenteeism is closely monitored and acted upon as suggested, it should not be necessary to pay an attendance bonus. That money would be better spent on rewarding good performance – not merely being at work which is every worker's duty any way.

Bruniquel & Associates (Pty) Ltd
Office: +27 31 309-4627
Web: Bruniquel & Associates

- HEALTH PROMOTION

Health promotion is the process of enabling people to increase control over, and to improve, their health. It moves beyond a focus on individual behaviour towards a wide range of social and environmental interventions.

WORLD HEALTH ORGANISATION

Workplace health promotion is the combined efforts of employers, workers, and society to improve the mental and physical health and well-being of workers. The term workplace health promotion denotes a comprehensive analysis and design of human and organizational work levels with the strategic aim of developing and improving health resources in an organization.

The Luxembourg Declaration on Workplace Health Promotion in the European Union of 1997 provides that health and well-being of workers at work can be achieved through a combination of:

- Improving the organization and the working environment

- Promoting active participation
- Enabling healthy choices
- Encouraging personal development.

Workplace health promotion combines alleviation of health risk factors with enhancement of health strengthening factors and seeks to further develop protection factors and health potentials. It also means employers actively helping their workers improve their own general health and wellbeing. Workplace health promotion is therefore complementary to occupational health and safety which consists of protecting workers from hazards.

Successful workplace health promotion strategies include the principles of participation, project management, integration, and comprehensiveness:

- Participation: all workers must be included in all program stages
- Project management: programs must be oriented toward the problem-solving cycle
- Integration: programs must be incorporated into organization management practices and workplace health-promotion strategies should influence corporate planning
- Comprehensiveness: programs must incorporate interdisciplinary individual-directed and environment-directed health strategies.

By making workers feel better and healthier, workplace health promotion leads to many positive consequences like reduced turnover and absenteeism, enhanced motivation and improved productivity, as well as improving the employer's image as a positive and caring organisation. Some workplace health promotion aspects are:

✓ Participation of workers in the process of improving work organisation
✓ Active involvement and consultation of workers in improving their work environment
✓ All measures aimed at enhancing wellbeing at work, for example, enabling flexible working hours or working from home especially after COVID-19
✓ Raising the topic of healthy eating at work, giving information on healthy nutrition as well as offering healthy canteen food or facilities to prepare own food
✓ Tobacco awareness, including the offer of free participation in smoking cessation programmes as well as declaring a comprehensive smoking ban within the entire organization through a 'No Smoking' policy
✓ Mental health promotion, offering courses for managers on how to deal with stress and tension within their team, providing the opportunity for anonymous psychological consultancy for all workers
✓ Exercises and physical activity, offering sport courses, encouraging physical activity, promoting an active and healthy culture at work
✓ Health monitoring, offering checks such as blood pressure or cholesterol level

• HEALTH RISK ASSESSMENT (HRA)

A health risk assessment which is also referred to as a health risk appraisal and health and well-being assessment is one of the most widely used screening tools in the field of health promotion and is often the first step in multi-component health promotion programs. A health risk assessment (HRA) is a health questionnaire, used to provide workers with an evaluation of their health risks and quality of life. Commonly a HRA incorporates key elements, namely:

- an extended questionnaire,
- a risk calculation or score, and
- some form of feedback in the form of a face-to-face consultation with a health advisor or an automatic online report.

The main objectives of a HRA are to:

✓ Assess health status
✓ Estimate the level of health risk
✓ Inform and provide feedback to participants to motivate behaviour change to reduce health risks

• BENEFITS OF CONDUCTING PERSONAL HEALTH RISK ASSESSMENTS

Health risk assessments:

✓ Provide workers with a snapshot of their current health status.
✓ Enable workers to monitor their health status over time.
✓ Provide workers with concrete information preparing them for lifestyle change.
✓ Help workers get involved with health coaching.

- ✓ Provide important information concerning the readiness of workers to change.
- ✓ Help employers measure and monitor population health status.
- ✓ Provide employers with important information that can help them build results-oriented health promotion programmes.
- ✓ Can provide employers with important information on productivity.
- ✓ Allow employers to evaluate changes in health behaviour and health risks over time.
- ✓ Engage both employers and workers in the health management process at work.

• PRIMARY HEALTH CARE

Primary health care involves the immediate treatment of minor/serious injury or illness, chronic diseases and minor ailments occurring at the factory, referral of the worker to the relevant doctor or hospital and follow-up treatment to ensure rapid recovery. It also covers aspects related to social welfare/health education and rehabilitation. The immediate treatment of injuries/occupational disease at work implies the administration of the Compensation Commissioner's procedures as outlined in COIDA.

• OFF-THE-JOB HEALTH AND SAFETY

It is equally important for an organization to preserve its most valuable asset, namely the labour force, both on and off the job. Even off-the-job incidents could result in business interruptions and lead to unwanted costs for the organization.

Motivations for off-the-job health and safety

There are two major reasons why every organization should be interested in off-the-job human reasons and business reasons for health and safety.

1. On the human side, no supervisor, manager or employer wants to see workers or members of their family killed, maimed or injured.
2. On the business side there are many reasons for being interested in off-the-job health and safety:
    - quality and production suffer when trained, skilled workers are off work,
    - job performance is affected when, for example, workers are worried about family members in hospital
    - high costs of medical treatment, production losses, absenteeism, waste and mistakes are unacceptable.

By focusing on off-the-job health and safety, the employer is also showing his interest in families of workers, which ensures that workers are committed to the organization. Health and safety awareness at work is transferred to the home environment and vice versa.

Creating awareness of off-the-job health and safety

Creating awareness of off-the-job health and safety is considered to be mainly a promotional activity. Special family competitions, posters and company magazines can be used to promote awareness at the plant or workplace. The following might be ideas that could contribute to creating awareness:

- ✓ Domestic health safety checklist for use by workers.
- ✓ Off-the-job injury reporting system for workers.
- ✓ Road and domestic health and safety items in house magazines and publications.
- ✓ Brochures and pamphlets.

• REVIEW QUESTIONS

1. What are the benefits of having an Employee Assistance Program (EAP)?
2. Develop an EAP in your particular workplace to support workers experiencing job-related stress.
3. What are some of the causes of absenteeism?
4. What are the principles that should guide a health promotion initiative or program?
5. What are the benefits of conducting a health risk assessment?
6. Why is off-the-job health and safety important?
7. Develop an EAP in your particular workplace to support workers experiencing job-related stress.

• REFERENCES

Aldana, Steven. G., 2001. Financial impact of health promotion programs: a comprehensive review of the literature. *American Journal of Health Promotion*, *15*(5), pp.296-320.

Anonymous (n.d.). Loose notes on safety, health and environment management

Canadian Center for Occupational Health and Safety (n.d.). (none). [online] Ccohs.ca. Available at: https://www.ccohs.ca/oshanswers/hsprograms/eap.html

EU-OSHA (2013). Workplace Health Promotion - OSHWiki. [online] Oshwiki.eu. Available at: http://oshwiki.eu/wiki/Workplace_Health_Promotion#What_is_workplace_health_promotion.3F

EU-OSHA (2013) Motivating workers to participate in workplace health promotion. Available at: https://www.ppe.org/motivating-workers-to-participate-in-workplace-health-promotion/

High Speed Training. 2018. Reducing absenteeism in the workplace. Available at: https://www.highspeedtraining.co.uk/hub/reducing-absenteeism-in-the-workplace/

HSA (n.d.). Managing Health and Safety. [online] Health and Safety Authority. Available at: https://www.hsa.ie/eng/Topics/Managing_Health_and_Safety/

Investopedia (2019). Causes and Costs of Absenteeism. Available at: https://www.investopedia.com/articles/personal-finance/070513/causes-and-costs-absenteeism.asp

Kirch, Wilhem. (2008). Encyclopedia of public health. New York: Springer.

National Automotive Roads Fuel Association (NARFA). 2020. Five Great Benefits of Employee Assistance Programs. Available at: https://www.narfa.com/five-great-benefits-of-employee-assistance-programs-eaps/

Ottawa Charter for Health Promotion, 1986, First International Conference on Health Promotion, WHO, Ottawa, Canada, 17–21 November 1986, WHO. Available at: http://www.euro.who.int/en/who-we-are/policy-documents/ottawa-charter-for-health-promotion

PDHI. 2021. What is a health risk assessment? Available at: https://www.pdhi.com/ncqa/what-is-a-health-risk-assessment/

Smallwood, John. and Haupt, Theodore. Conrad. 2003. “Occupational Health Risk Assessment in Construction” *Occupational Health Southern Africa*, 9 (1), pp. 22-25

Totalcare People Solutions (2018). *employers – Totalcare People Solutions*. [online] Totalcarepeoplesolutions.co.za. Available at: http://totalcarepeoplesolutions.co.za/tag/employers/

Wikipedia. 2021. Workplace health promotion. Available at: https://en.wikipedia.org/wiki/Workplace_health_promotion

PART 4

MANAGEMENT OF THE ENVIRONMENT

This section comprises of five chapters that cover the management of the environment in general industry in South Africa. Chapter 22 provides an overview and introduction to important aspects of environmental production management while Chapter 23 deals with the effects of various types of environmental pollution. Common environmental challenges are presented in Chapter 24. Given the recently declared national disaster in South Africa as a result of COVID-19, Chapter 25 addresses disaster management and aspects of dealing with the pandemic more effectively. The final chapter 26 covers environmental management systems (EMS) including their design and implementation in the workplace.

CHAPTER 22

INTRODUCTION TO ENVIRONMENTAL PRODUCTION MANAGEMENT

Environmental management consists of "actual decisions and action concerning policy and practice regarding how resources and the environment are appraised protected, allocated, developed, used, rehabilitated, remediated, and restored." Industry around the world has realised that the production of consumer goods and services does not take place in a vacuum. Raw materials, water and energy are used as inputs in production processes. By-products and waste are typically produced during manufacturing or other processes. The products that are eventually sold to consumers may also impact on the environment. Cement, for example, is washed down the drain or into the storm water system after wheelbarrows on a housing project are washed.

The management of the natural environment is becoming increasingly important within operations as customers, suppliers and the public demand that manufacturers minimize any negative environmental effects of their products and operations. Therefore, managers play a critical role in determining the environmental impact of manufacturing operations through choices of raw materials used, energy consumed, pollutants emitted and wastes generated. Organizations are realizing the need to take responsibility for the environmental impacts associated with their products and production processes.

The concepts of "*cradle –to-grave*", "*life cycle analysis*" or "*extended producer responsibility*" confirm that it is no longer possible for industry to merely manufacture products without taking account their impact on the environment. Manufacturers need to look at the raw materials they use, the processes involved in making the products, the possible impact of the use of their products, and how their products are eventually disposed of.

Reasons for implementing an environmental programme

There are many reasons to consider the implementation of an environmental programme which include:

- ✓ Environmental programmes, which focus on waste minimization or resource conservation can potentially save an organization money and other resources
- ✓ Organizations can avert possible costly litigation through their proactive efforts.
- ✓ Organizations may be compelled to introduce pollution control measures as a result of the complaints of neighbouring communities or non-compliance with government regulations.
- ✓ Organizations could also be forced to comply with the environmental standards of countries to which they are exporting their products.
- ✓ Others regard it is as their moral duty to conserve resources and prevent pollution.

Integrating the environmental programme with the health and safety programme

The safety and health system in the organization may be well established. This provides a perfect opportunity for integrating an environmental programme into the system. There have been many debates about the integration of safety, health and the environment given that they have many factors in common which include the need:

- ✓ For timely management instead of crisis control.
- ✓ To set clear policies and adhere to them.
- ✓ To check and audit the programme continually.
- ✓ To report progress to management.

By integrating the three aspects the SHE manager can save time and resources and can generally improve the efficiency of the overall programme.

Stakeholders in the environmental management process

While safety or health is predominantly the concern of the management of the organization and its workers, there are usually more stakeholders where the environment is concerned. The following groups may be interested in the environmental performance of an organization, namely:

- ✓ Top management.
- ✓ Workers.
- ✓ Shareholders.

- ✓ Financial houses.
- ✓ Customers.
- ✓ Suppliers.
- ✓ Surrounding communities.
- ✓ Non-governmental environmental organisations.
- ✓ Government departments.

As South African consumers become more aware of the possible environmental impacts associated with the products they buy, they will begin to ask questions about aspects such as pollution, use of raw materials and toxic substances, and waste management. Developing an environmental programme for the organization allows it to be well prepared for public attention to these matters when it arises.

• ENVIRONMENTAL ACTS, REGULATIONS AND GUIDELINES

The primary environmental management legislations are the National Environmental Management Act, 1998 (NEMS) and the National Environment Laws Amendment Act No 14, 2009. Other relevant legislations include:

- Atmospheric Pollution Prevention Act, 1965
- Environment Conservation Act, 1989
- National Environmental Management: Protected Areas Act, 2003
- National Environmental Management: Biodiversity Act, 2004
- National Environmental Management: Air Quality Act, 2004

The National Environmental Management System (NEMS) seeks to integrate the rather fragmented environmental laws already in place. Additional policy initiatives of relevance to industrial environmental managers include:

- ✓ Integrated Pollution Control (Sub directorate: Pollution Control).
- ✓ Environmentally Safe Management of Hazardous Chemicals (Sub directorate: Hazardous Substances).
- ✓ Use of Environmental Resource Economics in Environmental Management (Sub directorate: Environmental Resource Economics).
- ✓ Proposed Regulations on Preparation of Environmental Impact Reports (Sub directorate: Integrated Environmental Management).
- ✓ Environmental Self-regulation in South Africa (Sub directorate: Self-regulation).

The Act includes some environmental provisions. It is difficult to be familiar with all laws relating to the environment, as there are many. Local authorities are also empowered to issue byelaws, which may govern, for example, the quality of effluents entering their municipal water treatment works.

A list of important national Acts, Regulations and Guidelines are provided in the next table. They are provided as a guideline to be consulted while being involved in SHE management. It is by no means a complete list of environmental laws but encompasses some of the issues covered in the Acts.

Acts

ACTS RELEVANT TO ENVIRONMENTAL ISSUES

ACT	DESCRIPTION
Environmental Conservation Act, 73 of 1989	Controls various polluting activities, including littering, the disposal of solid waste, and noise.

ACT	DESCRIPTION
Water Act, 54 of 1956	Concerns the use of water, as well as the treatment and disposal of wastewater. The Act provides for regulations governing the use of water, as well as quality standards for effluent disposal. The Dumping at Sea Control Act, 73 of 1980 and the Seashore Act, 29 of 1935 provide for coastal environmental issues.

Hazardous Substances Act, 15 of 1973	Provides for the use and disposal of substances, which may be hazardous to human health due to their toxic, corrosive, irritant, strongly sensitizing or flammable nature. The Petroleum Products Act, 120 of 1977 provides for special requirements governing petroleum products.
Atmospheric Pollution Prevention Act, 45 of 1965	Provides for the prevention of pollution of the atmosphere, and for the implementation of regulations to achieve this end.

Regulations

The following table provides an overview of regulations relevant to environmental issues:

REGULATIONS RELEVANT TO ENVIRONMENTAL ISSUES

ISSUE	REGULATION	REFERENCE
General:	Environmental Regulations for Workplaces	GN R1593 GG 11458 of 12 August 1988, as amended
	Draft Regulations on Environmental Impact Assessments	GN R171 GG 15529 of 4 March 1994, and GN R1018 GG 16527 of 14 July 1995 (Provides for a procedure for identifying and addressing environmental issues during the planning process.)
	Identification of activities which may have a detrimental effect on the environment	GN R1017 GG 16527 of 14 July 1995 (in terms of the Environmental Conservation Act, Section 21(1)).
	Draft Major Hazard Installation Regulations	GN R1097 GG 16574 of 28 July 1995.
Hazardous materials:	Asbestos Regulations	GN R773 GG 10693 of 10 April 1987, as amended.
General:	Lead Regulations	GN R586 GG 13082 of 22 March
	Declaration of Group I hazardous substances	GN R2777 GG 9533 of 21 December 1984
	Declaration of Group II hazardous substances	GN R1382 GG 15907 of 12 August 1994
	Declaration of Group III hazardous substances	GN R689 GG11823 of 14 April 1989
	Declaration of Group IV hazardous substances	GN R246 GG 14596 of 26 February 1993
	Determination of policy on hazardous waste management	GN R1064 GG 15987 of 30 September 1994
Water	Requirements for the purification of waste water (or effluent)	GN R991 GG 9225 of 18 May 1984
	Regulations pertaining to measures aimed at the prevention of water pollution from the operations of mines and works	GN R287 GG4989 of 20 February 1976)

Guidelines

Guidelines do not have the force of law, but are often written as recommendations for good management practice. As the environmental management field is a new one, guidelines often provide extremely useful information on various topics.

GUIDELINES RELEVANT TO ENVIRONMENTAL ISSUES

ISSUE	REGULATION	REFERENCE
General	Integrated Environmental Management Guidelines Series	Department of Environmental Affairs and Tourism
Air Pollution	Internal Guidelines for Scheduled Processes; Internal Ambient Guidelines	Department of Environmental Affairs and Tourism (Air Pollution Control Directorate)
	Guidelines for the design, installation, and operation of incinerators	Department of Environmental Affairs and Tourism: (Directorate Air Pollution Control)
Water Quality	South Africa Water Quality Guidelines	Department of Water Affairs and Forestry
	SABS Water Standards	South African Bureau of Standards
Waste management	Minimum Requirements for Waste Disposal by Landfill, Minimum Requirements for the Handling and Disposal of Hazardous Waste, and Minimum Requirements for Monitoring at Waste Management Facilities	Department of Water Affairs and Forestry
Hazardous materials	SABS 087: The Handling, Storage and Distribution of Liquefied Petroleum Gas in Domestic, Commercial and Industrial Installations	South African Bureau of Standards
	SABS 0228: Identification and classification of dangerous substances and goods	South African Bureau of Standards

Monitoring compliance

The purpose of monitoring compliance to the requirements for a safe and healthy working environment is to ensure that workers in the workplace

- ✓ apply healthy and safe and environmentally friendly working practices
- ✓ wear the correct PPE
- ✓ conform to good housekeeping practices

To protect the environment, it is also necessary that workplace procedures be evaluated to ensure that the environment is not harmed, for example, when disposing of waste.

• ENVIRONMENTAL IMPACT OF TOURISM

Extensive damage to the environment is caused by recreation and tourism, including

- ✓ despoiling of coastlines by construction of tourist facilities;
- ✓ pollution of the sea;
- ✓ loss of historic buildings to make way for tourist facilities; and
- ✓ loss of agricultural land for airport development.

• ENVIRONMENTAL IMPACT OF ENERGY

Energy and environmental problems are closely related, since it is nearly impossible to produce, transport, or consume energy without significant environmental impact.' These challenges involve the production and consumption of energy and include air pollution, water pollution, thermal pollution, and solid waste disposal.

In general, environmental problems increase with energy use and combined with the limited energy resource base is the crux of the energy crisis. An energy impact assessment should compare these costs with the benefits to be derived from energy use.

• ENVIRONMENTAL IMPACT OF TRANSPORT

Impact of transportation-related activities on the environment, in particular, those impacts dealing with air pollution, noise, displacement of people and businesses, disruption of wildlife habitats, and overall growth-inducing effects.

• ENVIRONMENTAL IMPACT OF HOUSEHOLDS

Household impacts on the environment include

- ✓ domestic heating emissions such as hot air, carbon dioxide, carbon monoxide, water vapour and oxide of nitrogen, Sulphur and other trace gases;
- ✓ domestic sewage consisting of human bodily discharges,
- ✓ water from kitchens, bathrooms and laundries;
- ✓ the dumping of bulky wastes such as old washing machines, refrigerators, cars and other objects that will not fit into the standard dustbin and which are often dumped in uncontrolled open spaces.

• TRADE IMPACT ON ENVIRONMENT

Trade impacts on the environment can be direct, such as trade of endangered species, of natural resources, and of natural products such as tropical timber, or indirect, such as deforestation, loss of habitats, pollution from mining, from energy production, oil spills, and global warming, and increases in transport infrastructures.

• SOCIAL RESPONSIBILITY

Social responsibility is an ethical theory that an entity, be it an organisation or individual, has an obligation to act to benefit society at large. Social responsibility is a duty every individual has to perform so as to maintain a balance between the economy and the ecosystems. Corporate Social Responsibility (CSR) is the continuing commitment by businesses to behave ethically and contribute to economic development while improving the quality of life of the workforce and their families as well as the local community and society at large.

CSR is one of the newest management strategies where companies try to create a positive impact on society while doing business. There is no clear-cut definition of what CSR comprises. Every organization has different CSR objectives although the primary motive remains constant. Typically, CSR could be:

- environmental management, such as waste reduction and sustainability
- responsible sourcing, such as using only fair trade ingredients
- improvement of working standards and conditions
- contributing to educational and social programmes
- worker volunteering
- socially responsible investment
- development of worker and community relations

All organization generally aim:

- ✓ to improve qualitatively (the management of people and processes)
- ✓ and quantitatively (the impact on society).

The second is as important as the first and stakeholders of every organization are increasingly taking an interest in "the outer circle" which encompasses the activities of the organization and how these impacts the environment and society. Organizations, particularly "green" organizations have made social responsibility an integral part of their business models. Environmental CSR aims to reduce any damaging effects on the environment from business processes and activities may focus on:

- energy use
- water use
- waste management
- recycling
- emissions
- eco-friendly office and business travel policies
- reducing waste and pollution
- employing local people
- becoming involved in charity work
- encouraging education in the community

Green CSR can reduce business risk, improve reputation and provide opportunities for cost savings. Even the simplest energy efficiency measures can generate savings and make a difference to the business of the organization.

The impact on the environment can be reduced in many ways. For example, the organization can:

- create products that can be recycled
- optimise the product life cycle
- source responsibly such as, for example, using recycled materials and sustainable timber
- reduce packaging
- buy locally to save fuel costs
- create an efficient and fuel-efficient distribution network
- work with environmentally conscious suppliers and distributors

Some of the business benefits of good environmental and sustainable practices include:

- Reducing energy use can save the organization money
- Reducing the use of raw materials can cut costs, minimise waste and lower waste disposal costs
- Reducing water use can lower water bills and also lower waste water disposal costs
- Recycling equipment and materials could generate income and save on disposal costs – the organization may be able to sell old equipment or parts to subsidise the cost of buying new ones
- Taking steps to prevent environmental damage will reduce the risk of incurring costs and of having to carry out potentially expensive remedial work in the event of an environmental incident.

The organization could also enjoy other benefits. Many government departments and other large organizations will only work with businesses who take their environmental responsibilities seriously. Further, investors use the social responsibility - or lack thereof – of an organization as an investment criterion. For example, an investor who has a moral or other objection to smoking may not want to invest in a tobacco company or one that manufactures tobacco products. It might be possible to more easily attract investment from banks and other sources of finance.

As well as making financial savings, the organization can enhance its reputation and brand value by fostering customer loyalty and motivating workers. Environmental CSR can be used as an opportunity to encourage innovation, increase investment and open new markets.

• ENVIRONMENTAL TRAINING AND AWARENESS

Training and awareness need to take place at all levels of the organisation. It is not sufficient to simply inform top or senior management of strategic environmental issues in the hope that the awareness will filter down to all levels. Too frequently, it does not. In the same vein, it is difficult to expect results from well-trained workers, if the management structure does not provide an incentive by integrating environment into performance evaluations.

In South Africa, environmental issues are often seen as a "luxury", far removed from the basic needs of access to housing, services and personal safety. Many people make the mistake of confusing "environmental issues" with "conservation of the rhino". While conservation of endangered species is important, it is only one facet of environmental responsibility. Environmental issues in fact concern basic needs and resources, and having a healthy and safe world to live and breathe in.

Environmental awareness is primarily about a change in attitude and behaviour. To achieve this goal implies that training cannot be done overnight, or by conducting a once-off training course. The organization needs to create an atmosphere of co-operation and an understanding of common environmental goals shared by workers and management. This will only be achieved if workers and their representatives in partnership with management are directly involved in planning and deciding on the necessary training programmes.

Environmental training and awareness need to be part of any long-term programme for all workers. By combining environmental training with safety and health, workers will be able to identify common elements and an understanding of the integrated nature of safety, health and environmental management.

Possible elements of an environmental training and awareness programme are outlined in the table.

ENVIRONMENTAL TRAINING AND AWARENESS PROGRAMME

ELEMENT	EXPLANATION
Environmental awareness for all workers	Induction training and ongoing awareness promotions through displays, competitions, celebration of special environmental days, and focus weeks such as, for example, on reducing waste.
Training for SHE representatives	SHE representatives should undergo integrated SHE training by attending courses
Training for on-site environmental professionals	Professionally trained environmental managers should be encouraged to keep up with new developments through attendance of conferences, subscriptions to journals, and use of Internet access to environmental news
Management and union leadership training	Leadership seminars can be arranged on specific topics, such as environmental management systems, waste minimisation, pollution prevention and hazardous chemical management. A focus on new legal developments and possible liabilities and opportunities for cost savings will always be of interest.
Senior executive training	Executive seminars can be arranged to make senior executives aware of strategic issues, such as environmental compliance for international trade, new policy developments, and the corporate image benefits of improved environmental performance.

As with many industrial training programmes, one possible way to increase the involvement and participation of workers could be through the establishment of a well-thought out and designed incentive schemes. Usually, the person best placed to effect environmental improvements is the worker on the shop floor. These workers are keenly aware of what waste is being produced by which processes, of hazardous chemicals which are used indiscriminately, or of the extent to which water and energy are wasted on a day-to-day basis. By offering some tangible incentive to the worker who identifies opportunities for environmental improvement, many creative ideas might be developed. The incentive could be financial such as, for example, based on a percentage of cost saving waste reduction ideas implemented or may simply be in the form of recognition. This may be achieved through regular performance evaluation channels or by special mention in departmental meetings. However, a cautious approach to incentive schemes should be adopted.

• REVIEW QUESTIONS

1. How would an organization benefit from having an environmental management program in place?
2. Develop an environmental management plan (EMP) for your particular workplace that demonstrates commitment to sustainable and eco-friendly business practices with some emphasis on corporate social responsibility (CSR).
3. What are the advantages of treating safety, health and environment in an integrated manner?
4. What are some of the benefits of environmental CSR?
5. What elements could be considered in an effective training and awareness program?
6. Why should a cautious approach be followed with respect to incentive schemes to increase involvement and participation of workers in improving environmental management at work?
7. Develop an environmental management plan (EMP) for your particular workplace that demonstrates commitment to sustainable and eco-friendly business practices with some emphasis on corporate social responsibility (CSR).

• REFERENCES

Anonymous (n.d.). Loose notes on safety, health and environment management

Borkovskaya, Victoria., Degaev, Evgeniy. and Burkova, Irina., 2018. Environmental economic model of risk management and costs in the framework of the quality management system. In *MATEC Web of Conferences* (Vol. 193, p. 05027). EDP Sciences.

Camarda Domenico. (ed.), Grassini Laura. (ed.). Local resources and global trades: Environments and agriculture in the Mediterranean region. Bari: CIHEAM, 2003. p. 263-270 (Options Méditerranéennes : Série A. Séminaires Méditerranéens; n. 57)

Choi, Seok-Beom., Min, Hokey., Joo, Hye-Young. and Choi, Han-Byul., 2017. Assessing the impact of green supply chain practices on firm performance in the Korean manufacturing industry. *International Journal of Logistics Research and Applications*, *20*(2), pp.129-145.

Chun-sheng, Zhao., Shu-wen, Niu. and Xin, Zhang. (2012). Effects of household energy consumption on environment and its influence factors in rural and urban areas. Energy Procedia, 14, pp.805-811.

Copeland Brian. R. (2010) How Does Trade Affect the Environment?. In: Heal G. (eds) Is Economic Growth Sustainable?. International Economic Association Series. Palgrave Macmillan, London

DEAT (2004) Environmental Management Plans, Integrated Environmental Management, Information Series 12, Department of Environmental Affairs and Tourism (DEAT), Pretoria.

Department of Environmental Affairs (1965). Atmospheric Pollution Prevention Act No. 45 of 1965.

Department of Environmental Affairs (1989). Environmental Conservation Act No 73 of 1989.

Department of Environmental Affairs (1998). National Environmental Management Act No: 107 of 1998.

Department of Labour (1993). Occupational Health and Safety Act No: 85 of 1993.

Davy, A. (1999). *Environmental Assessment UPDATE Sourcebook*. 25th ed. Washington: Environment Department, The World Bank, pp.1-8.

Dubravská, Mariana., Marchevská, Martina., Vašaničová, Petra. and Kotulič, Rastislav., 2020. Corporate Social Responsibility and Environmental Management Linkage: An Empirical Analysis of the Slovak Republic. *Sustainability*, *12*(13), p.5431

European Environment Agency. 2004. Environmental Impact of Energy. Available at: https://www.eea.europa.eu/help/glossary/eea-glossary/environmental-impact-of-energy

General Multilingual Environmental Thesaurus. (n.d.) Environmental impact of Households. Available at: https://www.eionet.europa.eu/gemet/en/concept/2836

General Multilingual Environmental Thesaurus. (n.d.) Environmental impact of tourism. Available at: https://www.eionet.europa.eu/gemet/en/concept/2839

General Multilingual Environmental Thesaurus. (n.d.) Environmental impact of Transport. Available at: https://www.eionet.europa.eu/gemet/en/concept/2840.

Klassen R.D. (2000) Environmental Issues and Competitive Manufacturing. In: Swamidass P.M. (eds) Innovations in Competitive Manufacturing. Springer, Boston, MA. https://doi.org/10.1007/978-1-4615-1705-4_30

Law, Michelle. Man. Suet., Hills, Peter. and Hau, Billy. Chi. Hau., 2017. Engaging employees in sustainable development–a case study of environmental education and awareness training in Hong Kong. *Business Strategy and the Environment*, *26*(1), pp.84-97.

Lochner, P. 2005. Guideline for Environmental Management Plans. CSIR Report No ENV-S-C 2005-053 H. Republic of South Africa, Provincial Government of the Western Cape, Department of Environmental Affairs & Development Planning, Cape Town.

Makelane, Hlamulo., 2016. The Impact of Environmental Management Plans on Firms' Compliance with Governmental Regulations and Environmental Laws. *Editorial Team*, p.30.

NiBusinessInfo. (n.d.). Corporate Social Responsibility: Environmental Impact. Available at: https://www.nibusinessinfo.co.uk/content/corporate-social-responsibility-environmental-impact

Omer, A. (2009). Energy use and environmental impacts: A general review. *Journal of Renewable and Sustainable Energy*, 1(5), p.053101.

Rau, G. J. & Wooten, D. C., 1990. Environmental Impact Analysis Handbook, McGraw-Hill, U.S.A., 1990

Schaltegger, Stefan., Burritt, Roger. and Petersen, Holger., 2017. *An introduction to corporate environmental management: Striving for sustainability*. Routledge.

Sharma, R. and Kumar, S. (2012). Impact of Transportation System on Environment in Developing Countries "A Review". *IJRREST: International Journal of Research Review in Engineering Science and Technology*, 1(2), pp.61-66.

Severo, Eliana. Andrea., de Guimarães, Julio. Cesar. Ferro. and Dorion, Eric. Charles. Henri., 2017. Cleaner production and environmental management as sustainable product innovation antecedents: A survey in Brazilian industries. *Journal of Cleaner Production*, *142*, pp.87-97.

Shahzad, Mohsin., Qu, Ying., Javed, Saad. Ahmed., Zafar, Abaid. Ullah. and Rehman, Saif. Ur., 2020. Relation of environment sustainability to CSR and green innovation: A case of Pakistani manufacturing industry. *Journal of Cleaner Production*, *253*, p.119938.

Suganthi, L., 2019. Examining the relationship between corporate social responsibility, performance, employees' pro-environmental behavior at work with green practices as mediator. *Journal of cleaner production*, *232*, pp.739-750.

Union of South Africa (U of SA) 1956. Water Act, No. 54, in U. of SA, Statutes of the Union of South Africa 1956. Part II Nos. 47–73. Cape Times Ltd, under supervision of the Government Printer, Parow, pp. 1046–1305.

CHAPTER 23

ENVIRONMENTAL POLLUTION AND EFFECTS

Pollution may be defined as:

The residue of any activity in the environment, which may be a nuisance or cause damage to life or materials.

It refers to any substance, emissions, product or by-product, which could decrease the quality of air, water or ground resources in any way. Substances giving rise to pollution are referred to as 'pollutants'.

According to the Pollution Management and Environmental Health program of the WHO, air, land, and water pollution caused 9 million premature deaths in 2016, or 16% of all deaths worldwide. About 92% of all pollution-related mortality is seen in low-income and middle-income countries, with the poor, marginalized, and young hardest hit by the health effects of the contamination. The economic burden is immense: in 2016, ambient air pollution alone cost the global economy US$5.7 trillion—4.8 percent of global GDP. Responding to pollution is a challenge that is solvable if everyone acts now to reduce it in all its forms. An integrated pollution and waste management approach is needed to limit the damage of pollution to the environment by workplaces. Integrated pollution and waste management is a holistic and integrated system and process of management, aimed at pollution prevention and minimisation at source, managing the impact of pollution and waste on the receiving environment and remediating damaged environments.

Throughout the world there are various types of pollution that interfere with the quality of life for all living creatures and with the natural functioning of the earth's ecological systems. Although some environmental pollution is a result of natural causes such as methane emissions from cattle and toxic materials expelled from volcanoes, sadly most pollution is caused by human activities.

• HUMAN INDUSTRIAL ACTIVITIES

The greatest source of pollution is the industrial manufacturing community. An industry is a collection of organizations that operate in a related set of goods or services, which are eventually sold to purchasers and users. Generally, industries are divided into three groups, namely:

- ✓ *Primary industries* which are involved in the collection, utilizing, and harvesting of resources directly produced by physical processes such as, for example, mining and smelting.
- ✓ *Secondary industries* which deal with manufacturing as these industries take raw materials, convert them in various ways, and produce tangible goods such as automobile factories and assembly plants.
- ✓ *Tertiary industries* which produce services for individuals and groups such as advertising and information technology service providers.

Each of these groups of industries produce differing amounts of pollution during their operations:

- ✓ Sectors such as tourism because of its nature and types of activities have a close relationship with the environment
- ✓ The motor vehicle industry, for example, has become environmentally proactive by considering and implementing innovative design and technologies and alternative energy sources such as production of electric cars
- ✓ The transport industry continues to have a negative impact on the environment such as carbon exhaust emissions and waste generation.

The largest impact from pollution and associated waste products is produced within the secondary industries

The following industries have some of largest polluters, namely:

- ✓ chemical,
- ✓ pesticide,
- ✓ oil refining,
- ✓ construction,

- ✓ petrochemical,
- ✓ metal smelting,
- ✓ iron and steel,
- ✓ agriculture, and
- ✓ food processing.

These industries are major users of energy that produce large amounts of waste products and pollution.

Other industries impact the environment less through their activities but are still considered highly problematic and include the following industries, namely:

- ✓ textile,
- ✓ leather tanning,
- ✓ paint,
- ✓ plastics,
- ✓ pharmaceutical, and
- ✓ paper and pulp.

The effects on the environment connected with industrial activities are mainly related to the production of industrial wastes that can be divided into various types:

- ✓ solid waste, such as dust particles or slag from coal;
- ✓ liquid wastes from various processes, including radioactive coolants from power stations;
- ✓ gas wastes, largely produced by the chemical industry.

As a result of rapid developments and innovations due to 4IR, new technologies can intersect with biology to derive new bio-economy opportunities. An example in the "automated synthesis machine", which combines computational chemistry, robotics and artificial intelligence to rapidly create and analyse new biological assets. The environmental impact of these developments and activities by emerging industries will have to be managed effectively to ensure no unintended consequences occur despite their positive outcomes and effects on quality of life.

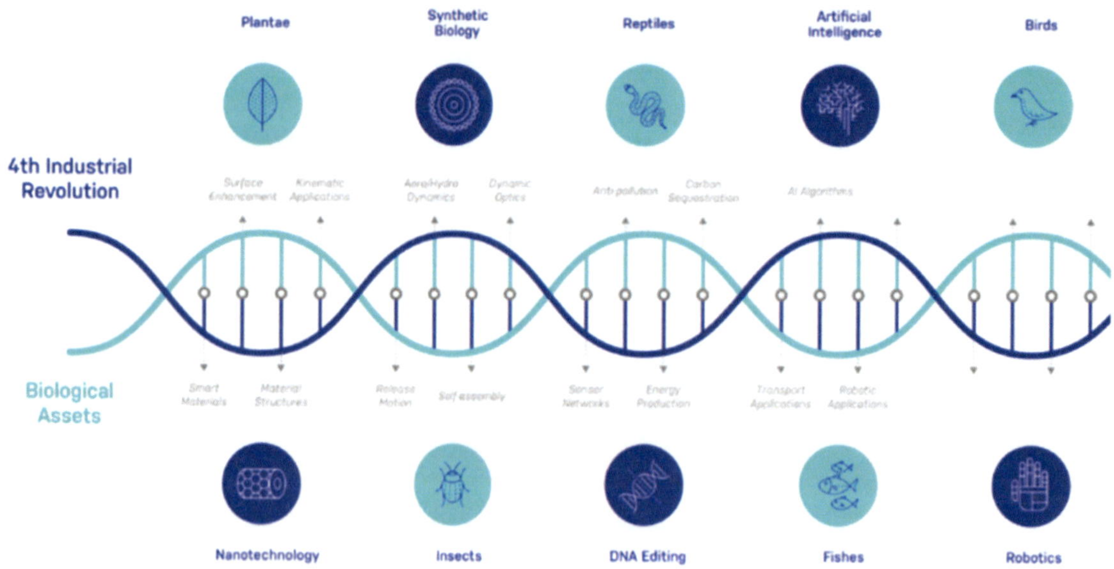

(Source: http://www3.weforum.org/docs/WEF_Harnessing_4IR_Life_on_Land.pdf)

• AIR POLLUTION

Air pollution is the contamination of the environmental air due to the introduction of unhealthy substances.

Air pollution is the contamination of the air by noxious gases and minute particles of solid and liquid matter or particulates in concentrations that endanger health. The major sources of air pollution are engines, power and heat generation, industrial processes and the burning of solid waste.

Air pollution is an undesirable change in the physical, chemical or biological characteristics of air.

Air pollutants are the substances which pollute the air.

Some of the most common industrial pollutants include:

- ✓ dust,
- ✓ soot,
- ✓ ash,
- ✓ carbon monoxide,
- ✓ excess of carbon dioxide,
- ✓ sulphur dioxide,
- ✓ oxides of nitrogen,
- ✓ hydrocarbons, chlorofluorocarbons (CFC),
- ✓ lead compounds,
- ✓ asbestos dust,
- ✓ cement dust,
- ✓ pollens
- ✓ radioactive rays

• SOURCES OF AIR POLLUTION

The pollution of air can be caused by natural processes or by human activities.

There are few human activities, which do not result in emissions. Industry, agriculture, burning of forests, and mining contribute in various ways to the problem of emissions and air pollution.

The combustion of petrol and other hydrocarbon fuels in motor vehicles, trucks and aircraft produces pollutants such as nitrogen oxides, gaseous hydrocarbons and carbon monoxide as well as quantities of lead particulates. In the presence of sunlight, nitrogen oxides combine with hydrocarbons when the sun shines to form photochemical oxidants which include ozone and the eye-stinging peroxyacetylnitrate (PAN). Nitrogen oxides react with oxygen in the air to form nitrogen dioxide which is a foul-smelling brown gas. Air pollutants may occur indoors or outdoors. Indoor air pollution poses a serious health hazard, as levels can build up more rapidly.

Air may be heavily polluted in urban areas by transportation but also by the burning of fossil fuels such as oil and coal in electricity power generating stations, factories, office buildings and homes and by the incineration of waste. The massive combustion produces tons of ash, fly ash, soot and other particulates along with quantities of sulphur oxides. The effects of these oxides include causing metal to rust or tarnish, nylon to be decomposed, and vegetation to be killed. Nuclear power plants pollute air by releasing radioactive rays.

Almost every industrial process pollutes the air. Petroleum refineries generate hydrocarbon and particulate pollution. Industries such as, for example, iron and steel plants, metal smelters, pulp and paper mills, chemical plants, sugar mills, cement and asphalt plants discharge vast amounts of various particulates. Uninsulated high-voltage power lines ionize the adjacent air, forming ozone and other hazardous pollutants. Airborne pollutants from other sources include insecticides, herbicides, radioactive fallout and dust from fertilizers, mining operations and livestock feedlots. Indiscriminate cutting of trees and clearing of forests increases the amount of carbon dioxide in the atmosphere and thereby pollutes it. Use of chlorofluorocarbons in refrigeration, fire extinguishers and aerosol sprayers pollutes air by depleting the ozone layer. Smoking pollutes air by emitting carbon monoxide and nicotine.

The sources of air pollution can be classified into two groups:

- ✓ Natural Sources
- ✓ Man-made sources

Natural sources include dust storms, forest fires, ash from smoking volcanoes, decay of organic matter and grains of pollen floating in air. Man-made sources on the other hand include the consequences of urbanization and population explosion, deforestation, and industrialisation. The following table provides a few examples of common pollutants and their sources:

COMMON AIR POLLUTANTS AND THEIR SOURCES

POLLUTANT	POSSIBLE SOURCE
Carbon Monoxide	Fuel combustion in motor vehicles
Hydrocarbons	Fuel combustion in motor vehicles, processing of fuels, use of solvents and petroleum products
Lead	Additives to petrol, mining, smelting, iron and steel production, battery manufacturing
Nitrogen Oxides	Fuel combustion, chemical processing industry, industrial heating
Particulate Matter, including asbestos dust	Combustion, grinding and spraying operations, wind erosion, wood combustion, mine dumps
Sulphur Oxides	Combustion of coal, smelting of sulphide ores
Volatile Organics (include benzene, aromatics, dioxins)	Use of solvents

Extensive reviews of the health effects of air pollution have been carried out in recent years. There is a huge body of literature and published research around the world that focuses on the negative impacts of all forms of pollution especially air pollution. In particular, in South Africa the relationships to respiratory illnesses and other health complications have been established in research carried out by, for example, the CSIR and Medical Research Council. The following table highlights the potential health effects of a selection of indoor air pollutants:

POTENTIAL HEALTH EFFECTS OF INDOOR AIR POLLUTANTS

POLLUTANT	MAJOR INDOOR SOURCE	POTENTIAL HEALTH EFFECTS
Tobacco smoke	Cigarettes, pipes, cigars	Respiratory irritation, bronchitis and pneumonia in children, emphysema, lung cancer and heart disease
Carbon monoxide	Unvented or malfunctioning gas appliances, vehicles or forklifts running indoors	Headaches, nausea, angina, impaired vision and mental functioning, fatal at high concentrations
Ozone, nitrogen oxides and iron	Welding fumes	Headaches, dryness of the eye membranes, impaired lung function, pulmonary congestion, haemorrhage
Asbestos	Damaged or deteriorating insulation, fire-proofing and acoustic materials	Asbestosis, lung cancer, mesothelioma, and other cancers
Lead	Sanding or open-flame burning of lead paint, dust industrial lead applications	Nerve and brain damage, anaemia, kidney damage, growth retardation
Radon	Soil under buildings, some earth-derived construction materials and groundwater	Lung cancer
Organic chemicals	Aerosol sprays, solvents, glues, cleaning agents, pesticides, paints	Eye, nose and throat irritation, headaches, loss of co-ordination, damage to liver, kidney and brain, various types of cancer

In South Africa, the National Environment Management: Air Quality Act 39 of 2004 (AQA), provides for the control of all air pollutants in South Africa. An atmospheric emission license is required to carry out any of the listed processes in the schedule under the AQA. The license is issued subject to certain minimum standard specifications. The polluting industry is expected to adopt the "best practicable means" of minimising the emission of noxious gases to the atmosphere. "Best practicable means" refers to measures which are "technically possible and economically feasible", bearing in mind the importance of the well-being of workers and the surrounding communities.

An atmospheric emission licence will reflect the following:

- the activity in respect of which it is issued;
- the premises in respect of which it is issued;
- the person to whom it is issued;
- the period for which the license is issued;
- the name of the licensing authority;
- the periods at which the license may be reviewed;
- the maximum allowed amount, volume, emission rate or concentration of pollutants that may be discharged in the atmosphere-
  - under normal working conditions; and
  - under normal start-up, maintenance and shut-down conditions;
- any other operating requirements relating to atmospheric discharges, including non-point source or fugitive emissions;
- point source emission measurement and reporting requirements;
- on-site ambient air quality measurement and reporting requirements;
- penalties for non-compliance;
- greenhouse gas emission measurement and reporting requirements; and
- any other matters which are necessary for the protection or enforcement of air quality.

There are several technologies and approaches that can be used to control pollution from various pollutants. Examples of these are shown in the following table:

OVERVIEW OF AIR POLLUTION CONTROL TECHNOLOGIES

POLLUTANT	TECHNOLOGY	CONTROL MEASURE
Particulate matter	Source controls (pollution prevention	Fuel substitution; process feed materials; change of process variables.
	Setting chambers	Enlarged section in exhaust chambers causes decrease in gas velocity. This permits coarser particles to fall out of suspension.
	Separators	Through centrifugal force, separators are used to remove heavier particles.
	Wet scrubbers	These devices wet particles in order to remove them from the gas stream.
	Fabric filters	Suspended tubes of woven or felted fabric are placed in a housing structure (baghouse) to filter dry particles.
	Electrostatic precipitators	Use electrical energy to charge and collect particles.
Gas controls (for example, VOCs, sulphur and nitrogen	Source controls (pollution prevention)	Fuel substitution; fuel cleaning; efficiency improvements through process modifications.
	Absorption	A chemical process in which the gas contaminant is dissolved into a relatively non-volatile liquid (for example, flue gas desulphurisation).
	Adsorption	A gas molecule is retained by either chemical or physical forces onto an adsorbing material (for example, activated carbon).
	Condensation	Pollutants are condensed out of the gas (turned to liquid phase) by cooling.
	Flaring	The burning of waste gases which are not economical to recover.
	Incineration	Incinerators use combustion to convert volatile gases and particulates to carbon dioxide, water and ash. External sources of fuel are needed.

Monitoring

Monitoring stations operated by local authorities, and industries such as chemical, refineries, energy generation and quarries typically continually assess levels of key air pollutants such as sulphur dioxide, particulate matter, dust and smoke.

• HARMFUL EFFECTS OF AIR POLLUTION

Air pollution has unhealthy effects on people, animals and plant life all across the world and South Africa is no exception. These pollutants can cause short-term effects such as eye and throat irritation. The long-term effects include cancer and damage to the immune, neurological, reproductive and respiratory systems of the body. The mechanisms by which air pollutants can cause serious adverse health effects are not yet fully understood. However, many possible mechanisms have been proposed. The possible mechanism related to lung disease, for example, is that small particles penetrate the alveolar epithelium and cause lung inflammation. If a worker already has damaged lungs or lung disease, then increased inflammation as a result of air pollution will worsen their condition. Two potential mechanisms have been proposed for causing heart attacks, namely:

- Lung inflammation can lead to the release of chemicals from macrophages that cause changes in the clotting mechanism of the blood.
- A neural reflex may be initiated by the irritant effects of air pollutants in the lung leading to changes in heart rhythm and rate.

A summarized list of the harmful effects of air pollution includes the following:

- Affects the respiratory system causing breathing difficulties and diseases such as bronchitis, asthma, lung cancer, tuberculosis and pneumonia.
- Affects the central nervous system causing carbon monoxide poisoning. CO_2 has more affinity for haemoglobin than oxygen and forms a stable compound carboxy haemoglobin (COHb), which is poisonous and causes suffocation and death.
- Causes depletion of the atmospheric ozone layer resulting in ultraviolet radiations reaching the earth and causing skin cancer, damage to eyes and immune system.
- Causes acid rain which damages crop plants, trees, buildings, monuments, statues and metal structures and also makes the soil acidic.
- Causes the greenhouse effect or global warming which leads to excessive heating of the atmosphere, which contributes to climate change, weather variability and rising sea levels. A global average temperature rise of only 1°C could have serious implications The increased temperature may cause melting of ice caps and glaciers, resulting in floods in various parts of the world, increases in precipitation and severe weather events like hurricanes, tornadoes, heat waves, floods and droughts. Indirect effects include increases in infectious disease, weather-related deaths and food and water shortages.
- Causes serious ailments from the kidney, liver, circulatory system and nervous system from the air pollution from certain metals, pesticides and fungicides.
- Causes anaemia, brain damage, convulsions and death from lead pollution.
- Causes nerve damage and death from fungicides and pesticides such as DDT (Dichloro diphenyl trichloroethane) which enter the food chain and cause kidney disorders and problems of brain and circulatory system.
- Affects many different ecosystems where pollutants remain toxic in the affecting dams, lakes, ponds, streams, fields and forests.

• PREVENTION AND CONTROL OF AIR POLLUTION

Air pollution control technologies

An emissions survey may be conducted to ascertain the need for air pollution control. This survey will locate sources and define quantities of all air contaminants. Before adoption and implementation, control technologies need to be assessed in terms of their:

- ✓ Suitability for the removal of pollutants.
- ✓ Organizational capital and operating costs.
- ✓ Operating efficiencies in the plant.
- ✓ Waste disposal practices.

Options that result in pollution prevention, rather than control, should be sought. Different techniques are used for controlling air pollution caused by gaseous and particulate pollutants respectively.

Methods of controlling gaseous pollutants:

The air pollution caused by gaseous pollutants like hydrocarbons, sulphur dioxide, ammonia, and carbon monoxide can be controlled by using the following methods, namely:

1. Combustion: This technique is applied when the pollutants are organic gases or vapours. The organic air pollutants are subjected to 'flame combustion or catalytic combustion' when they are converted to less harmful product carbon dioxide and a harmless product water.
2. Absorption: In this method, the polluted air containing gaseous pollutants is passed through a scrubber containing a suitable liquid absorbent. The liquid absorbs the harmful gaseous pollutants present in air.
3. Adsorption: In this method, the polluted air is passed through porous solid adsorbents kept in suitable containers. The gaseous pollutants are adsorbed at the surface of the porous solid and clean air passes through.

Methods of controlling particulate emissions:

The air pollution caused by particulate matter like general and saw dust, soot, and fly ash can be controlled by using fabric filters, wet scrubbers, electrostatic precipitators and certain mechanical devices.

1. Mechanical devices work on the basis of following:
   - Gravity in terms of which the particulates settle down by the action of gravitational force and get removed.
   - Sudden change in the direction of air flow that brings about the separation of particles due to greater momentum.
2. Fabric filters where the particulate matter is passed through a porous medium made of woven or filled fabrics.
   - The particulate present in the polluted air are filtered and gets collected in the fabric filters, while the gases are discharged.
   - The process of controlling air pollution by using fabric filters is called 'bag filtration'.
3. Wet scrubbers are used to trap SO_2, NH_3 and metal fumes by passing the fumes through water.
4. Electrostatic precipitators where the polluted air containing particulate pollutants is passed through an electrostatic precipitator and induces an electric charge on the particles resulting in the aerosol particles being precipitated on the electrodes.

Other methods of control include:

- Better designed equipment and smokeless fuels in homes and industries.
- Renewable and non- polluting sources of energy like solar energy, and wind energy
- Properly maintaining and adhering to emission control standards of motor vehicles.
- Planting wherever possible more trees.

• WATER POLLUTION

Water pollution occurs when contaminants enter the natural groundwater, rivers, streams, lakes, dams, reservoirs or oceans. Most water pollution is caused by humans from industrial waste, fertilizers, untreated sewage, underground storage leaks, marine dumping and oil pollution. Not only does water pollution cause the death of many organisms, it also disrupts entire ecosystems. It has been estimated that more than half of over 300 freshwater species have declined because of water pollution, while 32% of the salt water and land species have declined.

While over 70% of the earth is covered by water only less than 3% is freshwater. Only about 0.3% of fresh water is found in lakes, rivers, and swamps. More than 99 percent of all the water on Earth is unusable by humans and many other living things which leaves a really small percentage of accessible fresh water. This small percentage is what makes the development of water pollution in rivers, lakes, and oceans worldwide so problematic. Fresh water is used for irrigation (70%), industry (22%), and domestic use (8%).

One half of the global population have health difficulties due to water-based diseases which amounts to an estimated 250 million people per year, resulting in between 5 and 10 million deaths. In developing countries, water pollution is a leading cause of death. The most harmful and toxic elements are heavy metals which enter the water system such as cadmium, lead, iron and mercury which hinder the transportation of fluids through the cell wall. They immobilize enzymes as they have strong affinity for sulphur bonds present in enzymes. These substances can produce physiological poisoning by accumulating over the tissues of water creatures. The toxic effects of mercury are neurological damage, birth defects, and chromosome breakage. Arsenic is one of the most important water polluting agents as it is a carcinogen or cancer causing agent. Apart from cancer, it also leads to circulatory problems and respiratory problems. Cyanide exists in water as hydrogen cyanide, and can cause nerve damage and thyroid problems. Ammonia is formed during microbial degradation of biomass and organic matters. The caustic nature of ammonia, damages gills of fish. Hydrogen sulphide which evolves from geothermal waters, and wastes from paper and textile mills, causes nervous system poisoning.

Sources of water pollution

Sewage Waste

According to the UN the amount of wastewater produced annually is six times more than exists in all the rivers of the world. Lack of adequate sanitation contaminates water and is one of the most significant forms of water pollution. Daily 2 million tons of sewage and industrial and agricultural waste are discharged globally into water, the equivalent of the weight of the entire human population of 6.8 billion people. It has been estimated that, for example, a single cruise ship can generate 3.8 million litres of greywater, 796 thousand litres of sewage, 150 thousand litres of oily bilge water, 8,128 kgs of solid waste, and other toxic wastes every week. Tragically, worldwide, 2.5 billion people live without improved sanitation.

Land-based activities

From plastic bags to pesticides, most of the waste produced on land by humans eventually reaches the oceans, either through deliberate dumping or from run-off through drains, streams and rivers. Unfortunately, for centuries and centuries, the oceans have been a convenient dumping ground for waste generated on land by mankind including toxic material such as pesticides, chemical weapons, and radioactive waste. Regularly, activist groups mop up and clean the washed up plastics and litter from the beaches and coastlines of South Africa.

Toxic Material Dumping

Asbestos, metallic elements, lead, and mercury are non-biodegradable and difficult to dispose. Lead and mercury cause problems with human health and development as well animals, and the health and development of ecosystems.

Fertilizers and pesticides

Fertilizers used in agriculture are one of the leading contributor of nitrates and phosphates. 2,4,5-T (trichlorophenoxy acetic acid, which contains the impurity dioxin), 2,4-D (dichlorophenoxyaceticacid) are the two common pesticide components that cause water pollution. There is a significant drive to use more natural and organic fertilizers.

Oils

Pollution from oil kills thousands of animal species with one spill. Runoffs and dumping of oil into the water system happen every day. Oil spills only count for 12% of the oils that enter the ocean however the concentrated and larger amounts cause a great deal of harm. They can destroy whole ecosystems and are not easy or cheap to clean up. Several major oil spills such as the recent incident off the coast of Mauritius create significant threats to marine life for decades after the incident. The following table provides some examples of water pollutants and their sources:

SUMMARY OF KEY WATER POLLUTANTS AND PREDOMINANT SOURCES

POLLUTANT	PREDOMINANT SOURCES
Organic material	Sewage, feedlots, industrial waste
Nutrients	Run-off from fertilised farms, feedlots, sewage
Toxic chemicals	Industrial waste, fertilisers, leachate
Acids	Mine drainage
Colour and foam	Industrial waste
Solids	Erosion, mining effluents, agriculture, building debris
Oils and grease	Industrial waste
Litter	Urban storm run-off
Hot water	Power plants, industries

Water resources

South Africa has a well-established and documented water problem with an average rainfall of less than 500 mm per annum. Although water supplies are renewable, they are not unlimited. Nearly half of the water demand in South Africa comes from irrigation in the agricultural sector. As the South African population increases, the demand for water places increasing strain on rivers, dams, reservoirs and groundwater. Most recently, the Western Cape and Cape Town in particularly experienced unprecedented water shortages that resulted in major restrictions of the use of water. The economic consequences of these restrictions were widespread. The Department of Water Affairs and Forestry has indicated that the quality of many water sources in South Africa is declining.

Impact of water pollutants

The impact of pollutants in water is extremely complex:

- ✓ Fertiliser, which is made up of nutrients to increase plant yield, can be washed off the soil into rivers. Once in the water, these nutrients are just as effective in increasing plant growth, resulting in blooms of algae and pest plants, clogging river systems and dams. This process is called eutrophication.
- ✓ Micro-organisms in decaying organic matter in the case of sewage use up the oxygen in the water, killing normal riverine species.
- ✓ Solid wastes, such as silt and litter, smother animals and can kill their eggs, destroying life cycles.
- ✓ Pesticides from agricultural run-off kill insects whose natural function is to clean the rivers.
- ✓ Toxic chemicals can build up in water supplies, affecting humans and animals. An excess of dissolved mineral salts in water can result in dehydration and renal failure.
- ✓ Radioactive elements such as radon, and radium-226 can be fatal if found in water and consumed. Radon is the most significant element that can cause lung cancer when consumed in a dissolved state or inhaled in its gaseous form. Strontium-90 and tritium are radioactive elements that get dumped into water during nuclear testing. They cause genetic mutations and pose destructive effects on our genes that can pass on to future generations also.

Most diseases that affect human beings are water-borne and include typhoid, cholera, paratyphoid fever, bacillary dysentery, infectious hepatitis, and poliomyelitis.

Legal requirements

The National Water Act 36 of 1998 as amended and National Water Amendment Act 45 of 1999, govern water pollution control. The Act requires that water used in industrial enterprises be purified to a specified standard before being discharged. Various authorities, with most industries falling under a local authority, set standards for discharge. While changes in government will modify the overly complicated and divided administrative structure, the general trend is towards more stringent and integrated control of pollution.

Water quality

Water quality is measured in terms of:

Physical factors:

- ✓ Conductivity.
- ✓ Odour.
- ✓ Taste.
- ✓ Colour.
- ✓ Temperature.
- ✓ Turbidity.

Biological factors:

- ✓ Animal life.
- ✓ Bacteria.
- ✓ Viruses.

Chemical factors:

- ✓ Inorganics (metals, salts, hardness).
- ✓ Organics (oil and grease, pesticides, organo-chlorides).

Biochemical and Chemical Oxygen Demand (BOD or COD) are commonly used measures for assessing total organic materials in the water. These measurements show how much the effluent will lower the oxygen content of a river into which it is discharged. Many local bye-laws stipulate that permission must be obtained in writing from the local authority before any industrial effluent may be discharged to the sewer system.

In obtaining such permission, the authority will typically require:

- ✓ A description of the processes which give rise to the effluent.
- ✓ The nature and quantity of the effluent.
- ✓ The rate of discharge.

In terms of most local authority bye-laws, industrial water users are charged both according to the quantity of waste water discharged, and a "treatment charge formula", based on pollution loading. Such a formula typically includes measures of:

- ✓ Chemical oxygen demand (COD).
- ✓ Conductivity.
- ✓ Suspended solids (SS).
- ✓ Ammonia concentration.
- ✓ Ortho-phosphate concentration.
- ✓ pH.
- ✓ Metal concentration.
- ✓ Permanganate value (PV).

An example of local authority requirements for discharge to the sewerage system:

EXAMPLE OF LOCAL AUTHORITY WASTE QUALITY STANDARDS

PARAMETER	REQUIRED STANDARD
Chemical Oxygen Demand	Not to exceed 5 000 mg/I
Permanganate Value	Not to exceed 1 400 mg/I
pH	6,0 — 10,0
Conductivity	Not greater than 500 mS/m at 35°C
Caustic Alkalinity (as CaCo)	2 000 mg/I
Substances not in solution (fat, grease)	.500 mg/I
Substances soluble in petroleum ether	500 mg/I
Sulphides	50 mg/I
Substances which may liberate hydrogen cyanide	20 mg/I
Formaldehyde (as HCHO)	50 mg/I
Sugars, starch (as glucose)	1 500 mg/I
Chlorine (as Cl)	100 mg/I
Sulphates (as SO4)	1 800 mg/I
Fluorine-containing compounds (as F)	5 mg/I
Non-organic solids in suspension	100 mg/I
Anionic surface active agents	500 mg/I
Total metals	50 mg/I
Any individual metal	20 mg/I

Where water is discharged directly to a stream or other natural watercourse, far more stringent standards are applied. A further legal stipulation is that no water from any process shall be allowed to enter the storm water drains, which are intended for clean storm water run-off only

Control Technologies

If there is any chance that effluent from an industrial source may contribute to water pollution, a plant-wide wastewater survey should be conducted. This survey should identify wastewater flows and outflows in order to:

- ✓ Ascertain the need for wastewater control and treatment.
- ✓ Identify opportunities for conserving water and preventing pollution.
- ✓ Separate incompatible wastewater streams and sewer systems.

Most industries will employ a simple means of separating grit, oils and greases from their wastewater. Oil interceptors or grease traps work by settling out solid material, such as sand and metal grits, and floating out oils and grease. They may also provide temporary containment in the event of an accidental spill. The oil interceptor is generally located outside, between the operational area and the sewer connection. It should be accessible and identifiable by a metal cover and vent. Design specifications for an oil interceptor are available from local or regional authorities.

Considerations for oil receptors

- ✓ Restroom sanitary wastes should never pass through the oil interceptor.
- ✓ A certified engineer should sign off any extensions or changes in design.
- ✓ Oil interceptors need to be cleaned regularly to ensure correct functioning and compliance with local authority bye-laws.

- ✓ A monitoring and maintenance schedule should be drawn up, based on the water quality and quantity flowing through the trap.
- ✓ Sludge and waste oil from oil interceptors are classified as hazardous waste. Unless recycled, these wastes must be disposed of on a Class 1 landfill, with a certificate of disposal being filed each time the interceptor is cleaned.
- ✓ All oil interceptors should be inspected at regular intervals (usually every 3 months) by the local authority.
- ✓ If wastewater is found to exceed the water quality parameters, the authority may impose a fine or charge a higher rate for effluent disposal.
- ✓ The oil interceptor and all gullies and drainage channels should be checked weekly for blockages and the accumulation of dirt.

An oil interceptor is not designed to remove all contaminants from wastewater. Dissolved solids and heavy metals such as lead will not be removed by the system. A conventional oil interceptor is also unable to reduce a high chemical oxygen demand, which may be a result of sewage or chemicals in the water. Such polluted water may need additional treatment before being discharged to drains.

EXAMPLE OF A CHECKLIST FOR MONTHLY MONITORING OIL INTERCEPTOR PERFORMANCE

MONTH:			
OIL INTERCEPTOR	WEEK	STATUS (clear/blocked;	ACTION TAKEN
Check for correct	1		
operation of	2		
oil interceptor	3		
	4		
Inspection of gullies			
	1		
and drainage channels	2		
	3		
	4		
Date of last service/cleaning: Servicing Company: Date of last inspection: Was effluent in compliance with regulations? Yes / No If No, attach copy of inspection outcome and details of action taken.			

Similar to control of air pollution, numerous technologies are available for treating polluted water. These may be used to control pollution. They may also be used in pollution prevention systems, for example, reverse osmosis may be used to recycle heavy metals back into the operational system. The following table provides an overview of various common treatment technologies:

OVERVIEW OF WATER TREATMENT TECHNOLOGIES

POLLUTANT	TECHNOLOGY	DESCRIPTION
Suspended particles	Screens Evaporation Sedimentation Flotation	Filter water by capturing solid particles. Water is evaporated, leaving the solids behind. Settling basin allows solids to settle out, after which they can be pumped from the base of the tank. Fine gas bubbles are introduced into the water. These attach themselves to the particles which rise to the surface to be removed.
Organic material	Activated sludge	Micro-organisms are introduced into the water which feed on the organic material available.
Nutrients (for example, phosphate)	Flocculation	Chemicals are added to increase the size of particles, which are then removed by sedimentation or filtration.
Gas controls (for example, VOCs, sulphur and nitrogen	Reverse osmosis	A semi-permeable membrane through which only certain components of a solution may pass.

- SOIL POLLUTION

Despite its importance, South African soil is deteriorating due to poor management practices. Erosion, compaction, acidification, salinization and infestation by weeds and pathogens pose serious threats to the soil resource.

Soil pollution takes place when waste or other harmful materials are disposed of into the soil thereby damaging its fertility.

Soil pollution, also known as soil contamination, is caused by man-made, harmful chemicals penetrating the earth and causing deterioration. There are many health risks associated with soil pollution, through direct contact with the soil or from air contaminants. Knowing the causes and the potential solutions to soil pollution can help to mitigate some of the risks. The potential causes of soil pollution include industrial chemicals, lead contamination, radioactive metals, leaking underground fuel tanks, improper disposal of hazardous materials, agricultural chemicals, salts, sewage. Ground pollution effects vary based on the specific type of soil pollution as each one causes a wide variety of harmful environmental effects.

Effects of soil pollution

Soil pollution has several other consequences such as, for example:

- Development of human diseases from contact with polluted soil or water
- Climate change leads to a disruption in ecosystems that causes harm to certain animal species
- Decreased soil fertility and yield leads to a lack of nutrients and an increase in diseases

Ground pollutants and their sources

The activity associated most directly with soil degradation is agriculture. Chemicals as part of agriculture are used to increase the production of crops or restrict the growth of harmful fungi. As a consequence, chemicals such as herbicides and pesticides which are not naturally produced and cannot be broken down in nature end up seeping into the ground, stripping away the fertility of the soil and harming the ecosystems. In other cases, plants absorb these chemicals and eventually die off.

In other industries, accidental spills of chemicals or contaminated waste waters, during transportation, storage, handling or disposal, can seriously pollute localised soils. The following table identifies some of the more common sources of soil pollution:

COMMON SOURCES OF LAND DEGRADATION

DEGRADATION	PREDOMINANT SOURCES
Erosion and compaction	Poor land management
Acidification	Excessive use of fertilisers; acid rain
Salinization	Over-irrigation
Biological invasion	Introduction of alien species or noxious weeds
Pollution	Accidental spills or illegal dumping of chemicals; irrigation with polluted water; windblown litter

Common causes of soil pollution:

- Industry
- Human waste
- Deforestation
- Acid rain when pollutants in the air such as sulphur dioxide and nitrogen oxide mix with rain
- Fertilizers - burned roots, stunted growth and mineral deficiencies
- Salt contamination in soil - caused by irrigation in arid and semi-arid climates
- Groundwater pollution effects
- Heavy metals - such as radioactive metals such as uranium, radium and plutonium and other heavy metals such as lead, mercury, arsenic, and chromium as a result of mining activities, industrial activity, and motor vehicle use
- Pesticides poison plants and decrease the fertility of the soil, making it less able to support plant life

Mitigating the effects of soil pollution

- Address farming practices by avoiding over-cropping and over-grazing.
- Recycle by using paper, plastic, aluminium and glass products and reduce the use of products that are not recyclable. Reuse items such as containers rather than throwing them in the trash.

(Source: https://www.seekpng.com/ipng/u2q8q8e6a9o0i1t4_reduce-reuse-recycle-reuse-reduce-recycle-symbol/)

- Limit pesticides to using more environmentally-friendly substances when appropriate.
- Reforestation by planting new trees where others have been cut down to help curb soil erosion.
- Bioremediation by adding microorganisms into the soil to break down contaminants which is a safe and environmentally-friendly way to clean up polluted soil.

- OPT FOR BIODEGRADABLE

- Use biodegradable products which will break down naturally and not cause any toxic chemicals to go into the soil.
- Use soil additives such as lime to adjust the pH levels of the soil and limit soil pollution.
- Soil washing also known as soil separation is a water-based process for scrubbing contaminants off soils and has been proven to be effective on soils contaminated by heavy metals and organic contaminants.

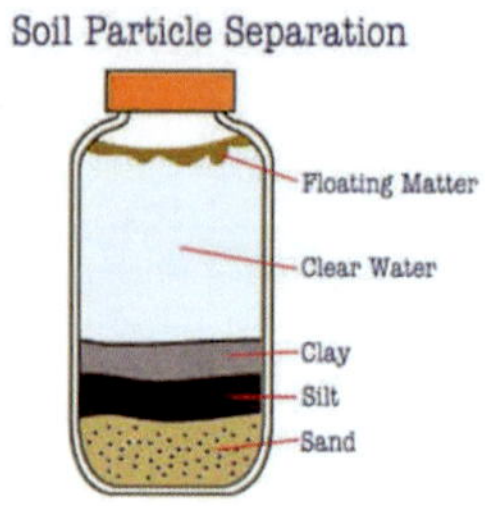

Legal Requirements

Numerous Acts exist to encourage the sound management of agricultural land. These include:

- National Environmental Management: Waste Act 59 of 2008
- National Environmental Management: Protected Areas Act 57 of 2003
- Environment Conservation Act 50 of 2003

Controls for ground pollution

Planning, proactive management and the development of appropriate contingency plans provide a safeguard against accidental spills. Storage tanks containing chemicals should be monitored and building walls of sufficient capacity (110% of the volume of the tank) constructed to prevent spillage and leaks onto the ground. Underground tanks should be registered and levels monitored carefully. The leakage from underground tanks can go undetected for long periods of time, resulting in serious pollution of soils and eventually groundwater.

Overfill prevention measures, such as automatic shut-off devices, overfill alarms or ball float valves, should be installed. An accurate leak detection system must be employed, and planned corrective actions available for every tank on site. A leaking underground storage tank may be the result of one or more of four factors, namely galvanic corrosion, faulty installation, piping failures and spills and overfills. The following table highlights some possible causes of underground leakages:

POSSIBLE CAUSE OF LEAKING IN UNDERGROUND STORAGE TANKS

FACTOR	EXPLANATION
Galvanic corrosion	Corrosion of the steel as a result of site-specific factors (for example, soil conductivity).
Faulty installation	This may involve inadequate backfill, allowing movement of the tank and separation of pipe joints.
Piping failures	Poor workmanship can frequently lead to corrosion of unprotected steel.
Spills and overfills	These are usually the result of human error, and can intensify the corrosiveness of the soils around the tank.

Every plant should have on record a recent survey of all chemical storage tanks on site, indicating:

- ✓ Location
- ✓ Contents (with environmental implications of accidental release to air, ground or water)
- ✓ Leak detection system
- ✓ Emergency plan in the event of a leak being detected

Companies have to be committed to rehabilitating land after construction, excavation, accidental spillage or any activity that upsets the land. This may take simple forms, such as the removal and storage of topsoil for reintroduction after excavation. Topsoil is rich in organic matter and will contain seeds to encourage the re-growth of natural vegetation.

Soils polluted with organic compounds, such as petroleum products, may need to be removed entirely or can be bio-remediated on site through the introduction of micro-organisms into the soil. These procedures are invariably extremely costly, and provide an incentive for sound planning and control.

In addition to pollution control, industrial environmental programmes need to encourage the organization to adopt responsible policies towards vegetation management such as, for example,

- ✓ The protection of endangered plant and animal species on site.
- ✓ The promotion of indigenous species.
- ✓ The promotion of irrigation systems that conserve water.
- ✓ The need to eradicate noxious weeds.
- ✓ The control of invasive alien species.
- ✓ Control of herbicide and pesticide use.

• NOISE POLLUTION

Noise is one of the most enveloping health hazards in the workplace and can be defined as undesirable or a physiologically impairing sound. Noise pollution takes place when there is either an excessive amount of noise or an unpleasant sound that causes a temporary disruption in the natural balance. This definition is usually applicable to sounds or noises that are unnatural in either their volume or their production. In urban settings the most common source of noise pollution is the sound generated through vehicles. There are many other sources of noise pollution such as aircraft, industrial machines, traffic, events, and construction sites. Noise pollution results in adverse psychological effects like mental stress and depression and also deafness or partial deafness in humans. Noise-induced hearing loss is one of the most common occupational diseases and the second most self-reported occupational illness or injury.

Sound is a variation in pressure occurring in the air, detected by the ear and interpreted by the brain. How the brain interprets the sound it hears depends on the amplitude and frequency of the sound. The amplitude is the loudness measured in decibel ampere (dB(a)). The frequency of the sound is also referred to as the pitch, which is a high or low sound, measured in Hertz. According to the Act, the maximum safety limit a person can be exposed to without damaging the ear is 85dB(a) for a period of 8 hours.

Causes of noise pollution

- ✓ Industrialization and mechanisation have contributed to a more widespread problem of noise generation from various equipment like compressors, generators, exhaust fans, grinding mills
- ✓ Poor urban spatial planning
- ✓ Transportation in the form of motor vehicles, trucks, aircraft and trains
- ✓ Construction activities such as mining, construction of roads, bridges, dams and buildings

Effects of noise pollution

- Hearing problems and damage to eardrums and loss of hearing from exposure for long periods to jackhammers, hooters, machinery, aircraft and motor vehicles
- Psychological and chronic health issues such as the occurrence of aggressive behaviour, disturbance of sleep, constant stress, fatigue, and hypertension.
- Sleeping disorders leading to fatigue and poor work performance
- Cardiovascular issues such as high blood pressure, increased heartbeat rate and stress-related heart problems
- Troublesome communication leading to misunderstanding work instructions, severe headaches and emotional imbalance
- Wildlife because of higher dependence on sound and hearing

Classification of noise

Noise can be classified based on its characteristics into three main groups, namely continuous noise, interrupted noise and impulse noise.

Continuous noise occurs on a continuous basis for an extended period of time. An example of a continuous noise is such as when an electric motor is running in the back ground. It has a monotonous continuous sound with little variance in the frequency and intensity of the noise.

Interrupted noise occurs when there are variations in sound levels due to the on and off switching of the machine during a production cycle. The nature of the sound is mostly the same, but alternates between sound and silence.

Impulse noise consists of a very intense sound lasting only a short period of time, such as a telephone ring tone or a gun shot.

The body responds differently to different exposures of sound. As hearing is such a valuable sense that a person has, it needs to be protected. After a loud exposure of sound, a temporary threshold shift in hearing occurs and quiet sounds cannot be heard for moments up to hours.

It is very difficult to control noise, but it can be managed through engineering controls such as, for example, modification of the machines design, insulated mountings, and gears, enclosing the source or operator or by making use of PPE for noise control.

Organizations can adopt a hearing conservation program by doing the following, namely:

1. Do a risk assessment including a noise survey
2. Apply engineering measures such as isolating equipment or the operator or both
3. Re-survey to determine effectiveness of interventions
4. Demarcate the area which is generating high levels of noise
5. Identify workers or the area
6. Take administrative measures such as education and training, a baseline audiogram and annual audiogram. Do exit audiograms.
7. Make use of PPE
8. Report noise levels to the relevant authorities.

According to the COID Act 130 of 1999, employers may not permit a worker to work in an area that has noise levels exceeding 85dB(A). Where this is the case, effort must be made to lower noise levels to below 85 dB(A), and if this is not possible, all efforts should be made to reduce the noise level as much as possible. Where it cannot be reduced, the area must be demarcated and no person should be allowed to enter the demarcated area without ear protectors.

• ILLUMINATION OR LIGHT POLLUTION

Natural light entrains or regularizes basic and fundamental biological activities across species from plants to humans. The effects of light pollution on plants and animals in the environment are numerous and are becoming more known over time. In general, the most common action is that light pollution alters and interferes with the timing of necessary biological activities. Good lighting is essential for the accurate processing of information given that 80% of information is absorbed through eyesight. Good and sufficient lighting is aimed at promoting productivity, safety, health, well-being and pleasant working conditions at an economical cost. Reasonable contrast is required between the object viewed and the background to enhance ease of vision. Compromises in lighting system conditions are required because as the level of lighting improves viewing, the negative effect of glare might increase. Well balanced levels of illumination are essential to establish safe working conditions. Industrial lighting involves a variety of tasks, operating conditions and economic considerations.

The purpose of industrial lighting is to help provide a safe working environment, provide efficient and comfortable seeing and to reduce losses in visual performance.

Poor lighting can add to:

- Headaches
- Tired, sore, dry eyes
- Incidents and injuries
- Blurred vision and double vision

Risk factors that should be addressed that include:

- Illumination or the level of lighting
- Colour rendering qualities
- Distribution of tight
- Eyesight of the workers
- Contrast in illumination
- Glare and undesirable reflections

SABS 0014 outlines a method that a survey of light conditions can be conducted. The guidelines address lighting levels such as intensity in respect of different types of working areas as well as emergency lighting and a few other general aspects mentioned in the Environmental Regulations, SANS 10400-0 and National Building Regulations Part O.

Good illumination is influenced by the quality or distribution of brightness, its colour, direction, diffusion, and degree of glare and quantity or amount of light. These factors contribute to good illumination, because it is also these factors that influence the ability of the worker to work safely and correctly.

Poor illumination includes direct and reflected glare, dark shadows, visual fatigue and delayed eye adaptation or coming from a bright to a dark area.

Lux is the metric unit to calculate illuminance. 1 Lux= 1,0 lumen/m2. It is the quality of light measured as it falls on an object. Illuminance levels as required by law, are specified in the environmental regulations for workplaces. Illuminance meters use photocells based upon silicon and they incorporate colour correcting filters to match the sensitivity of the human eye. These meters should be calibrated once a year and should be zero checked occasionally.

Visible light is a type of electromagnetic radiation and classified as a "non-ionising" radiation. Other non-ionising radiation lights include ultraviolet, infra-red, microwaves, lasers and ELF fields. Of these non-ionising types of radiation, microwaves and laser beams form the most dangerous to health types of radiation. Ultraviolet radiation can cause skin burns, but microwaves can penetrate into the skin and cause damage to deeper tissue.

Ionising radiation contains large amount of energy and are capable of penetrating into living tissue and can only be shielded by high density materials such as lead or concrete. Ionising radiation is divided into four main types: Alpha particles, Beta rays, Gamma and X-rays, and, neutrons. Alpha particles can penetrate human tissue to 1/10th of a millimetre to the immediate vicinity of the source. Beta rays can penetrate a few millimetres into human tissue before being absorbed. Gamma and X-rays are electromagnetic radiations of varying levels of energy that have a high penetration potential and can irradiate all parts of the body uniformly. Neutrons are uncharged particles with a wide range of energy and penetration power.

Ionising radiation can have the following biological effects on the human body when exposed to it. Physical chemical reaction takes place after exposed to ionization followed by a chemical reaction that may last for seconds, causing a biological manifestation of the cell damage. Cell damage may include the destruction of cells, the delay in the cell division and/or alteration of the inherent cell structure. The amount of cell tissue damage will depend on the amount of energy absorbed into the tissue per mass of tissue, as well as the dose and the time of exposure.

The International Commission of Radiological Protection set the recommendations followed by South Africa as the basis for radiation protection. These norms stipulate that the exposure of individual to radiation must be justified in terms of the benefits secured when the work is performed. Should exposure be justified, it must be optimized and must be kept as low as reasonably possible in the work situation.

• OTHER FORMS OF POLLUTION

Other forms include

- ✓ radioactive pollution that is caused by the disintegration of radioactive metals releasing dangerous rays that can cause diseases like cancer.

• REVIEW QUESTIONS

1. What are some of the major causes of air pollution?
2. How can the effects of water pollution be effectively mitigated?
3. What impact does soil pollution have on the environment?
4. What are the elements of an effective hearing conservation program?
5. What are the consequences of poor lighting?

• REFERENCES

Abdul-Rahman, Fahzy. 2014. Reduce, reuse, recycle: alternatives for waste management.

Air pollution: Sources and effects of air pollution. https://www.yourarticlelibrary.com/air-pollution/air-pollution-sources-and-effects-of-air-pollution/4688. Accessed May 28, 2020

Alam, S., Akter, S.T., Sheikh, M.S. and Bahar, I., 2016. The Impact of noise pollution on workers health in selected industries of mirzapur industrial area, Tangail, Bangladesh. *Journal of Environmental Science and Natural Resources*, *9*(2), pp.155-160.

Anonymous (n.d.). Loose notes on safety, health and environment management

Apte, Komalkirti. and Salvi, Sundeep., 2016. Household air pollution and its effects on health. *F1000Research*, *5*.

Arthur, W. and Koth, P. 2021. Pollution Issues. Available at: http://www.pollutionissues.com/Ho-Li/Industry.html

Atkins, W. and Koth, P. (n.d.). Industry - water, effects, environmental, pollutants, United States, history, types, causes, impact, EPA, pesticide, chemicals, industrial, toxic, world, human, power, disposal, use. [online] Pollutionissues.com. Available at: http://www.pollutionissues.com/Ho-Li/Industry.html

Azam, Muhammad., Alam, Md. Mahmudul. and Hafeez, Muhammad. Haroon., 2018. Effect of tourism on environmental pollution: Further evidence from Malaysia, Singapore and Thailand. *Journal of cleaner production*, *190*, pp.330-338.

Beltrán-Esteve, Mecerdes. and Picazo-Tadeo, Andres. J., 2015. Assessing environmental performance trends in the transport industry: eco-innovation or catching-up?. *Energy Economics*, *51*, pp.570-580.

Campbell, Fiona. (2019). Human Factors: The Impact on Industry and the Environment. Natural Resources Management and Biological Sciences [Working Title].

Chowdhary, Pankaj., Yadav, Ashutosh., Kaithwas, Gaurav. and Bharagava, Ram. Naresh., 2017. Distillery wastewater: a major source of environmental pollution and its biological treatment for environmental safety. In *Green Technologies and Environmental Sustainability* (pp. 409-435). Springer, Cham.

Conservation Institute (n.d.) Soil pollution: Causes, effects and what to do about it. https://www.conservationinstitute.org/soil-pollution/. Accessed May 28, 2020

Conserve Energy (n.d.) What is Noise Pollution? https://www.conserve-energy-future.com/causes-and-effects-of-noise-pollution.php. Accessed May 28, 2020

Department of Environmental Affairs (1965). Atmospheric Pollution Prevention Act No. 45 of 1965.

Earth Reminder (n.d.) Effects of noise pollution. https://www.earthreminder.com/effects-of-noise-pollution/. Accessed May 28, 2020

Effects of Water Pollution On The Environment. (2019). In *ScienceAid*. https://scienceaid.net/Effects_of_Water_Pollution_On_The_Environment Retrieved May 28, 2020

EU (2006). Environment fact sheet. [Luxembourg]: [Office for Official Publications of the European Communities].

Florida Atlantic University (n.d.) Light pollution harms the environment. Available at: http://cescos.fau.edu/observatory/lightpol-environ.html. Accessed May 28, 2020

Greally, T., Cuddihy, J., Leane, G. and Cowley, H. (2011). [online] Epa.ie. Available at: https://www.epa.ie/pubs/reports/air/quality/HSE%20Report%20to%20the%20Air%20Quality%20Health%20Information%20Group%2023rd%20Dec%202011.pdf.

Help Save Nature. (n.d.) Long term effects of water pollution on the environment. https://helpsavenature.com/long-term-effects-of-water-pollution-on-environment. Accessed May 28, 2020

Hill, Marquita. K., 2020. *Understanding environmental pollution*. Cambridge University Press.

Inyinbor Adejumoke, A., Adebesin Babatunde, O., Oluyori Abimbola, P., Adelani Akande Tabitha, A., Dada Adewumi, O. and Oreofe Toyin, A., 2018. Water pollution: effects, prevention, and climatic impact. *Water Challenges of an Urbanizing World*, *33*.

Jariwala, Hiral. J., Syed, Huma. S., Pandya, Minarva. J. and Gajera, Yogesh. M., 2017. Noise Pollution & Human Health: A Review. *Noise and Air Pollution: Challenges and Opportunities*, *29*.

Jayaswal, Komal., Sahu, Veerendra. and Gurjar, B.R., 2018. Water pollution, human health and remediation. In *Water Remediation* (pp. 11-27). Springer, Singapore.

Keller, S. (n.d.). Water Pollution. Sustainable Sanitation and Water Management Toolbox, Available At: Https://Sswm.Info/Sswm-University-Course/Module-8-Water-And-Sanitation-Future-Challenges/Further-Resources-Phosphorus/Water-Pollution.html

Mutumi, Prue. P. J., 2018. *For profit or the environment?: an analysis of the relationship between the motor vehicle manufacturing industry and the green economy in South Africa.* Unpublished PhD submitted to University of the Witswatersrand.

Ngarambe, J., Lim, H.S. and Kim, G., 2018. Light pollution: Is there an Environmental Kuznets Curve?. *Sustainable cities and society*, *42*, pp.337-343.

OSHA (2011). [online] Osha.gov. Available at: https://www.osha.gov/Publications/3430indoor-air-quality-sm.pdf.

Rodríguez-Eugenio, Natalia., McLaughlin, Michael. and Pennock, Daniel., 2018. *Soil pollution: a hidden reality*. FAO.

Sampson, Ian. 2001. *Introduction to a Legal Framework to Pollution Management in South Africa* WRC Report No TT 149/01 March 2001

Sciencing (n.d.) The effects of soil pollution on the environment. https://sciencing.com/the-effects-of-soil-pollution-on-the-environment-13406897.html. Accessed May 28, 2020

Sheppard, Charles. (2019). World Seas. [Place of publication not identified]: Academic Press.
Smolensky, Michael. H., Sackett-Lundeen, Linda. L. and Portaluppi, Francesco., 2015. Nocturnal light pollution and underexposure to daytime sunlight: Complementary mechanisms of circadian disruption and related diseases
Sun, Zhuanlan. and Zhu, Demi., 2019. Exposure to outdoor air pollution and its human health outcomes: A scoping review. *PloS one*, *14*(5), p.e0216550.
Thermax Group (2021). Techniques of air pollution control. Available at: https://www.thermaxglobal.com/techniques-of-air-pollution-control.html
Vallero, Daniel. (2014). Fundamentals of Air Pollution. 5th ed.
World Bank. (n.d.). Pollution Management and Environmental Health. Available at: https://www.worldbank.org/en/programs/pollution-management-and-environmental-health-program.html
World Economic Forum. 2018. *Harnessing the Fourth Industrial Revolution for Life on Land.* World Economic Forum's System Initiative on Shaping the Future of Environment and Natural Resource Security in partnership with PwC and the Stanford Woods Institute for the Environment
WWF (n.d.) Pollution: Ocean Pollution. Available at: https://www.environment.co.za/pollution/ocean-pollution.html and sourced from: http://wwf.panda.org/our_work/oceans/problems/pollution/

CHAPTER 24

COMMON ENVIRONMENTAL CHALLENGES

• VENTILATION

Ventilation is the provision of fresh air and the removal of contaminated air from an area concerned. Poor ventilation is a hazard which a risk to the health of workers. Not only does poor ventilation affect the physical health of workers, but psychological effects like stress also rise when workers are constantly exposed to the effects of poor ventilation.

Where there are low light conditions, there will normally also be bad ventilation. It is important to control dust, fumes, gases and vapours. It is used to control climatic comfort factors and entails the provision of fresh air from outside the building to keep interior building air circulated or the removal of contaminated air. Proper ventilation removes unpleasant smells and excessive moisture, and to prevent stagnation of the indoor air.

Ventilation requirements for health reasons include:

- Oxygen for respiration
- Prevention of accumulation of bacteria
- Removal of toxic dust, fumes and vapours

Thermal comfort ventilation control is applied to:

- Remove excess heat
- Body cooling
- Structural cooling
- Moisture control

There are two types of ventilation, namely

- General ventilation which uses forces of nature and features of buildings and decision to remove gases and dust; and
- Local exhaust ventilation which creates air velocity by mechanical means through fans or air conditioner. Can be used to blow in fresh air and/or extract old air.

Effects of poor ventilation

1. High levels of carbon dioxide and low levels of oxygen can cause fatigue and affect the ability of workers to concentrate.
2. Build-up of chemical and biological contaminants that cause poor indoor air quality. Poor indoor air quality can lead to workers suffering from headaches, fatigue, hypersensitivity and allergies, sinus congestion dizziness, shortness of breath, coughing and nausea.
3. Extreme temperatures cause fatigue, discomfort and distraction and can consequently increase accidents in the workplace.
4. Low humidity can cause a dry throat, dry skin and static electricity build-up. On the other hand, high humidity contributes to bacterial and mould growth.
5. Excessive and irritating workplace odours cause discomfort and affect concentration such as, for example, ammonia and chlorine.
6. Poor ventilation causes Sick Building Syndrome (SBS), the symptoms of which include irritation of the eyes, nose and throat, headaches, fatigue, and a susceptibility to colds and flu.

• THERMAL STRESS

The work environment may either be comfortable or extremely cold or hot and consequently uncomfortable. A common physical hazard in most industries is heat. Extremely hot temperatures occur where workers are working in foundries or in industries where they use open fires for energy such as, for example, soap factories and where extreme heat is used to mould iron or process other materials.

Sources of heat in the workplace

Heat can be transferred in various ways including:

- Muscular activity from the work where heat is released as the body performs work and consumes oxygen.
- Radiation from surrounding hot working machines such as ovens, fires or other equipment that radiate heat into the surrounding air. If the worker works in an environment where there are machines that radiate extreme heat into the air, the bodies of workers will absorb the heat radiated because their bodies is cooler than the radiated heat.
- Conductive and convective heat from working in hot air transfers to the bodies of workers as they handle equipment hotter than themselves, or when working in hot air, the bodies of workers will also have the strain of the hot air heating up their bodies, and their bodies trying to cool themselves down.
- Evaporation is the cooling effect that takes place as a result of perspiration from the surface of the skin. The evaporation of the perspiration is encouraged through air movement or convection that flows over the skin. In a very humid environment evaporation is hampered, and the sweat cannot evaporate fast enough to cool the body down properly. When working in a very hot environment, workers should be wearing loose clothes to ensure that sweat can evaporate easily. Natural fibres, such as cotton or woollen clothing ease the process of evaporation.

The human body starts to experience physiological heat strain that can cause different diseases depending on the degree of heat stress experienced.

The following are major health conditions from heat in order of the least to the most stressful, namely:

- Heat Stress
  Fluctuation in the core temperature of the body about 2°C below or 3°C above the normal core temperature of 37.6°C impairs performance markedly and a health hazard exists. When this happens the body attempts to counteract by Increasing the heart rate, dilation of the capillaries in the skin to bring more blood to the surface so that the rate of cooling is increased, and sweating to cool the body.
- Heat Oedema
  The swelling of the feet or ankles because fluid balance of the body changes as an initial response to heat
- Heat Rash
  Or Prickly heat develops when a worker perspires a lot. The rash occurs because of pores that are obstructed and result in inflammation of the skin tissue, causing a swollen red, itching and burning skin. If skin is simultaneously exposed to chemicals and heat the situation will be aggravated.
- Heat Cramps
  Muscles start cramping when a worker is exposed to heat for many hours, while the muscles are used and when the worker is unacclimatised to the heat. When this happens a temporary blood shortage occurs, and oxygen, electrolytes and other metabolites will be the main reason for the muscle contractions
- Heat exhaustion
  Heat exhaustion develops when the body can no longer cope with the heat, and the body loses too much fluid or salt or both. This is characterized by an accelerated heart rate, dizziness, weakness, nausea and headache. The worker can recover with rest and lots of fluids. Ensure that the worker is acclimatised and has regular access to water at work.
- Heat Syncope/ collapse
  After experiencing heat exhaustion and the worker is still exposed to heat, he/she will faint as a result of insufficient blood circulation to the brain. Immediately remove the worker from the hot environment and give plenty of water.
- Heat Stroke
  Heat stroke is potentially life-threatening state and can result in irreversible damage. A predisposing factor for heat stroke is excessive physical exertion in extreme heat conditions. If a worker has a temperature greater than 40,6°C a total collapse of the defence mechanisms of the body against the high environmental heat occurs. This result in hot dry skin, molted or cyanotic or blush and can have causes including over exertion, loss of fluids, cardiac disease. The body fails to sweat, so evaporative cooling does not function. The body temperature rises uncontrollably, accompanied by convulsions and loss of consciousness. Delayed treatment may be fatal. Rapid cooling by splashing with chilled water or cooling with fans and wet sheets is necessary. Seek urgent medical assistance. This emergency is characterized by: a dry warm skin, absence of sweating, muscle contractions, aggression, delirium, dilated pupil and unconsciousness.

Heat stress can practically be measured either through the WBGT (Wet bulb globe temperature) Index, or the thermal environment. There are legal requirements that need to be adhered to when a WBGT Index is higher than a certain minimum amount. The WBGT Index means a number, which characterizes the thermal conditions in the environment. It is calculated by adding seven tenths of the reading in degrees Celsius obtained with a wet bulb thermometer (WB), to one 5th for the reading. in Celsius obtained with a globe thermometer (GT), and adding the sum to one tenth of the reading in degrees Celsius obtained with a dry bulb thermometer (DB):

WBGT= WB (0. 7) + GT (0,2) + DB (0, 1)

Where in °C
WB= Wet Bulb Temp
GT= Globe Temp
DB= Dry Bulb Temp

The index reading may also be obtained by using a heat stress monitor, which is an electronically integrated direct reading instrument that has been designed, built and calibrated specifically for that purpose.

Exposure to cold working environments

Efforts to prevent health problems in cold environments are directed at the maintenance of body heat.

The human body responds to cold environments by constricting blood vessels of the skin and shivering to try and keep itself warm, as a signal is sent from the hypothalamus to create more heat. Severe cold slow nerve reactions and hands may fumble, causing additional safety hazards. The effects of cold stress can be any of the following:

- **Cold stress** is the effect of external working environments in very low temperatures such as less than 6°C, and the resultant inability of the body to maintain a constant internal body temperature. In cold temperatures, a high air flow will increase cold stress effect considerably, called the wind chill factor.
- **Chilblains** occur when insufficient blood reaches the extremities and the fluids around cells actually freeze. This causes tissue damage and occurs on fingers, toes, nose, cheeks and ears.
- **Frostbite** occurs by handling very cold objects, which is a sensation of initial coldness, followed by numbness. It occurs in 3 degrees: freezing, freezing with blistering and freezing with dying of tissue.
- **Hypothermia** is a condition that occurs when the body loses heat faster than it can produce it, Blood vessels in the skin constrict to conserve internal heat. Hands and feet are first affected. If the body continues to lose heat, involuntary shivering begins (the first signs of impending hypothermia). Further heat loss causes speech difficulty, forgetfulness, loss of manual dexterity, collapse, cardiac arrest and finally death.
- **Trench foot is** an injury which results from long exposure of the feet to wet conditions at temperature of 10°C with little movement which causes changes in the circulation of blood in the feet. The result if untreated in time will be loss of toes or part of the feet. Treatment includes keeping the foot dry and warm and doing exercise for good circulation.
- **Immersion foot results** from immersion of the foot in water that is below 10°C, for a prolonged time, usually in excess of 24 hours

The temperature of hands and feet can drop to as much as 23-28°C below normal body temperature without lasting harm. A temperature drop of only about 1.5°C of the core body temperature produces shivering. As temperature continues to drop, the brain becomes less efficient and the worker becomes confused and disoriented. Consciousness is lost at a core temperature of 29-31 °C. When the core temperature drops to below 25°C it becomes life threatening.

Cold stress can be prevented through acclimatisation, decrease of air flow, suitable dry clothing such as semi-permeable clothing with the ability to counteract wind chill factors, layered clothing, avoidance of direct skin exposure to cold, correct selection and placement, carry out jobs in pairs and create a reliable communication system and always train exposed workers in PPE, first aid, and rescue signals.

• ENERGY OPTIMISATION

Energy demand and costs are rising dramatically while financial constraints increase. There is a need to find ways to do more with less which could include measures to lower overall energy usage or just finding a way to manage demand penalties and other charges that drive up the per-unit cost of the energy that is required.

Energy optimization means using energy wisely

Typically, there are two types of energy efficiency, namely, Passive Energy Efficiency and Active Energy Efficiency, with active energy efficiency being dependent upon passive energy efficiency.

• PASSIVE ENERGY EFFICIENCY

This method includes installing and using efficient devices and so providing a baseline for optimized energy efficiency. The following is included:

✓ Installing low consumption devices
✓ Well-insulated building envelope
✓ Roof insulation
✓ Sealing
✓ Well-insulated hot water pipes
✓ Walls and fenestration

• ACTIVE ENERGY EFFICIENCY

This method is needed to optimize and sustain the gains from passive energy efficiency and to provide additional saving. This can be achieved by system automation and regulation such as, for example,

✓ Lighting Control Systems
✓ Heating, ventilation, air-conditioning
✓ Building management systems

The National Building Regulations Act Amendments stipulate that buildings complying with the green building codes of the Green Building Council of South Africa must generate 50% of their hot water using renewable energy, either through solar water heating or heat pump technology.

• WASTE MANAGEMENT

Management often overlooks the potential costs associated with producing waste. Many different types of waste are generated, including municipal solid waste, agricultural and animal waste, medical waste, radioactive waste, hazardous waste, industrial non-hazardous waste, construction and demolition debris, extraction and mining waste, oil and gas production waste, fossil fuel combustion waste, and sewage sludge. Waste can be in solid, liquid or gaseous form. Liquid wastes may be watery effluent that leaves the factory through the drains, or used chemicals. Gaseous wastes may leave the factory as emissions, or may be stored and treated on site. Waste can also refer to the excess energy or water used by the operation of an organisation.

Waste generation, generally represents inefficient use of materials. For example, solid wastes may be generated from:

✓ Raw materials that are not used (for example, off-cuts).
✓ Packaging removed from components used in production (for example, corrugated cardboard boxes, drums or wooden pallets).
✓ Unused by-products of the production process (for example, dewatered sludge).
✓ The office environment (for example, paper waste).
✓ Canteens and restaurant facilities (for example, waste food).
✓ Contractor operations (for example, builders' rubble).

Once generated, waste must be managed through reuse, recycling, storage, treatment, and/or disposal. Most municipal solid waste and hazardous waste are managed in land disposal units. For hazardous wastes, land disposal includes landfills, surface impoundments, land treatment, land farming, and underground injection. Modern landfill facilities are engineered with containment systems and monitoring programs.

The way waste management is handled evolved primarily due to health concerns and the need to control unwelcome odours. Disposing of waste has huge environmental impacts and can cause serious problems. In the past, waste often was deposited on land just outside developed areas. However, more recently, excavation of land specifically for deposition of wastes became common, often accompanied by burning of wastes to reduce volume, a practice eventually determined to be a contributor to degraded air quality in urban areas. Burning of wastes occurred at multiple levels, from backyard burning to large, open-burning dumps of municipal solid wastes to onsite burning of commercial and industrial wastes. Land disposal also created problems such as ground water contamination, methane gas formation and migration, and disease vector hazards. Leachate produced as waste decomposes may cause pollution. Badly-managed landfill sites may attract vermin or cause litter.

In the past few decades, the paradigm of waste management has shifted to one of sustainable materials management which focuses on resource, environmental, and human health impacts over the entire life cycle of materials.

Impact of waste

The effects associated with waste vary widely and are influenced by the substances or chemicals found in waste and how they are managed.

- Hazardous waste, by definition, has the potential to negatively affect human health and the environment, which is why it is so strictly regulated. Hazardous wastes are either specifically listed as hazardous, or exhibit one or more of the following characteristics:
  - ignitability,
  - corrosivity,
  - reactivity, or
  - toxicity.

  Generation and management of hazardous wastes can contaminate land, air, and water and negatively affect human health and environmental conditions.
- Municipal solid waste landfills are a major source of human-related methane emissions. Methane is one of several non-CO_2 gases that contribute to global climate change. Methane gas is released as wastes decompose.
- Gas emissions can be affected by recycling and changing product use, which will also create positive environmental benefits, such as reductions in energy consumption and greenhouse gases.
- Diarrhoea, dysentery, cholera, typhoid, salmonellosis, plague, hepatitis, chronic respiratory diseases can spread due to poor management of solid waste.

There are many costs associated with waste. Upfront costs include the direct cost of disposal into a landfill and the price paid for the raw materials that were not used. But waste also imposes other "costs" on society. These include the environmental problems associated with landfills such as leaching and bad smells and the use of scarce land for landfill operations. By using and disposing of products, resources that are required to produce them are being unnecessarily being used up. For these reasons, waste minimisation has become an important aspect of environmental management in industry.

Waste minimisation

The best way to manage waste is by not producing it in the first place. Avoiding waste generation is the ultimate goal of a waste minimisation programme.

According to the National Environment Management Act of 1998 as amended, waste minimisation can be described as:

All the activities, and processes, to the extent feasible, that are required to reduce the amount of waste that is generated or subsequently treated, stored or disposed of from the point of waste collection to recycling and monitoring and includes any treatment that reduces the volume, results in recycling or reuse of waste, or that reduces the hazard rating of hazardous waste. Waste minimisation also involves the redesigning of products, goods and processes to reduce the amount of waste produced.

In the past, waste management focused on ways to handle waste after it had been produced. Technologies for land filling, treatment and disposal at sea were developed. As costs associated with waste management, and legislation preventing improper dumping increase, these options are becoming less feasible. The focus is shifting to waste prevention technologies.

Waste minimisation was originally developed to reduce the toxicity and volume of hazardous wastes, but can readily be applied to all wastes generated in production. The basic principles of waste minimisation are as follows:

- ✓ Consideration should be given to ways of avoiding waste at the initial planning stages of all projects.
- ✓ Waste minimisation considerations should be incorporated into the design of products and production systems.
- ✓ Existing sources of waste should be identified and opportunities for the reduction or recycling evaluated.
- ✓ Modification of processes to reduce waste should be evaluated.
- ✓ Substitution of a material to eliminate a waste source should be considered.

Aim to "*reduce, re-use and recycle*" wherever possible.

REUSE
REDUCE
RECYCLE

Disposal of waste

Achieving the objective of zero waste is likely not to be achieved in the very near future. Waste will need to be disposed of appropriately. A small amount of waste such as medical waste is incinerated, but most waste disposal is done through land filling. Sanitary land filling can be defined in an environmental thesaurus as:

"an engineered method of disposing of solid waste on land in a manner that protects the environment, by spreading the waste in thin layers, compacting it to the smallest practical volume, and covering it with compacted soil by the end of each working day or at more frequent intervals if necessary."

The following table illustrates how following the Reduce, Reuse and Recycle approach results in savings when applied to office waste paper:

SAVINGS FROM OFFICE WASTE PAPER RECYCLING

APPROACH	EXAMPLE	SAVING/GAIN
Reduce	The use of paper can be eliminated by: using electronic forms of communication (E-mail) and computer faxes) posting notices instead of copies to all workers establishing a computer ordering system for your customers installing double-sided copying machines and enforcing their use	Time-saving and increased efficiency; Save paper (and limit plantation expansion); Energy and water (in paper production);
Re-use	Ensure paper is re-used as much as possible by: cutting scrap paper which is used only on one side to use as refills for desk memos using scrap paper for internal memos	Avoid purchase of desk memos; Save paper; Save energy and water (in paper production);
Recycle	Recycle waste paper by contacting a waste recycling company and setting up an office recycling scheme. The paper recycling industry is well developed in South Africa and numerous organisations will purchase high quality waste paper.	Potential income from sale of paper; Save paper; Less energy and water used than in virgin paper production;

Environmental concerns associated with land filling are:

- ✓ Short-term impacts in the form of flies, odour, unsightliness, and windblown litter.
- ✓ Long term impacts in the form of pollution of the water regime from leaching through the landfill site.

Leachate is a very strong wastewater generated as rain falls onto the landfill, or as off-site storm water flows across the landfill. All landfills will generate leachate, as it is impossible to protect the entire site from rainfall or runoff throughout its existence. All waste disposal facilities should therefore have a method for addressing leachate problems, where they develop.

Consideration should be given to the following regarding waste disposal, namely:

- ✓ The reputation of the organisation that is removing waste;
- ✓ Disposal of removed waste at the appropriate class of waste site;
- ✓ Soundness of the means of waste transportation;
- ✓ Execution of an EIA of the waste disposal facility, and availability on request;
- ✓ Provision by the waste disposal organisation provide of a description of their operational guidelines, including:
  - ✓ Waste specification (what will and won't be accepted).
  - ✓ Leachate management system (including leachate disposal) and water balance.
  - ✓ Monitoring control of the waste site (analyses and records)

It is the duty of the waste generator to ensure that its waste is adequately transported, treated and disposed of. Finally, the results of the environmental performance indicators should be displayed clearly in each department. This will serve to increase 'awareness, and even a sense of challenge to meet the stated target. Competitions may be arranged between departments to encourage greater commitment and participation.

• ARTIFICIAL INTELLIGENCE AND ENVIRONMENTAL CHALLENGES

In order to meet the listed critical challenges there are opportunities for using artificial intelligence to help unlock solutions. In this list are priorities for six of the world's most pressing environmental challenges and the priority action areas for successfully addressing them:

Climate change	Biodiversity and conservation	Healthy oceans	Water security	Clean air	Weather and disaster resilience
• Green power • Smart transport options • Sustainable production and consumption • Sustainable land use • Smart cities and homes	• Habitat protection and restoration • Sustainable trade • Pollution control • Invasive species and disease control • Realising natural capital	• Fishing sustainability • Preventing pollution • Protecting habitats • Protecting species • Impacts from climate change (including acidification)	• Water supply • Catchment control • Water efficiency • Adequate sanitation • Drought planning	• Filtering and capture • Monitoring and prevention • Early warning • Clean fuels • Real-time integrated, adaptive urban management	• Prediction and forecasting • Early warning systems • Resilient infrastructure • Resilience planning

• REVIEW QUESTIONS

1. What are the effects of poor ventilation?
2. In the workplace what are some of the sources of heat?
3. What are examples of health conditions that workers could experience due to exposure to extreme heat and cold?
4. How could exposure to the effects of extreme cold be mitigated?
5. What are the primary concerns about the impact of waste on the environment and how can these be effectively managed?

• REFERENCES

Anonymous. Nd. Loose notes on Safety, Health and Environment Management

Boukabara, Sid-Ahmed., Krasnopolsky, Vladimir., Stewart, Jebb. Q., Maddy, Eric. S., Shahroudi, Narges. and Hoffman, Ross.N., 2019. Leveraging modern artificial intelligence for remote sensing and NWP: Benefits and challenges. *Bulletin of the American Meteorological Society*, *100*(12), pp.ES473-ES491.

Burger,S. 2013. Law changes seen spurring increase of Solar Heating in Commercial Buildings. Available at: https://www.engineeringnews.co.za/article/Commercial-Buildings-to-use-more-large-scale-solar-water-heating-2013-11-01.html

Canadian Center for Occupational Health and Safety (n.d.). *Cold Environments - Working in the Cold.* [online] Ccohs.ca. Available at: https://www.ccohs.ca/oshanswers/phys_agents/cold_working.html.

Department of Environmental Affairs (2012). National Waste Information Baseline Report. Department of Environmental Affairs. Pretoria, South Africa.

European Environment Agency (n.d). Sanitary Landfill. Available at: http://www.eionet.europa.eu/gemet/aliss_scripts/concept/7438

Environmental Protection Agency (n.d.). What are the trends in wastes and their effects on human health and the environment. https://www.epa.gov/report-environment/wastes. Accessed May 29, 2020

HSE (2013). *Heat stress in the workpace: A brief guide.* [online] Hse.gov.uk. Available at: https://www.hse.gov.uk/pubns/indg451.htm

Makgae, Mosidi. 2011. Key Areas in Waste Management: A South African Perspective, Integrated Waste Management - Volume II, Sunil Kumar, IntechOpen, DOI: 10.5772/18023. Available from: https://www.intechopen.com/books/integrated-waste-management-volume-ii/key-areas-in-waste-management-a-south-african-perspective

National Building Regulations and Building Standards Act 103 of 1977 (1977). [online] Gov.za. Available at: https://www.gov.za/sites/default/files/gcis_document/201504/act-103-1977.pdf.

OSHA (2017). *OSHA Technical Manual (OTM) | Section III: Chapter 4 - Heat Stress | Occupational Safety and Health Administration.* [online] Osha.gov. Available at: https://www.osha.gov/dts/osta/otm/otm_iii/otm_iii_4.html.

OSHA (n.d.). *Emergency Preparedness and Response | Safety and Health Guides - Cold Stress Guide | Occupational Safety and Health Administration.* [online] Osha.gov. Available at: https://www.osha.gov/SLTC/emergencypreparedness/guides/cold.html.

OSHA (n.d.). *OSHA Technical Manual (OTM) | Section III: Chapter 3: Ventilation Investigation | Occupational Safety and Health Administration.* [online] Osha.gov. Available at: https://www.osha.gov/dts/osta/otm/otm_iii/otm_iii_3.html.

Solarzone (n.d.). Six ways poor ventilation can affect your workers' health. http://www.solazone.com.au/poor-ventilation/ Accessed May 29, 2020

Sundell, J., Levin, H., Nazaroff, W.W., Cain, W.S., Fisk, W.J., Grimsrud, D.T., Gyntelberg, F., Li, Y., Persily, A.K., Pickering, A.C. and Samet, J.M., 2011. Ventilation rates and health: multidisciplinary review of the scientific literature. *Indoor air*, *21*(3), pp.191-204.

United States Environmental Protection Agency. (2017). Report on the Environment. Available at: https://www.epa.gov/report-environment/wastes

Ye, Zhiping., Yang, Jiaqian., Zhong, Na., Tu, Xin., Jia, Jining. and Wang, Jiade., 2020. Tackling environmental challenges in pollution controls using artificial intelligence: A review. *Science of The Total Environment*, *699*, p.134279.

CHAPTER 25

DISASTER MANAGEMENT

The United Nations defines a disaster as

a serious disruption of the functioning of a community or a society

Other definitions include:

a sudden accident or a natural catastrophe that causes great damage or loss of life

an event or fact that has unfortunate consequences

A **disaster** is a sudden, calamitous event that seriously disrupts the functioning of a community or society and causes human, material, and economic or environmental losses that exceed the community's or society's ability to cope using its own resources. Though often caused by nature, **disasters** can have human origins.

Unfortunately, all disasters typically involve widespread human, material, economic or environmental impacts, which exceed the ability of the affected community or society to cope using its own resources and external assistance on a large scale is required.

Disaster Management can be defined as the organization and management of resources and responsibilities for dealing with all humanitarian aspects of emergencies, in particular preparedness, response and recovery in order to lessen the impact of disasters.

The World Health Organisation defines a disaster as:

Any occurrence that causes damage, ecological disruption, loss of human life and deterioration of health and health services on a scale sufficient to warrant an extraordinary response from outside the affected community

A disaster, therefore is

A sudden, calamitous event which results in widespread human loss, loss of livelihood, property and life.

A disaster may also be described as the product of a hazard such as earthquake, flood, landslide or windstorm coinciding with a vulnerable situation which might include communities, cities, towns or villages both rural and urban. There are two main components in this definition, namely hazard and vulnerability**.** Without vulnerability or hazard there is no disaster. A disaster occurs when hazards and vulnerability meet. Disasters are different from accidents. The loss of a sole breadwinner in a motor vehicle accident may be a disaster to the affected family, but only be an accident to the rest of the community.

Variables determine the difference such as, for example,

- ✓ Causes,
- ✓ Frequency,
- ✓ Duration of the impact,
- ✓ Speed of onset,
- ✓ Scope of the impact,
- ✓ Destructive potential,
- ✓ Human vulnerability.

Emergency management or disaster management is the discipline of dealing with and avoiding risks.

Emergency management is a discipline that involves

- ✓ preparing for disaster before it occurs,
- ✓ responding to a disaster such as emergency evacuation, quarantine, mass decontamination, and national lockdowns,

✓ and supporting, and rebuilding society after natural or human-made disasters have occurred.

It is an emergency management system which is a continuous process involving individuals, groups and community to manage hazards and:

✓ Deals with and avoid risks and impact of a disaster.
✓ Involves in preparing for disaster before it happens.
✓ Responds immediately to the outcome prior to and following a disaster.
✓ Deals with the process of supporting and rebuilding society after a disaster.

There are four phases in managing disasters, namely mitigation or prevention, preparedness, response, and recovery as depicted in the following diagram:

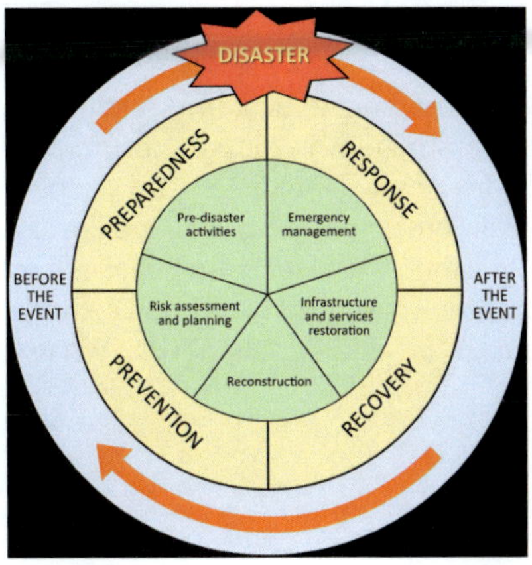

The Disaster Management Cycle
(Source: Erdeli, Krol and Natalizio, 2017)

• TYPES OF DISASTERS

There are four main types of disaster. These are:

✓ Natural disasters: including floods, hurricanes, earthquakes and volcano eruptions that have immediate impacts on human health and secondary impacts causing further death and suffering from, for example, floods, landslides, fires, and tsunamis.
✓ Environmental emergencies: including technological or industrial accidents, usually involving the production, use or transportation of hazardous material, and occur where these materials are produced, used or transported, and forest fires caused by humans.
✓ Complex emergencies: involving a break-down of authority, looting and attacks on strategic installations, including conflict situations and war.
✓ Pandemic emergencies: involving a sudden onset of contagious disease that affects health, disrupts services and businesses, brings economic and social costs. COVID-19 has been the latest challenge and will likely not be the last such threat.

Any disaster can interrupt the provision of critical essential services, such as health care, electricity, water, sewage/garbage, refuse and waste removal, transportation and communications. This interruption in service delivery can have serious effects on the health, social and economic networks of the affected local communities and countries. It is also true that disasters have a major and long-lasting impact on both directly and indirectly affected people long after the immediate effect has been mitigated. Further, poorly or inefficiently planned relief activities can have a significant negative impact not only on the disaster victims but also on donors and relief agencies.

• PRINCIPLES OF DISASTER MANAGEMENT

Preparation

Preparation is the early key to reducing the loss of life and destruction to physical property after a disaster. This includes training personnel in disaster protocols, moving necessary equipment into threatened areas, predicting potential dangers and developing collaborative systems between emergency personnel and even local communities. The goal of preparation is to decrease emergency response time and ensure that necessary equipment is available and on-site after a disaster. The lessons learnt from the recent COVID-19 pandemic and the level of preparedness of South Africa when the pandemic hit the country will be invaluable when faced with a similar emergency challenge of such magnitude.

Mitigation

Mitigation is the collaborative attempt to reduce the overall cost in human life and property resulting from a disaster. This includes flood plans, evacuation strategies, floodwalls and trained emergency personnel who are on-site and working to minimize the effects of the disaster. Additionally, as part of the mitigation principle, emergency personnel are responsible for responding to potential dangers that could pose long-term hazards within a community, such as corrosion of the land, disease from poor conditions or severe risks to the wildlife in the area. The declaration of a state of national disaster by the South African government is an example of a mitigative effort to contain the spread of COVID-19 and save lives. Other strategies such as a national lockdown consisting of 5 phases were implemented as part of this effort. Supporting frontline emergency healthcare workers across the country and operationalizing the South African National Defence Force (SANDF) were critical elements in this endeavour.

Protection

Protection is a two-step procedure that includes relaying critical information to people in the affected area, and teaching them how to reduce their individual risks. Information is the key to protecting human lives and reducing potential property loss. During a disaster situation, emergency alerts help warn people about status of a situation, and instruct them on how to respond. This may include an order to evacuate an area, or a set of preventive measures citizens can take within their own homes. In the face of a national disaster being declared as has been seen with the recent COVID-19 pandemic repeated and frequent communications using all media and communication platforms in multiple languages about correct protective measures to be taken are critical. Helplines, WhatsApp numbers and national information sources were operationalized.

Response

Active response plans are responsible for addressing the moment-to-moment needs of individuals during a disaster. This includes providing food, emergency needs, shelter and access to emergency personnel throughout a disaster situation. Response teams act quickly, addressing the changing needs during a disaster to reduce the loss of human lives and property damage. A massive team of health screeners and trackers was mobilized in response to COVID-19. Many non-governmental organizations (NGOs) mobilized to provide emergency supplies such as food parcels and meals to the most vulnerable of South African society. Government introduced many extraordinary relief initiatives and programs to alleviate the effects of COVID-19.

Recovery

Recovery begins when the disaster is finished and serves the community, establishing long-term medical care for those in need after a disaster, rebuilding and working to reduce the chance of future similar disasters from occurring. The recovery of an area requires long-term effort and a strict analysis of a disaster in order to avoid future recurrences of the situation. At the time of writing South Africa is quite a long way from this phase relative to COVID-19.

• THE ROLE OF MUNICIPALITIES IN DISASTER MANAGEMENT

Every municipality must have a disaster management plan as part of its Integrated Development Plans, according to the Municipal Systems Act. Many municipalities have been severely criticized for their unpreparedness and poor management of the COVID-19 pandemic despite having been given clear guidelines and protocols.

Structure and Mechanism: This plan must set up the structure and mechanisms for dealing with disasters and it must anticipate future disasters. Plans must be developed to deal with disasters that occur regularly - for example, flooding of informal settlements and roads.

Protection Services Department: In each municipality, the Protection Services department is responsible for disaster management. The department usually deals with traffic policing, fire brigades, law enforcement, and sometimes ambulances on an agency basis for provincial government. The role of the disaster management team is to coordinate the response to disasters and emergencies, ensuring that resources are applied effectively, whatever it may be. Fire services, ambulance services, emergency medical services, engineers and traffic services can all become involved in the disaster management effort.

Capacity: When a disaster exceeds the capacity of a local authority, the district, province or national can become involved, coordinating and facilitating the response and efforts of various local authorities. Other parties such as the SANDF as well as volunteer organizations such as the Red Cross, St John's and the National Sea Rescue Institute can also be drawn in if needed. In the case of COVID-19, the SANDF were brought in to assist the response and efforts to enforce provisions of the lockdown and support initiatives to contain the spread of the virus.

Disaster Management Activities: These activities include the co-ordination of disaster response agencies, the compilation and exercising of contingency plans, and disaster management education and training.

Funding: Following the finalisation of the Act, the national government will announce on a funding mechanism for provinces and municipalities to finance their comprehensive disaster management plans.

• IDENTIFYING POTENTIAL LOCAL DISASTER HAZARDS

These can include all or some of the following:

- ✓ Mass-event situations such as concerts, sport, other social gatherings such as funerals and weddings;
- ✓ Storms and storm damage;
- ✓ Flooding;
- ✓ Fires: Domestic, mountain and veld;
- ✓ Oils spills, at sea, on land;
- ✓ Transport accidents;
- ✓ Hazardous material spills such as, for example, spilling of chemicals from factories and trucks on highways

• DISASTER MANAGEMENT IN THE ORGANISATION

Every organization should have in place a strategy to respond to a disaster should it occur. The following would be some of the actions to be taken, namely:

- Create a disaster planning team involving key functional managers from all operations of the organization as well as human resources and health and safety or risk management professionals. Officials from local fire, police and emergency-management offices as well as city, municipality or provincial officials are also important members of this team.
- Decide who will be authorized to make decisions for the planning team and the extent of the authority to make decisions. If additional authority is needed for important decisions, this may be requested or additional decision makers invited to join the team.
- Develop a mission statement and plan structure. The team needs to decide the purpose and scope the project to be undertaken so that there are no misunderstandings and the end product of the team can be implemented. The President and members of his cabinet repeatedly emphasized that the mission during COVID-19 was to save lives by all means. A timetable, deadlines and budgets are useful tools in helping this process along.
- Establish what personnel should be in the Emergency Management Group (EMG). This group can be considered a crisis team who will coordinate all activities related to the emergency or disaster. An incident commander or leader must be chosen out of this group. The incident commander is the leader of the EMG and the EMG activities support the decisions, leadership and activities of the incident commander.
- Install an incident command system (ICS). An incident command system is a protocol used by municipal fire services, police organizations, emergency response units and business organizations to organize a coordinated response to an emergency or disaster. Although incident command systems are standardized, there is some flexibility to customize an ICS to fit special circumstances.
- Set up an emergency operations centre (EOC). The emergency operations centre is a centralized area that serves as the base of communications, decisions and emergency operations. It is essential to set this up so that decision makers gather in a known central location in an organized fashion to immediately coordinate disaster response. Set up the EOC in an area unlikely to be directly affected by the disaster or emergency but accessible for all responders. Ensure the EOC has adequate communication and transportation resources.
- Coordinate assistance from outside resources. If required ensure that mutual aid agreements are in place, as well as provisions to transfer authority for incident command to emergency responders such as police, fire or health officials.

• EMERGENCY PREPAREDNESS AND RESPONSE

Emergency preparedness and response in South Africa is governed by the Disaster Management Act No. 57 of 2002 and is designed to provide for-

- ✓ an integrated and coordinated disaster management policy that focuses on preventing or reducing the risk of disasters, mitigating the severity of disasters, emergency preparedness, rapid and effective response to disasters and post-disaster recovery;
- ✓ the establishment of national, provincial and municipal disaster management centres;
- ✓ disaster management volunteers

An emergency is any unplanned event that threatens workers, customers, or the public; that shuts down business operations; or that causes physical or environmental damage.

In order to be defined as an emergency, the situation should:

- ✓ Threaten life, health, property or environment
- ✓ Have already caused a loss of life, put the health of people in danger, damaged property or the environment
- ✓ Have a high probability to escalate and causing danger or loss of life, loss of property and danger to the environment

Incident: An event that may cause or causes an interruption or a crisis, such as an accident. An incident could lead to an emergency.

• WHEN A LOCKDOWN IS NECESSARY – A RECENT EXAMPLE

• WHAT IS A LOCKDOWN?

A lockdown is an emergency protocol that requires all South Africans to stay at home except for essential purposes. Under various levels of lockdown, the protocols and allowed activities and economic engagements will differ depending on the level being enforced. Under total lockdown or Level 5 only grocery stores, pharmacies, banks and other essential industries will remain open. All non-essential activities are suspended. The intention is to slow down the spread of the virus and flatten the curve of escalation of infections.

• PURPOSE OF THE LOCKDOWN

In the case of the COVID-19 pandemic, the nation-wide lockdown was primarily necessary to fundamentally disrupt the chain of transmission and prevent the spread of the virus and save the lives of South Africans. The lockdown was for a period of an initial period of 21 days which was extended depending on the need to do so – Level 5. Toll-free WhatsApp and Emergency numbers were provided in the notice of the lockdown. For example, 0600 123 4567.

• RESTRICTIONS ON WORKING?

A list of essential services was published in the Government Gazette and includes pharmacies, hospitals, banks, supermarkets, petrol stations and health care providers. Companies that are essential to the production and transportation of food, basic goods and medical supplies will remain open under Level 5.

• PERMISSABLE ACTIVITIES

In terms of the level 5 level of lockdown, South Africans were able to:

- Seek medical care;
- Buy groceries;
- Visit the pharmacy;
- Access banking services;
- Get petrol; and
- Collect a social grant.

Special powers were granted to the SANDF, SAPS and other enforcement agencies to ensure that the lockdown was enforced. Violators were criminally prosecuted. Any additional measures were communicated after broad consultations by a Central Command led by the President. It is important to become knowledgeable about the provisions of the Disaster Management Act of 2002 as amended in terms of Section 27(2) and special powers given to the President and his Cabinet.

As capacity and conditions improved the lockdown was eased through staggered phases.

Workplace plans

In terms of Regulation 16 (6) (b) a COVID-19 workplace plan must be developed prior to reopening of an organization that employs workers or serves the public. In the case of small businesses, the plan is expected to be basic reflecting the size of the business, whereas for medium and larger business a more detailed written plan should be developed given the larger numbers of workers at the workplace. This detailed plan must include the following, namely:

1. The date the business will open and its operating or opening hours;
2. The timetable setting out the phased return-to-work of workers, to enable appropriate measures to be taken to avoid and reduce the spread of the virus in the workplace;
3. The steps taken to get the workplace COVID-19 ready;
4. A list of workers who can work from home; workers who are 60 years or older; and workers with comorbidities who will be required to stay at home or work from home;
5. Arrangements for workers in the organization:
   a. Sanitary and social distancing measures and facilities at the entrance and exit to the workplace;
   b. Screening facilities and systems;
   c. The attendance-record system and infrastructure;
   d. The work area of workers;
   e. Any designated area where the public is served;
   f. Canteen and bathroom facilities;
   g. Testing facilities for establishments with more than 500 workers;
   h. Worker rotational arrangement for establishments where fewer than 100% of workers will be permitted to work.
6. Arrangements for customers or members of the public, including sanitation and social distancing measures.

• PREPARATION FOR AN EMERGENCY

The first step is to find out what the emergency telephone numbers are for the area that is being worked in and ensure that every health and safety representative has the numbers available. The numbers should also be given to all the department and section heads and should be displayed in prominent places all over the workplace.

THE NATIONAL EMERGENCY TELEPHONE NUMBER IS 10177. COVID-19 NUMBER IS 0600 123 4567

Employers should therefore

- ✓ Post fire escape plans in prominent locations such as notice boards, toilets and change rooms on every level of the facility
- ✓ Ensure all workers are familiar with exit locations; escape routes and fire extinguisher locations.
- ✓ Conduct regular fire drills.
- ✓ Post the emergency numbers on or near all telephones.
- ✓ Make provisions for safe evacuation of workers with disabilities by appointing someone to assist them usually trained Marshalls.
- ✓ Appoint and train a fire warden on each level of the building to ensure safe evacuation and ongoing health and safety programs.

(Source: https://blog.ansi.org/2018/11/nfpa-72-2019-fire-alarm-signaling-code-smoke/#gref)

Workers should

- ✓ Know the location of all building exits.
- ✓ Know the location of the nearest fire alarms and how to use them.
- ✓ Count the doors or desks between their work area and the nearest exit. During a fire, exit signs may not be visible due to smoke or a power failure

(Source: https://za.pinterest.com/pin/542754192572280547/)

Emergency procedures must be developed to be followed in the case of an emergency. Management and all workers must know what these procedures are, so that they are able to:

- ✓ Identify the nature of the emergency in order to take the appropriate action
- ✓ Prioritise and grade the emergency in order to determine evacuation drills and or emergency signals to be used

• EMERGENCY PLANS

Emergency plans contain the steps that have to be taken in an emergency, to ensure that everyone acts in an organised manner. The main purpose of these plans is to safeguard lives, protect property and to maintain essential services.

It must include the following:

- ✓ How to co-operate with external organisations such as the SAPS, ambulance services, and hospitals.
- ✓ If there are local emergency response units in the area, establish how the efforts of the organisation can be integrated with these units during emergencies.
- ✓ What actions to take during specific emergencies.

It is important that the emergency plan

- ✓ Is simple so that everybody involved can understand it.
- ✓ Is practical and can be activated quickly and without causing confusion or panic
- ✓ Is coordinated with the local security and emergency response units, where possible
- ✓ Is exercised periodically.

Everyone involved must know exactly what their responsibilities and tasks are.

• CO-ORDINATION

The emergency plan must be controlled from a central point. A chief coordinator must be appointed. For example, in the case of COVID-19 the President was the chief co-ordinator of the National Command Council. It is of the utmost importance that when an emergency situation arises every member of the emergency team knows exactly where to report to and who to report to. Instructions from the chief coordinator must be strictly adhered to and enforced.

Source: https://images.news18.com/ibnlive/uploads/2013/12/disaster-management-officers-deserve-salary-hike-for-their-good-work-says-vijay-bahuguna_191213040222.jpg

• CO-ORDINATING ACTIONS

The following actions must be co-ordinated during an emergency, namely:

- ✓ First aid and medical services
- ✓ Fire fighting
- ✓ Access and egress control
- ✓ Evacuation of personnel
- ✓ Rescue of personnel
- ✓ Isolation of electricity, gas and other fuel supplies
- ✓ Engineering requirements

• TESTING THE PLAN

Practical exercises must be held to test the practicability and effectiveness of the emergency plan and to make necessary changes. It must be done discretely so as not to cause panic. The objective is that the execution of the plan must become second nature to those involved. In the case of the COVID-19 pandemic which was an unprecedented and arguably an unexpected emergency plans had to be rapidly drawn up with assistance of the WHO and other countries with more experience together with South African health experts. However, the lessons learnt through this exercise will enable better preparedness and emergency plans in the future which will be able to be tested based on recent past experiences.

• LOCATION OF EMERGENCY EQUIPMENT

The location of emergency equipment is vital to the emergency team in all emergency situations. Every member of the team and staff should know the location of the following: -

- ✓ Fire Alarm.
- ✓ Emergency alarm (robbery).
- ✓ First-aid equipment.
- ✓ Fire-fighting equipment.

(Source: https://za.pinterest.com/pin/792000284446138488/)

(Source: http://www.k-direct.biz/Personal-Protective-Equipment/Fire-Fighting-and-Emergency-Response/fire-hosereel-25m-with-6kg-extinguisher)

• SEQUENCE OF AN EVACUATION

The sequence of an evacuation where required can be divided into the following phases:

- ✓ detection
- ✓ decision
- ✓ alarm
- ✓ reaction
- ✓ movement to an area of refuge or an assembly station
- ✓ transportation

It is critical that every member of the emergency team knows the following, namely:

- ✓ The warning signal that will be broadcasted to indicate that an evacuation will take place i.e. siren, hooter, vocal instructions or combination.
- ✓ The location of emergency exits and secondary exits or special exits for handicapped persons.
- ✓ The location of lifts.
- ✓ The location of assembly points.
- ✓ The location of emergency equipment such as fire extinguishers and first aid kits.
- ✓ The means of by-passing electronic access control such as locks, gates, doors, turnstiles.
- ✓ The elderly, infirm or handicapped be allowed out first.
- ✓ All other persons, preferably women and children first with males second and staff third.
- ✓ Obviously, the nature of the emergency will dictate in what order, if any, people leave the store.
- ✓ It is vital that security officers prevent panic at all cost. Panic is an absolute killer. Security officers must not allow people to run, shout, scream, or otherwise lose control.

When the evacuation signal is heard, everyone should follow the emergency evacuation procedure

• GENERAL AND PARTIAL EVACUATION

Evacuation normally causes a certain amount of panic. The decision to order an evacuation must be taken with great care. In most instances only a partial evacuation may be necessary.

Partial Evacuation: This action is necessary where a suspicious object is found or when a specific place was indicated by a person making a bomb threat. It will also be ordered where a fire is localised and can be easily contained.

Advantages of Partial Evacuation

- ✓ the risk of injuries is minimised
- ✓ workers are available to effect searching of the area and to assist
- ✓ essential services can continue
- ✓ workers must remove all personal belongings such as handbags, laptops, cell phones and lunch boxes

General Evacuation: The decision to order a general evacuation is a difficult one to take because of the accompanying responsibility, and the problems it may cause.

Disadvantages of General Evacuation

- ✓ workers can be injured in the rush to evacuate, particularly if they are not well trained.
- ✓ in the event of a bomb threat the possibility exists that another similar device could be placed outside the building.
- ✓ workers and supervisors are not available after the evacuation to help search the premises.
- ✓ loss of production can have an adverse financial effect.

• ENSURING SAFETY FOR DRIVERS IN PROTESTS

In recent years the phenomenon of public protests has become a regular feature of South African life which have the tendency to be disruptive to normal operations and sometimes become violent. It is important to know what steps to take to ensure personal safety and that of workers. The following steps should be considered while driving when being confronted with protest action in streets and motorways as advised in a News24 article in 2019:

- The simplest way to remain safe is to avoid the situation altogether by using alternative routes and points of access;
- Listen to the news and look out for alerts on social media and apps before choosing a route;
- Keep an eye for people gathering at the side of the road or on bridges;
- Drive watching 12 seconds ahead of the vehicle so that any volatile situations can be identified immediately;
- If forced to stop or slow down because of a protest, identify an escape route;
- If the situation is becoming dangerous, unbuckle the seatbelt so that the vehicle can be exited quickly while being aware that the vehicle is probably the safest place to be;
- Do not engage with protesters negatively;
- If forced to get away from an area quickly, remain calm;
- Be aware that taking aggressive evasion action that does not account for other drivers or people on the road could be more dangerous than acting calmly and slowly;
- Listen to authorities who are trained to defuse the situation and protect all citizens from harm;
- The primary consideration is personal safety and that of passengers if there are any;
- While it is preferable that the vehicle does not get damaged, if there is a need to choose between the two, personal safety should be prioritized; and
- Keep someone posted on where and what is happening.

Finally, the objective is to avoid the situation completely. Never be caught by surprise by a potentially dangerous situation or be blocked in when things can turn violent. Drivers should always be prepared for every situation.

These rules also apply to drivers of emergency vehicles and those delivering much needed emergency supplies such as during the recent COVID-19 pandemic.

• HAZARDS EVALUATION AND CONTROL

The Act defines

a hazard as "a source of or exposure to danger" and a risk as "the probability that injury or damage will occur."

Each hazard has risk(s) involved. The degree of risk involved varies according to the situation. When using a sharp knife to peel an apple, for example, a two-year-old toddler will run a much greater risk of cutting himself than an adult performing the same exercise.

- PHYSICAL HAZARDS

These hazards can harm the body and/or influence or stop the effective functioning thereof. Examples include:

✓ Fire and explosion – The hazard is the fire, where the workers is exposed to danger, while the risk is that the fire can cause severe burns and injuries.
✓ Extreme heat – working in a steel mill or very deep underground. The hazard is the extreme heat and the risk is injury due to heat exhaustion
✓ Extreme cold – working outdoors in winter or in a cold room. The extreme cold is the hazard while the risk is injury due to frost bite.
✓ Noise – working with pneumatic or powered by compressed air tools and equipment such as drills and jackhammers. The hazard is the loud noise, while the risk is damage to hearing due to the loud noise.
✓ Radiation – from exposure to arc welding, X-rays or radioactive material. Being exposed to radiation is a hazardous condition and the risk is damage to health due to the radiation
✓ Water – drowning and flooding. Water is the hazard and the risk is drowning.
✓ Extreme physical conditions – continuous vibration experienced by drivers of heavy vehicles, the effect of extreme g-forces on fighter- and aerobatic pilots, and the effect of high pressure on divers. The physical conditions are the hazards and the risks are injury due to the hazards.

- CHEMICAL HAZARDS

There are many categories of chemicals, each with unique properties and potential hazards. Examples include:

✓ Acids – sulphuric acid, "spirits of salts", pool acid, etc. are highly corrosive and can cause severe burns. The nitrous fumes caused when nitro-glycerine based explosives like dynamite explode contain enough nitric acid to burn and cause blistering inside the lungs if inhaled, often with fatal results if not treated timely, as the blisters burst and fills the victim's lungs with liquid and he drowns. So, the hazard would be the presence of acid and the risk is injuries caused by inhaling the acid.
✓ Alkalis – caustic soda and soda ash are also very corrosive and can cause severe burns. Caustic soda is a hazard and the risk is injury if it comes into contact with the human body.
✓ Air pollution – harmful gases and dust is often released into the atmosphere and breathed in by living creatures or carried over great distances and deposited as acid rain, affecting life.
✓ Water pollution – toxic chemicals are released into river systems with disastrous effect on life and the environment.

- BIOLOGICAL HAZARDS

Biological hazards involve, for example, living organisms, like bacteria, spores, and fungi. Examples include:

✓ Infections caused by bacteria, fungi, and parasites. The bacterium is the hazard and the risk is infection when it enters the human body.
✓ Farm workers and animal handlers get infections from animals. The hazardous condition exists when working with farm animals and the risk is infection due to handling the animals.
✓ Health workers get infected through exposure to used needles, inhaling or swallowing bacteria or viruses. A used needle is a hazard that becomes a risk when it is used.
✓ Sewage or waste disposal workers are continuously exposed to bacteria and viruses. Sewage is a hazard and the risk is infection when contact is made with it.

- ERGONOMIC HAZARDS

These hazards relate to how "worker friendly" the worker's working environment is. Working under unfavourable conditions can cause discomfort and injury. Factors like having to stand for long periods, working in cramped spaces, not having adequate ventilation, etc. can all cause discomfort and eventual injury. The hazard is present in the working condition and the risk is the injury that could happen as a result of the hazard.

• PSYCHOSOCIAL HAZARDS

This normally goes hand in hand with stress and tension in the workplace. A disgruntled worker might be so upset that it affects his ability to concentrate properly. If he works with dangerous machinery or equipment his lack of concentration may cause him to overlook safety precautions, resulting in injury or damage to property. Working with machinery is hazardous and the risk is injury or damage to property if the machine is not operated safely.

• RISKS

Risks arise when a hazardous condition exists. This means that health and safety precautions must be undertaken:

- ✓ work according to laid down safety procedures
- ✓ wear protective clothing
- ✓ be alert to the hazard and the risk involved

• PRACTICAL HAZARD IDENTIFICATION

In order to effectively manage risks, it is necessary to classify hazards identified during the health and safety inspections and allocate points to different hazards according to severity, frequency and possible exposure. The effect of this action will be that incidents can now be prioritised and dealt with according to priority, namely the most serious incident will be dealt with first and then the others in order of priority. Any classification system may be used as long as it works effectively and reflects the real state of affairs.

The first and most important stage in the risk assessment process is to identify hazards.

All the mining activities have to be assessed. If one leaves out some activities, the risk increases.

When an inspection is done to identify hazards, the risk assessment must be based as follows:

- ✓ geographically by looking at shafts, haulages, and workshops;
- ✓ functionally – linked to work activities such as trucking, blasting, and winding operations;
- ✓ Hazards or issues, such as the hanging wall, electricity, and machinery;
- ✓ Type of work such as engineering, environmental, and metallurgical.

It is best to use a mixture of all these approaches or at least the first two, to make sure that something important is not overlooked.

The first step is to identify the purpose and extent of the investigation.

The aim of an inspection is not to identify a culprit to punish for something that might be wrong. On the contrary, it is a valuable tool to identify potential hazards and prevent them from becoming real hazards. The information gathered during an inspection is analysed and recorded. It is then available for reference or research purposes.

• CLASSIFICATION OF HAZARDS

Normally three factors are used for this mathematical classification exercise namely severity *(how serious it is)*, frequency *(how often it occurs)* and exposure *(how many people will be exposed)*. Points are allocated based on the rating of each factor and the overall outcome or score based on the mathematical calculation is indicative of the classification or category of the exposure.

• SEVERITY

Class A – (Major, catastrophic, high priority, 4 points) – an incident that causes death or permanent disability or extensive damage to property or the environment.

Class B – (Serious, medium priority, 3 points) – such incidents may cause serious injury or disease or sufficient property damage to cause disruption, but without permanent far-reaching effects.

Class C – (Minor, low priority, 2 points) – these incidents normally cause minor injury or disease, if any, and minor disruption in the workplace.

Class D – (Negligible, 1 point) – no injury or loss of property involved.

- FREQUENCY

Class A – Likely to happen immediately. (4 points)

Class B – Probably will happen in time. (3 points)

Class C – Possible to happen in time. (2 points)

Class D – Unlikely to happen. (1 point)

- EXPOSURE

The number of people likely to be exposed to the hazard:

Class A – More than 50. (4 points)

Class B – 10 – 49. (3 points)

Class C – 5 – 9. (2 points)

Class D – Fewer than 5. (1 point)

Example:

A major hazard (4 points), which is likely to become a major incident immediately (4 points) and likely to involve more than 50 people (4 points) will score a total of 16 points, the worst case scenario.

By repeating the same exercise on other incidents the total points scored will determine the priority of the hazard.

Hazards with a high score (12 – 16 points) will take high priority, demanding immediate corrective action; while hazards with a low score (1 – 5 points) will not be treated with the same urgency.

- REVIEW QUESTIONS

1. Describe the principles of disaster management
2. What is the role of the municipality in a disaster management crisis?
3. What are the elements of a disaster management strategy in an organization?
4. Why would a partial evacuation be considered before a general evacuation?
5. What should drivers do to protect themselves when confronted with violent protests?
6. Develop the key aspects of a disaster management plan for your particular organization.

- REFERENCES:

Akter, Shahriar. and Wamba, Samuel. Fosso., 2019. Big data and disaster management: a systematic review and agenda for future research. *Annals of Operations Research*, *283*(1-2), pp.939-959.

Atkinson, Christopher., 2019. Theme-Based Book Review: Disaster Management and Environmental Policy.

Anonymous. Nd. Loose notes on Safety, Health and Environment Management

Department of Labour (1993). Occupational Health and Safety Act No: 85 of 1993.

Disaster Management. (2019, October 2). Physiopedia, . Retrieved 11:30, February 26, 2020 from https://www.physio-pedia.com/index.php?title=Disaster_Management&oldid=223951.

Education and Training Unit for Democracy and Development (ETU). (n.d.). Disaster Management. Available at: https://www.etu.org.za/toolbox/docs/government/disaster.html

Erdelj, Milan. and Natalizio, Erdelj., 2016, February. UAV-assisted disaster management: Applications and open issues. In *2016 international conference on computing, networking and communications (ICNC)* (pp. 1-5). IEEE.

Erdelj, Milan., Krol, Michal., and Natalizio, Enrico. 2017. Wireless Sensor Networks and Multi-UAV Systems for Natural Disaster Management. *Computer Networks,* 124

International Federation of Red Cross and Red Crescent Societies (IFRC). (n.d.). What is a disaster? Available at: https://www.ifrc.org/en/what-we-do/disaster-management/about-disasters/what-is-a-disaster/

Kirschenbaum, Alan. ed., 2019. *Chaos organization and disaster management*. CRC Press.

Makrosafe Holdings. (n.d.) Security Industry Emergency Preparedness. Available at: https://www.makrosafe.co.za/Security-Industry-emergency-preparedness.html

News24 (2019). Clashes in Gauteng: what to do when encountering protesters on SA's roads. Available at: https://www.news24.com/wheels/news/guides_and_lists/clashes-in-gauteng-what-to-do-when-encountering-protesters-on-sas-roads-20190830

OSHA (2016). HAZARD COMMUNICATION Hazard Classification Guidance for Manufacturers, Importers, and Employers. [online] Osha.gov. Available at: https://www.osha.gov/Publications/OSHA3844.pdf
Oxford Dictionaries (n.d.). Meaning of Disaster in English. Available at: https://www.lexico.com/definition/Disaster
Ray, Partha. Pratim., Mukherjee, Mithun. and Shu, Lei., 2017. Internet of things for disaster management: State-of-the-art and prospects. *IEEE Access*, *5*, pp.18818-18835.
The International Water Association Publishing (n.d.). Disaster Management. Available at: https://www.iwapublishing.com/news/disaster-management
UN (2009). Terminology on Disaster Risk Reduction. [online] Unisdr.org. Available at: https://www.unisdr.org/files/7817_UNISDRTerminologyEnglish.pdf
WHO (2007). Risk reduction and emergency preparedness. [online] Who.int. Available at: https://www.who.int/hac/techguidance/preparedness/emergency_preparedness_eng.pdf.
Zawawi, E.M.A., Yusof, Nor. Syazwani. and Ismail, Zulhabri., 2018. Adoption of post-disaster waste management plan into disaster management guidelines for Malaysia. *Journal of Material Cycles and Waste Management*, *20*(1), pp.223-236.

CHAPTER 26

ENVIRONMENTAL MANAGEMENT SYSTEMS (EMS)

Care for the environment includes concern about how industry manufactures products and the impact that such products and manufacturing processes could have on the environment. Controlling this requires a documented environmental management system (EMS) so that everyone understands how they must work to protect the environment from pollution and degradation. There are many different types of environmental management systems, which can be used by companies of differing sizes.

In 1996, the South African Bureau of Standards (SABS) published a code of practice, SABS ISO 14001 that provides companies with guidelines and requirements for adopting an environmental policy and implementing an environmental management system. This code helps companies to set their own environmental policies and objectives. The code is an international standard and provides the basis for a uniform EMS, which will conform to wider international standards and requirements.

Environmental management has been defined as:

"actual decisions and action concerning policy and practice regarding how resources and the environment are appraised protected, allocated, developed, used, rehabilitated, remediated, and restored."

Environmental management has over time developed into a multi-disciplinary process requiring careful preparation and planning. Planning for the implementation of environmental management should be carried out in a participative manner to ensure that all stakeholders and interested and affected parties who will be subject to it, will understand its purpose and the manner and form in which it is to be applied. Information on natural and human activities, processes and systems must be gathered, plans and procedures decided upon, documents drawn up and systems implemented. The results add environmental perspectives to existing planning processes. The process will result in the identification of new issues and concerns, but in most cases, the matters identified are well known in concept, if not in environmental jargon.

An Environmental Management System (EMS) is a set of processes and practices that enable an organization to reduce its environmental impacts and increase its operating efficiency. EMS is a framework that helps an organization achieve its environmental goals through consistent review, evaluation, and improvement of its environmental performance. The assumption is that this consistent review and evaluation will identify opportunities for improving and implementing the environmental performance of the organization. Although formal environmental management systems such as ISO 14001 are desirable, for most smaller businesses a simple environmental management system is all that is needed and, in most cases, it can be linked to the existing operational management systems. ISO 14001 standards are of interest particularly to South African organizations, which are currently exporting or considering exporting their products to countries in Europe, or to the United States or Japan, as an internationally recognised EMS may become a prerequisite for supplying certain organizations in these countries. National standards for EMS may also be consulted in the development of a programme. In South Africa, the SABS Code 0251 describes the recommended aspects of an EMS.

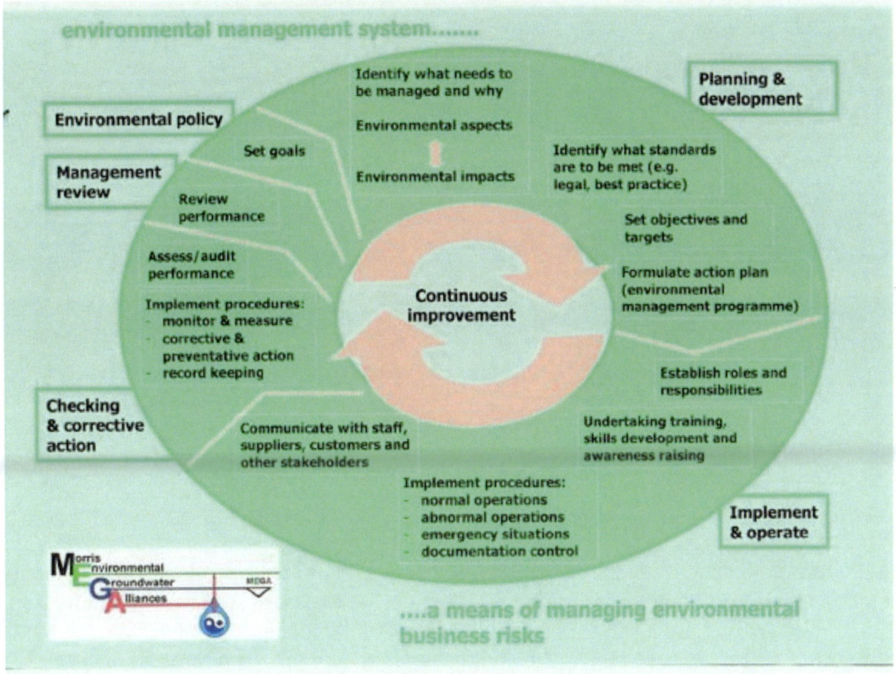

Source: http://enviropaedia.com/topic/default.php?topic_id=98 Accessed May 27, 2020)

Basic elements of an EMS include the following:

- Reviewing the environmental goals of the organization;
- Analysing environmental impacts and legal requirements of the organization;
- Setting environmental objectives and targets to reduce environmental impacts and comply with legal requirements;
- Establishing programs to meet these objectives and targets;
- Monitoring and measuring progress in achieving the objectives;
- Ensuring the environmental awareness and competence of workers; and,
- Reviewing progress of the EMS and making improvements.

Waste, air, water and ground pollution; and hazardous chemicals are all key aspects of an industrial environmental programme, although certain industries will need to focus in more detail on some issues. The use of resources, such as energy in the form of electricity, heavy furnace oils, and LP gas and water should form part of the EMS programme.

- BENEFITS OF AN EMS

The benefits of implementing an EMS in an organization include the following, namely:

- Improved environmental performance
- Enhanced compliance
- Pollution prevention
- Resource conservation
- New customers and markets
- Increased efficiency and reduced costs
- Enhanced worker morale
- Enhanced image with the public, regulators, financiers, and investors
- Worker awareness of environmental issues and responsibilities

An EMS encourages an organization to continuously improve its environmental performance. The system follows a repeating cycle as shown in the figure.

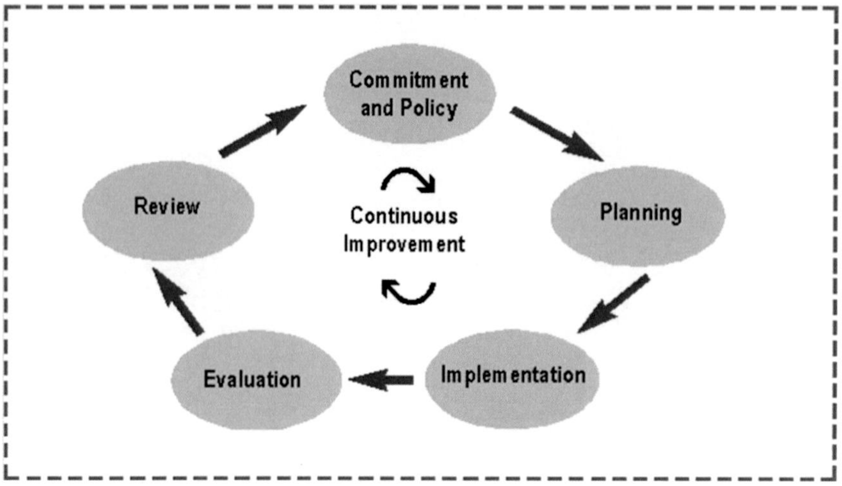

Source: https://www.epa.gov/ems/learn-about-environmental-management-systems#what-is-an-EMS. Accessed May 30, 2020

The organization first commits to an environmental policy, then uses its policy as a basis for establishing a plan, which sets objectives and targets for improving environmental performance. The next step is implementation. After that, the organization evaluates its environmental performance to see whether the objectives and targets are being met. If targets are not being met, corrective action is taken. The results of this evaluation are then reviewed by top or senior management to determine whether the EMS is actually working in the organization. Management reviews the environmental policy and sets new targets in a revised plan. The organization then implements the revised plan. The cycle repeats, and continuous improvement occurs.

The basic requirements of an EMS in terms of ISO 14 001 are:

- **Commitment and Policy** - Top or senior management commits to environmental improvement and establishes the environmental policy of the organization and which is the foundation of the EMS.
- **Planning** - An organization first identifies environmental aspects of its operations such as air pollutants or hazardous waste, that can have negative impacts on workers, people outside the organization and the environment. Legal and other considerations are considered. Criteria considered most important by the organization are chosen which could include, for example, worker health and safety and environmental compliance. Objectives and targets are set. The final part of the planning stage of establishing an environmental management program is devising an action plan for meeting the targets which includes setting up a structure designating responsibilities, establishing a schedule or timetable, and outlining clearly defined steps to meet the targets.
- **Implementation** - An organization implements the action plan using the necessary allocated resources. A training and awareness programme for workers is developed. Other steps in the implementation stage include documentation, following operating procedures, and setting up internal and external communication lines.
- **Evaluation** - An organization monitors its operations to evaluate whether targets are being met. If not, the organization takes corrective action. To achieve this objective, the organization establishes procedures for monitoring and measurement, for corrective action, for maintaining records, and conducting environmental audits.
- **Review** - Top or senior management at regular intervals reviews the results of the evaluation to determine the effectiveness of the current EMS in the organization and whether it is consistent with the values of the organization. The plan is then revised to optimize the effectiveness of the EMS. The review stage creates a loop of continuous improvement for the organization.

Elements of an Environmental Management Program Work Plan

Element	Comment
	Required by ISO 14001
Objective	By implication, the standard requires that the objective be incorporated into the environmental management program. Reference should be made to the associated activity, its environmental aspect, the environmental impact, and possibly even the element of environmental policy that the environmental management program is related to.
Target	The environmental target is also required by implication. Where appropriate, the organization should include the performance indicators and time frame associated with the target. An explanation can also be included about why the particular target was selected, how it fits into a sequence of targets, and how it relates to other environmental targets of the enterprise.
Personnel	This refers to who is responsible for carrying out the overall program and its principal elements, and for achieving the target and objective. These responsible personnel should not be confused with workers who may carry out specific tasks under their supervision.
Tasks	A description of the basic tasks, in sequence including, for each task, who does what, the time period in which the task has to be done, reports and other documents such as procedures and work instructions that need to be generated, and the expected results.
Schedule	All the standard requires is a timeframe for achieving the target and objective. But the only way to establish the timeframe is to estimate, task-by-task, the time required. Moreover, a task-by-task schedule will be needed to make revisions in the work plan later, so this information, summarized from the "Tasks" section, might as well be included. Schedule information could be represented as a bar chart or critical-path graphic.
	Not Required by ISO 14001
Communications	Who is responsible for ensuring full communication, the e-mail or WhatsApp group and who controls its membership, reports to be generated and how they are to be circulated, and specific reporting events, such as various meetings and whether they will be physical meetings or virtual meetings.
Coordination	Who is responsible for overall coordination of environmental management program implementation activities, the schedule for program team meetings both physical and virtual, and the schedule for regular progress presentations to top or senior management.
Resources	Time, personnel, equipment, and money committed to the program, and how they are to be utilized. Commitments of administrative and clerical personnel should be explicit.
Approach	A brief overall description of the program, with an explanation or justification of why the program is designed the way it is and perhaps how it represents implementation of the environmental policy of the organization. Here other comments may be included about the program, its work plan, and its aims that do not fit elsewhere but are important to understanding what the program is really all about.
Baseline data	Additional baseline data that will help the organization to evaluate the progress of the program and the progress of the organization toward the target and objective.

• ENVIRONMENTAL IMPACT ASSESSMENT

Environmental impact assessment (EIA) is one of the environmental assessment tools being used worldwide and is a systematic analysis of projects and work activities to determine their potential environmental impacts and the significance of such impacts and to propose measures to mitigate the negative impacts. It is the process of predicting, identifying, evaluating and mitigating the biological, social and other relevant effects of developmental proposals prior to major decisions being taken and commitments made. It is an important procedure for ensuring that the likely effects of new developmental activities on the environment are fully understood and taken into account before the development is allowed to go ahead.

Various tools exist for identifying the environmental impact of activities, products or services of an organization. The pollution matrix is one possible tool. Once the possible environmental impacts of an operation have been identified, a procedure should be established to ensure that any new developments such as, for example, a change in product line, construction, or major process change undergo a formal assessment of possible environmental impacts. Most organizations would use an environmental checklist to encourage planners to incorporate issues such as pollution prevention, resource conservation, waste minimisation and alternatives to hazardous substances into their designs.

• SETTING UP PERFORMANCE INDICATORS

It is important to promote and support environmental management efforts by accurately measuring and evaluating the impacts or burden of activities of any organization on the environment and the outcomes of environmental actions or environmental performance. What is required for measuring and evaluate environmental performance are environmental performance indicators. Environmental performance indicators promote environmental efforts of organizations for internal evaluation and decision making.

ISO14031 is a dedicated Environmental Performance Evaluation International Standard which has been developed to provide "guidance on the design and use of environmental performance evaluation, and on identification and selection of environmental performance indicators, for use by all organizations, regardless of type, size, location and complexity".

Having set targets and performance indicators invariably create a sense of progress, direction and meaning to the programme. A realistic measure of environmental performance must be established that is linked to output. If an organization was successful in meeting their target of a 20% reduction in total waste generated, but that reduction was only achieved because production fell by 20%, there really has been no improvement in environmental performance at all. Likewise, if production increased by 20%, and the total amount of waste generated remained the same, the organization has improved their environmental performance.

Other indicators may be developed which focus on specific targets, such as recycling which could measure the ratio between waste, which is recycled and that which is land filled; or the ratio of recycled water used, to water consumed from external sources.

If measuring the relationship between environmental performance and units of production is difficult, it may be advisable to start off with simpler performance indicators, such as total water usage, total electricity consumption and volume of waste removed by hazardous waste contractors.

Examples of performance indicators linked to production units include:

- ✓ Hazardous waste generated per unit of production.
- ✓ Waste generated per unit of production.
- ✓ Volume of pollutant emissions to air per unit of production.
- ✓ Water consumption per unit of production.
- ✓ Energy consumption per unit of production which includes electricity, coal, or furnace oil.

Performance indicators may also be linked to compliance with regulations or standards, both voluntary and mandatory. These may include:

- ✓ Number of times waste quality standards are exceeded.
- ✓ Number of unscheduled water releases.
- ✓ Results of environmental audits.

Targets need to be established for each department to meet. The target would usually refer to a percentage reduction in waste generation, water or energy use, or a score achieved in an audit. The key to a successful environmental management system is the setting of performance indicators and targets, which help the organization to achieve specific objectives. Results need to reported on a regular basis and communicated to workers, top or senior management, unions, and contractors and suppliers. Once these targets have been established, an auditing system should be implemented to provide a regular physical check on environmental performance.

• ENVIRONMENTAL AUDITING

The environmental audit is designed to check on the progress of the system. As such, it forms an important part of the EMS. Environmental auditing is a systematic, documented, periodic and objective process in assessing an organization's activities and services in relation to:

- Assessing compliance with relevant statutory and internal requirements
- Facilitating management control of environmental practices
- Promoting good environmental management
- Maintaining credibility with the public
- Raising worker awareness and enforcing commitment to departmental environmental policy
- Exploring improvement opportunities
- Establishing the performance baseline for developing an Environmental Management System (EMS)

Conducting an environmental audit is no longer an option but a sound precaution and a proactive measure. Evidence suggests that environmental audit has a valuable role to play, encouraging systematic incorporation of environmental perspectives into many aspects of an organisation's overall operation, helping to trigger new awareness and new priorities in policies and practices.

The environmental auditors will need to collect data that enable them to evaluate the performance of each division or department in the organization, and to provide a report, which can be justified by documented findings. South African law currently does not require organizations to conduct specific environmental audits. Recent policy developments, however, suggest that mandatory environmental audits are considered as a tool to ensure industry compliance with environmental regulations. There are many types of environmental audits. All audits provide information used to guide decisions, but audits may have a specific focus or objective to achieve.

SOME TYPES OF ENVIRONMENTAL AUDIT

TYPE OF AUDIT	EXPLANATION
Compliance audits	Establishes the extent to which an organization complies with environmental legislation, such as water quality requirements, and permits.
Environmental risk assessment	Establishes the likely frequency and consequences of environmentally damaging events. This is becoming a regular requirement of insurance agencies.
Management systems audit	Focuses on the procedures established in the EMS, usually checking them against a specific standard, such as the ISO 14 001 guidelines.
Waste audits	Identifies and quantifies waste streams of an organisation. Would usually be carried out as a first step in a waste minimisation programme.
Site audits	Evaluate the full environmental performance of the organisation, by assessing progress towards objectives. This would be carried out by a combination of documentary review and site visits.

A site audit is usually considered essential to maintaining continuous improvement in environmental performance. The level of depth at which a site audit is conducted would depend on the objectives to be achieved. The audit sheet would need to be prepared in consultation with those parties who may have a stake in the outcome. Depending on the objectives of the audit, interested parties may include:

- ✓ Workers.
- ✓ SHE Department.
- ✓ Management.
- ✓ Unions.
- ✓ Local communities.

• GUIDELINES FOR ENVIRONMENTAL AUDITING

These are several considerations when conducting an environmental audit of the EMP and EMS of the organization. The following are important guidelines to be followed, namely:

- Eliminate underlying environmental problems - Auditing can be used to identify the root causes of environmental problems and allow program managers to take steps to eliminate them rather than continuing to rely on temporary stop-gap or control measures.
- Identify systemic environmental problems - audit results can be used to identify systemic environmental problems that must be resolved. Strategic planning may be needed to address these systemic environmental problems.
- Forecast future compliance - Audits provide an understanding of the current state of compliance, but also can be used to determine what activities are necessary to remain in compliance with upcoming or anticipated future regulations.
- Evaluate effectiveness of internal environmental management program - This review may identify issues such as insufficient resources, lack of vision, or poor training that may compromise future compliance.

An environmental audit should include the following aspects, namely:

- On site review of relevant documentation
- Interviews / discussions with relevant stakeholders
- Operational sampling
- The audit can be carried out in two stages or alternatively in a "one off" initial compliance audit visit.

Stage one audit: This is primarily a document review of the environmental management system to ensure the requirements of the standard are in place. Dependent upon the outcome of this stage, the auditor will recommend and agree the schedule and date for the stage two audit. During this step the following will be examined, namely:

- ✓ Environmental policy.
- ✓ Legal requirements.
- ✓ Objectives, targets and programme.
- ✓ Organisation, training and communication.
- ✓ Operational control and records.
- ✓ Maintenance and modification procedures.
- ✓ Emergency procedures and response plans.
- ✓ Environmental impact assessment.
- ✓ Purchasing, suppliers and contractors.

Stage two audit: This will involve a review and completion of any non-conformances identified within the stage one process, together with a sampling of effective implementation of the systems in accordance with the ISO 14001 standard including examination of the following, namely:

- ✓ Air quality management.
- ✓ Water management.
- ✓ Land management.
- ✓ Hazardous materials management.
- ✓ Waste management.
- ✓ Noise, odours, radiation, vibration.
- ✓ Energy management.

• ENVIRONMENTAL SPECIFIC TRAINING AND AWARENESS

Environmental training and awareness need to be part of a long-term programme for all workers in the organization. By combining environmental training with safety and health training, workers will be able to identify common elements and an understanding of the integrated nature of safety, health and environmental management.

ENVIRONMENTAL TRAINING AND AWARENESS PROGRAMME

ELEMENT	EXPLANATION
Environmental awareness for all workers	Induction training and ongoing awareness promotions through displays, competitions, celebration of special environmental days, and focus weeks such as, for example, on reducing waste.
Training for SHE representatives	SHE representatives should undergo integrated SHE training by attending courses
Training for on-site environmental professionals	Professionally trained environmental managers should be encouraged to keep up with new developments through attendance of conferences, subscriptions to journals, and use of Internet access to environmental news
Management and union leadership training	Leadership seminars can be arranged on specific topics, such as environmental management systems, waste minimisation, pollution prevention and hazardous chemical management. A focus on new legal developments and possible liabilities and opportunities for cost savings will always be of interest.
Senior executive training	Executive seminars can be arranged to make senior executives aware of strategic issues, such as environmental compliance for international trade, new policy developments, and the corporate image benefits of improved environmental performance.

The results of the environmental performance indicators should be displayed clearly in the workplace.

• ENVIRONMENTAL REPORTING

Environmental reporting is to promote communication of organizations, to fulfil its accountability regarding environmental efforts in their activities, and to provide useful information to decision making of interested parties. Notably, in South Africa very few organizations prepare and issue environmental reports, either internally or to the public. These reports may remain internal for management or worker awareness benefits or include other parties, such as shareholders, financial institutions, the media, local communities, customers, government and non-governmental organisations. Items to be considered in an environmental report should the organization wish to produce one include, for example:

- ✓ Corporate environmental policy.
- ✓ Information on air emissions and effluent releases.
- ✓ Information on waste generated and waste minimization achievements.
- ✓ Results and description of environmental auditing procedures.
- ✓ Description of environmental management system.
- ✓ Extent to which applicable laws and standards are met.
- ✓ Possible liabilities associated with poor environmental performance such as, for example, waste disposal.
- ✓ Environmental goals of the organization
- ✓ Programmes to encourage worker training and engagement.
- ✓ Resource use such as energy and water consumption.
- ✓ Expenditure on environmental programme.
- ✓ Research and development programmes.
- ✓ Verification of report findings by an external third party.

Internal environmental reporting will be crucial to maintain the momentum of the environmental programme. Reporting environmental progress to workers and management allows improvements to be made to existing environmental management objectives, plans and management procedures.

• USE OF CONSULTANTS IN ENVIRONMENTAL MANAGEMENT

Consultants may be usefully contracted to assist in various areas of environmental management.

They are commonly used to:

- ✓ Identify strategic environmental needs of an organization.
- ✓ Conduct waste minimisation audits.
- ✓ Conduct compliance audits.
- ✓ Identify environmental impacts of the activities of the organization.

- ✓ Establish environmental management systems.
- ✓ Review internal environmental audits.
- ✓ Conduct external environmental audits.
- ✓ Compile external environmental reports.
- ✓ Assist in training.

Give consideration to:

- ✓ Obtaining referrals from other organization engaged in similar business operations.
- ✓ Examining the knowledge of and prior experience in the industry.
- ✓ Defining the scope of work explicitly.
- ✓ Ensuring all relevant workers, supervisors and line managers get to meet with the consultant.
- ✓ Indicating whether the information should be kept confidential.

• REVIEW QUESTIONS

1. What is an environmental management system (EMS)?
2. What are the benefits of implementing an EMS?
3. What are the key elements of an environmental management system?
4. What are the basic steps in an EMS using ISO 14001?
5. What are some items that could be included in the external environmental report of an organization?
6. Develop a SHE management system (SHEMS) for your organization. This exercise will demonstrate whether you have fully understood and applied the key elements of the manual.

• REFERENCES

Anonymous. Nd. Loose notes on Safety, Health and Environment Management

British Safety Council (n.d.). ISO 14001 Environmental Audit. Available at: https://www.britsafe.org/audit-and-consultancy/health-safety-and-environmental-audit/iso-14001-environmental-audit/

de Oliveira Neves, Fabio., Salgado, Eduardo. Gomes. and Beijo, Luiz. Alberto., 2017. Analysis of the Environmental Management System based on ISO 14001 on the American continent. *journal of environmental management*, *199*, pp.251-262.

Department of Environmental Affairs, 2014, Environmental Impact Assessment and Management Strategy for South Africa (Draft).

Environmental Protection Agency (EPA) (2017). *Environmental Management Systems | US EPA*. [online] US EPA. Available at: https://www.epa.gov/ems.

Environmental Protection Department (n.d.). Environmental Audit – A Simple Guide. Available at: https://www.epd.gov.hk/epd/english/how_help/tools_ea/audit_2.html

Enviropaedia (2007). Environmental Management Systems (EMS). Available at: http://www.enviropaedia.com/topic/default.php?topic_id=98

Richard N. Foster, Andrea R. Nagy, and Alexandra Barton-Sweeney, "Polaroid," *Yale SOM Case 08-037,* December 9, 2008

Hengky K Salim, Rory Padfield, Sune Balle Hansen, Shaza Eva Mohamad, Ali Yuzir, Khadijah Syayuti, Mun Hou Tham, Effie Papargyropoulou., 2018. Global trends in environmental management system and ISO14001 research. *Journal of cleaner production*, *170*, pp.645-653.

Kuhre, W. Lee., 2018. *ISO 14001 Certification: Environmental Management System*. Prentice Hall.

Mazzi, Anna., Toniolo, Sara., Mason, Marco., Aguiari, Filippo. and Scipioni, Antonio., 2016. What are the benefits and difficulties in adopting an environmental management system? The opinion of Italian organizations. *Journal of Cleaner Production*, *139*, pp.873-885.

Ministry of the Environment (2003). Environmental Performance Indicators Guideline for Organizations. Available at: https://www.env.go.jp/policy/j-hiroba/PRG/pdfs/e_p_guide.pdf

Pahuja S. (2013) Environmental Audit. In: Idowu S.O., Capaldi N., Zu L., Gupta A.D. (eds) Encyclopedia of Corporate Social Responsibility. Springer, Berlin, Heidelberg. https://doi.org/10.1007/978-3-642-28036-8_692

SABS (2015). *Environmental Management System SANS/ISO 14001:2015*. [online] Sabs.co.za. Available at: https://www.sabs.co.za/Training-Academy/docs/SABS_ISO14001brochureA5.pdf.

Sciencedirect (2019). Environmental Management. Available at: https://www.sciencedirect.com/topics/agricultural-and-biological-sciences/environmental-management/pdfStruwig, Fredrika. Miemie. W. and Lillah, Riyaadh., 2017. South African small and medium-sized enterprise owners' intention to implement an environmental management system. *The Southern African Journal of Entrepreneurship and Small Business Management*, *9*(1), pp.1-8.

Tepaskoualos, Filippos. and Chountalas, Panos., 2017. Implementing an integrated health, safety and environmental management system: the case of a construction company. *International journal for quality research*, *11*(4), pp.733-752.

United States Environmental Protection Agency (EPA). (n.d.). Learn about Environmental Management Systems. Available at: https://www.epa.gov/ems/learn-about-environmental-management-systems

Western Cape Government Environmental Affairs and Development Planning (2015). *Environmental Impact Assessment (EIA): An Introduction*. [online] Westerncape.gov.za. Available at: https://www.westerncape.gov.za/eadp/files/atoms/files/EIA_2015.pdf.

SOME WEB SITES WITH HEALTH AND SAFETY INFORMATION

- British Occupational Hygiene Society www.bohs.org
- British Psychological Society www.bps.org.uk
- British Safety Council www.britsafe.org
- British Safety Industry Federation www.bsif.co.uk
- BSI Group www.bsigroup.com
- Chartered Institute of Environmental Health www.cieh.org
- Department for Environment, Food and Rural Affairs www.defra.gov.uk
- Department for Health www.dh.gov.uk
- Department for Transport https://www.gov.uk/government/organisations/department-for-transport
- Department for Work and Pensions www.dwp.gov.uk
- European Agency for Health and Safety at Work www.osha.europa.eu
- European Commission www.europa.eu
- Faculty of Occupational Medicine www.fom.ac.uk
- Hazards Campaign www.hazardscampaign.org.uk
- Hazards Forum www.hazardsforum.org.uk
- Health and Safety Executive www.hse.gov.uk
- Institute of Ergonomics and Human Factors www.ergonomics.org.uk
- Institute of Occupational Medicine www.iom-world.org
- Institution of Occupational Safety and Health www.iosh.co.uk
- International Institute of Risk and Safety Management www.iirsm.org
- International Labour Office www.ilo.org
- National Archives www.nationalarchives.gov.uk
- NCID. Frequency of cleaning and disinfection for items in health care workplace. https://ncid.ac.za/wp-content/uploads/2020/05/ipc-guidelines-covid-19-version-2-21-may-2020.pdf
- NEBOSH – National Examination Board in Occupational Safety and Health www.nebosh.org.uk
- Office of the Rail Regulator www.rail-reg.gov.uk
- Royal College of Nursing www.rcn.org.uk
- Royal Environmental Health Institute of Scotland www.rehis.com
- Royal Society for the Prevention of Accidents www.rospa.com
- Safework Australia. Frequency of cleaning and disinfection for items in non-health care workplace. https://www.safeworkaustralia.gov.au/doc/how-clean-and-disinfect-your-workplace-covid-19
- Trades Union Congress www.tuc.org.uk
- WHO. Frequency of cleaning and disinfection for items in health care workplace. https://www.who.int/publications-detail/cleaning-and-disinfection-of-environmental-surfaces-in the-context-of-covid-19

INDEX

C

D

E

I

J

K

L

M

T